International Conference on Screw Machines 2024

IOP Conference Series: Materials Science and Engineering
Volume 1322

Dortmund, Germany
3-5 September 2024

ISBN: 979-8-3313-1007-3
ISSN: 1757-8981

Printed from e-media with permission by:

Curran Associates, Inc.
57 Morehouse Lane
Red Hook, NY 12571

Some format issues inherent in the e-media version may also appear in this print version.

Institute of Physics
Dirac House, Temple Back
Bristol BS1 6BE UK

Phone: 44 1 17 929 7481
Fax: 44 1 17 920 0979

techtracking@iop.org

Additional copies of this publication are available from:

Curran Associates, Inc.
57 Morehouse Lane
Red Hook, NY 12571 USA
Phone: 845-758-0400
Fax: 845-758-2633
Email: curran@proceedings.com
Web: www.proceedings.com

TABLE OF CONTENTS

IOP Conf. Series: Materials Science and Engineering **1322** (2024) 011001 doi:10.1088/1757-899X/1322/1/011001

International Conference on Screw Machines 2024

Dear members of the screw machine community,

It is a pleasure to look back on the International Conference on Screw Machines 2024, which turned out to be a huge success. Held at TU Dortmund University, this year's event was particularly special as we celebrated forty years since the conference was founded by the late Professor Knut Kauder. His pioneering work at the Chair of Fluidics laid the foundation for what has become a key forum for advancing screw machine technology.

My team and I were excited to introduce several new elements to enhance the conference experience, including the shortened two-year cycle, which allowed us to spotlight the latest developments in the field. The option to present technical work without submitting a full paper opened the door for more people to share their insights. The continuation of the ICSM Science Update sparked lively conversations around ongoing projects and fresh ideas. Our lab tour and evening social event further added to the friendly, welcoming atmosphere, offering everyone a chance to connect in a more informal setting.

A key highlight of the conference, as always, is the collection of research papers presented here in the proceedings. These papers reflect the outstanding work being done across our field, from fundamental research to real-world applications. The diversity of topics covered showcases the continued evolution of screw machine technology and demonstrates the importance of collaboration within our community. Additionally, all papers from past conferences are available as open access via the conference homepage (www.icsm.tu-dortmund.de), providing an accessible archive of knowledge on all types of screw machines.

What truly stood out this year was the energy and engagement from everyone involved. The depth of expertise, constructive discussions, and collaborative spirit throughout the event reflect the strength of our community. Thank you to all participants, presenters, sponsors, and organisers for making the International Conference on Screw Machines 2024 a memorable event.

Sincerely,
Andreas Brümmer
Chair of Fluidics
TU Dortmund University

IOP Conf. Series: Materials Science and Engineering **1322** (2024) 011001 doi:10.1088/1757-899X/1322/1/011001

Conference committee

General chair

Andreas Brümmer, TU Dortmund University, DE

Programme committee

Thomas Dreifert, Leybold GmbH, DE

Hans-Ulrich Fleige, Aerzener Maschinenfabrik GmbH, DE

Eckhard A. Groll, Purdue University, US

Knut Kauder (†), TU Dortmund University, DE

Ahmed Kovacevic, City, University of London, UK

Johann Lenz, KÖTTER Consulting Engineers, DE

Ronald Sachs (retired), Busch Produktions GmbH, DE

Jack Sauls (retired), Trane, US

Contact

TU Dortmund University
Chair of Fluidics
Leonhard-Euler-Str. 5
44227 Dortmund
Germany

icsm.ft.mb@tu-dortmund.de

www.ft.mb.tu-dortmund.de

+49 (0) 231 755 5721

IOP Conf. Series: Materials Science and Engineering **1322** (2024) 011002 doi:10.1088/1757-899X/1322/1/011002

Peer Review Statement

All papers published in this volume have been reviewed through processes administered by the Editors. Reviews were conducted by expert referees to the professional and scientific standards expected of a proceedings journal published by IOP Publishing.

- **Type of peer review:** Double Anonymous
- **Conference submission management system:** ConfTool Pro
- **Number of submissions received:** 24
- **Number of submissions sent for review:** 23
- **Number of submissions accepted:** 21
- **Acceptance Rate (Submissions Accepted / Submissions Received × 100):** 87.5
- **Average number of reviews per paper:** 2
- **Total number of reviewers involved:** 29
- **Contact person for queries:**
  Name: Manuel Grieb
  Email: manuel.grieb@tu-dortmund.de
  Affiliation: TU Dortmund University - Chair of Fluidics

IOP Conf. Series: Materials Science and Engineering **1322** (2024) 012001 doi:10.1088/1757-899X/1322/1/012001

Experimental investigation and modelling of the noise and vibration in screw compressors

J F Willie and R Ganatra

Compressors and Vacuum Pumps Systems (CVS) Engineering, Germany

E-mail: james.willie@cvs-eng.de

Abstract. This article presents work on the extension of torsional models development for screw compressors. These compressors are common in industry and their vibration behavior and signature are very important. When neglected in the design, it can lead to torsional or structural resonance or lateral vibration in the machine during operation and in the worst case to catastrophic failure of the machine. The screw compressors presented here are used for bulk transport on trucks. They are designed and built by CVS Engineering GmbH. Measurements in the lab and field test showed that the torque peak in the drive shaft can fluctuate and show large values than normal when the first torsional mode is excited by 1 × the input speed. They also show that the nature of the excitation and hence the response of the machine depends on the drive and the excitation amplitude, which depends strongly on the prop-shaft angle. To predict this behavior, detailed free and forced torsional models are developed and validated with measurements. The latter enables the validation of the torque peaks obtained in measurement. The drive train configuration (branched or unbranched) is

1. Introduction

There are many sources of pulsation/vibration in compressors. Some include mechanical (torsional/structural/lateral), flow induced/process oscillations, speed/dynamic oscillations, reliability/structural integrity issues, to mention but a few. Specifically, sources of pulsation/vibration in screw compressors include the shape of the discharge port, the screw profile, the internal clearances, the internal volume ratio, and over compression, and under compression [1]. Pulsation and noise issues are encountered in practice almost on a daily basis. For example, Leader, et al. [2] did a review of torsional resonance cases encountered in industry. They discussed the practical implementation of torsional analysis in real machines. Various cases, for example, synchronous motors, reciprocating engines, compressors and pumps, variable frequency drives (VFDs) and gears are discussed. In [3], a VFD is identified as the source of high torsional vibration in a motor/ID fan system operating far away from the system natural torsional frequencies. Identification of this type of torsional resonance, which is not classical, during the design, is difficult. Noise and pulsation can also be flow induced as reported in [4].

Screw compressors are used in bulk transport in the mobile industry and due to the stringent environmental regulations the space envelope available for mounting them on the trucks is constantly changing. In most cases, the space is decreasing and this leads to large prop-shaft angles which causes fluctuation in the torque and this torque fluctuation can drive noise and vibration in these machines. Figure 1 shows a sketch of the layout of such trucks with the PTO, prop-shaft, gearbox, and compressor shown. To protect the PTO gearbox on the trucks, a thin shaft (not shown here) is usually used which limits the first torsional mode that is easily excited by the input speed of the compressor [9]. Many other sources of vibration and noise exist in such machines, for example, rotors contact, gear noise due to imperfect micro gear geometry or misalignment in the drive train, gear eccentricity, flow induced noise, and so on. It is therefore critical that the source of the vibration or torque peak be identified before attempting to resolve it. In previous work done by the authors the source of the torque peak is identified as the excitation of the first mode torsional frequency of the drive train. The critical element is identified as the elements between the driver (engine, electric motor, and so on) and the compressor. The most critical element is the prop-shaft, including the coupling. The authors reported these findings in [8, 9, 10].

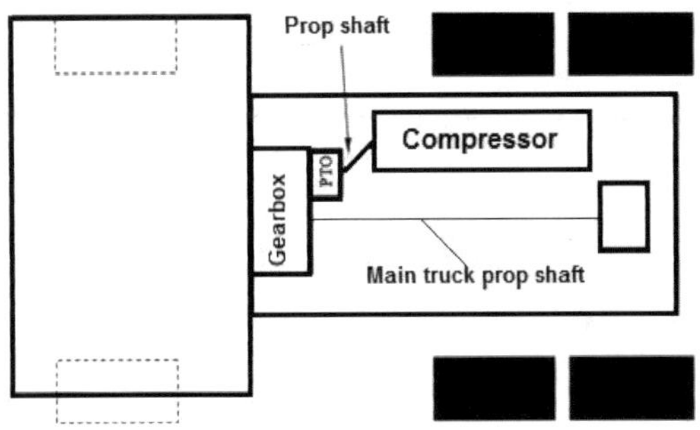

Figure 1. Sketch of truck with mounted compressor and PTO and prop-shaft

The aim of this work is to understand the sources of the torque peak fluctuation measured at the compressor input shaft and the noise in the drive train of a screw compressor. In previous work, the focus was on modeling the natural torsional modes of screw compressors that are operated using electric motors with VFDs in the lab and operated on trucks [8, 9, 10]. In other words, free torsional models were developed and validated with measurements. In this paper, the focus is mainly on forced torsional analysis. The paper will begin with the measurement setup followed by the torsional vibration over-view and then the model development will be presented. After this, the results and discussions will be presented followed by the conclusions and future work.

2. Measurement setup

The set up used in the torque measurement is shown in figure 2. Torque is measured by mounting the HBM torque flange onto the compressor input shaft and the signal input as a voltage to the Brüel & Kjaer (B & K) pulse noise and vibration analysis system. Microphone and accelerometers are used to correlate noise and vibration to the torsional response synchronized by a tachometer signal on the prop-shaft drive coupling. The compressors are driven by a 90 kW 4 pole asynchronous electric motor with moment of inertia of 1.15 kgm^2.

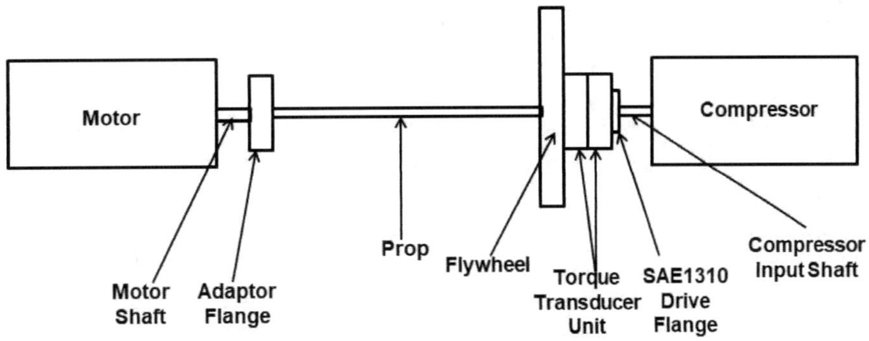

Figure 2. Schematic of the set up used in torque measurements

2.1. Motor torque

Time series data of the electric motor torque is required to enable the implementation of forced torsional models. This will in turn help with the validation of the motor torque being determined from measurement at the 1× input speed. For this motor, between 1000 rpm to 1500 rpm, the torque is steady with some fluctuation about the rated torque and from 1500 rpm, the torque decreases as shown in the schematic shown in figure 3. In the sketch t_1 and t_2 represent the times at which the motor torque is steady with fluctuation about the nominal or rated torque. In general, $\omega = 2\pi f$ and $f = \frac{np}{60}$, where n is the rpm and p is the number of pole pair. For four poles motor, p=2. We get $\omega = \frac{4\pi n}{60}$. This gives us $T(t) = T_0 + T_1 sin(\frac{4nt\pi}{60} \pm \alpha)$, where T_0 is the rated or steady torque, T_1 is the amplitude of the dynamic torque and P_0 is the rated power of the motor. The general torque equation can therefore be written as:

$$T = \begin{cases} T_0 + T_1 sin\left(\frac{4\pi nt}{60} \pm \alpha\right) & \text{for n} \leq 1500 \text{ rpm,} \\ \frac{60P_0}{2\pi n} & \text{for n} > 1500 \text{ rpm} \end{cases}$$

In reality, the relationship between torque and angular speed and time is a 3D one. n in the equation above is the critical speed and T_0 and T_1 are the torque amplitudes. In some special cases $T_0 = T_1$. To determine the motor torque, time series data of the electric motor is needed, which can be obtained by measurement as it was done for this work and presented under results and discussions.

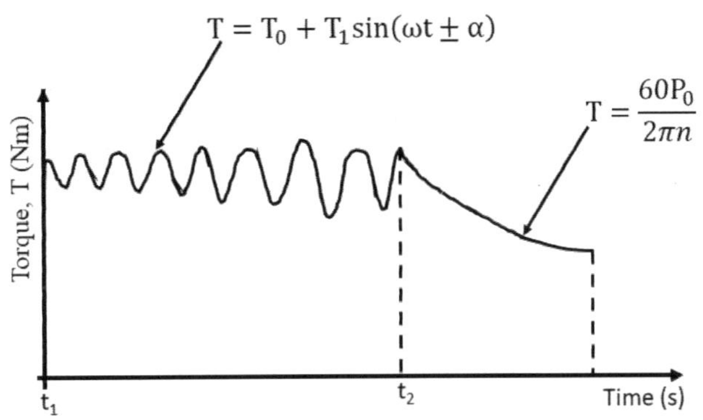

Figure 3. Schematic showing variation of motor torque with time

3. Torsional vibration

3.1. Background

To determine the torsional natural frequencies or forced response of a drive train the structural dynamic equation 1 is used.

$$[I]\{\ddot{\theta}\} + [C]\{\dot{\theta}\} + [K]\{\theta\} = \{T(t)\} \tag{1}$$

In this equation, $T(t)$ is the forcing torque or excitation, and C, which is the damping coefficient, is usually negligible in torsional systems and therefore is set to zero. I is the moment of inertia and K is the torsional stiffness. For this work and previous work presented in [8, 9, 10], lumped mass models, in which inertias are concentrated at gears and rotors, are developed with the shafts inertias as well as the stiffness of the gears neglected. Gears are assumed to be infinitely stiff. Each inertia is treated as a node, N, and the number of frequencies depends on N. For N nodes, we have N-1 natural torsional frequencies ω_1, ω_2,....ω_{N-1}, including rigid body modes. The corresponding relative angular deflection, θ, is a function of the angular frequencies. That is, $\theta_N = f(\omega_N)$.

In general, two drive trains are encountered in practice. In one, the torque is transmitted from the drive through all the rotating elements without branching and in the other case, two or more rotating elements are connected by parallel shafts connected to the same gear stage as shown in figures 4(a) and 4(b). In the latter, the torque is divided in the shafts at the branched point (Node 3 in figure 4(b)).

To determine the torsional natural frequencies and the mode shapes of a drive train, the forcing term $T(t)$=0 and this was the focus of the work done by the authors and reported in [8, 9, 10]. To determine the magnitude of the angular deflections, torques and hence the shear stress in the shaft elements, forced torsional models are required. In this case, $T(t) \neq 0$ and it is obtained by taking time series data of the torque from the drive and curve fitting it to obtain the analytical function, which is of the form $T(t) = T_0 sin(\omega t \pm \alpha) + \bar{T}$, where T_0 is the torque amplitude, $\bar{T}$ is the steady torque of the drive and α is the phase shift.

4. Torsional Models Development

Many methods exists for finding the torsional modes and frequencies of rotating machinery including screw compressors. One method that is commonplace is the eigenvalue method. Other methods like the Holzer method, the transfer matrix method and Finite Element method can be used [1, 6, 7]. In the work presented here, the Holzer method was selected because it is capable of predicting the angular deflection of the modes accurately in both unbranched and branched drives as well as the torsional frequencies, similar to the transfer matrix method. The eigenvalue method, on the other hand, can determine the torsional frequencies accurately of both drive trains but the angular deflection of branched drive trains cannot be determined accurately using this method. A detailed description of how this method was applied to determine the natural torsional modes has been presented by the authors in [8, 9, 10]. The same applies for forced torsional models (with the excitation torque included in the dynamic equation) that are used to determine the torque amplitude and the shear stress in the shafts [5]. Using the forced torsional models presented here the effect of prop-shaft angle on torque amplitude can be investigated. In the cases presented here, benchmarking was first done with sample cases for which the results exist in the literature followed by the extension of the models on screw compressors mounted on an electric motor in the laboratory [7]. The models developed and presented here are implemented in MathCAD.

In the cases presented here, the stiffness and the inertias were reduced to the input shaft. To determine the equivalent diameters of the male and the female rotors Finite Element method was used [9, 10].

IOP Conf. Series: Materials Science and Engineering **1322** (2024) 012001 doi:10.1088/1757-899X/1322/1/012001

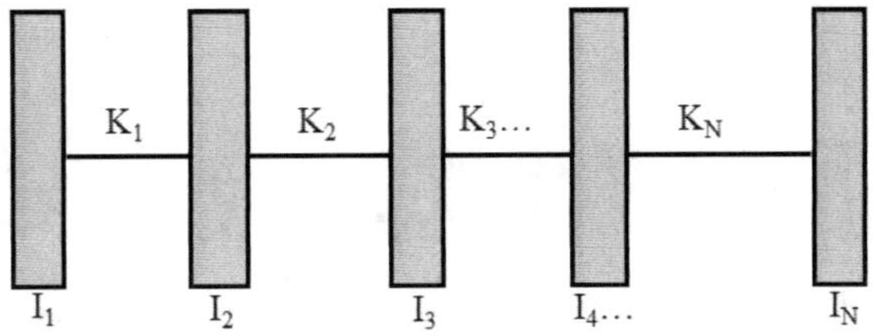

(a) Schematic, unbranched drive train

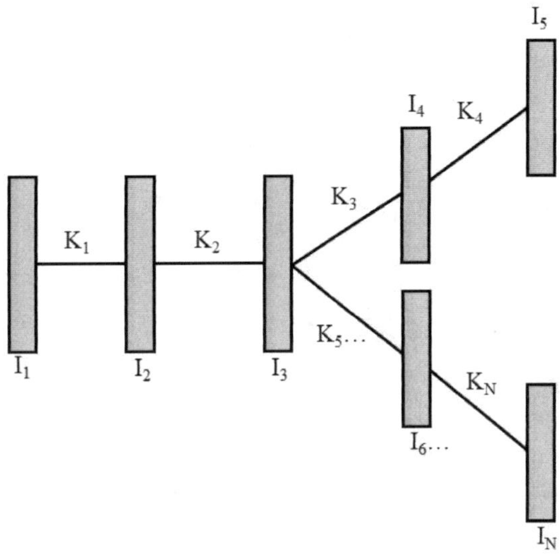

(b) Schematic, branched drive train

Figure 4. Schematic showing unbranched and branched drive trains

To illustrate the implementation of the forced torsional model the line diagrams of the unreduced and reduced models of a typical screw compressor is shown in figure 5 which is a two stage gearbox design followed by the synchronization gears which are located on the non-drive side. The direction of the arrow, denoted as 'a' in figure 5(b) is used to show the direction of torque flow as used in the Holzer method. In figure 5(a), N1 is the node representing the electric motor, N2 is the node for the input stage gears, N3 is the node for the intermediate stage gears, N4 is the node for the female rotor, N5 is the node for the synchronization gears and N6 is the node for the male rotor. G is used to denote the gears in the drive train with G1 the input stage gear wheel, G2 the input stage gear pinion and so on.

The motor torque is applied at the input gear assembly and this is done in the model by applying the forcing torque at node 2 depicted by the arrow.

4.1. Relationship between forced and natural torsional frequencies

If the natural angular torsional frequency is denoted by ω_n and the forced response angular torsional frequency by ω, then according to reference [7], we can write:

$$\frac{\theta}{\theta_0} \sim \frac{1}{1-(\frac{\omega}{\omega_n})^2} \tag{2}$$

From equation 2, we have three regions, namely, $\frac{\omega}{\omega_n} =< 1$, $\frac{\omega}{\omega_n} = 1$ or $\frac{\omega}{\omega_n} > 1$. When the ratio

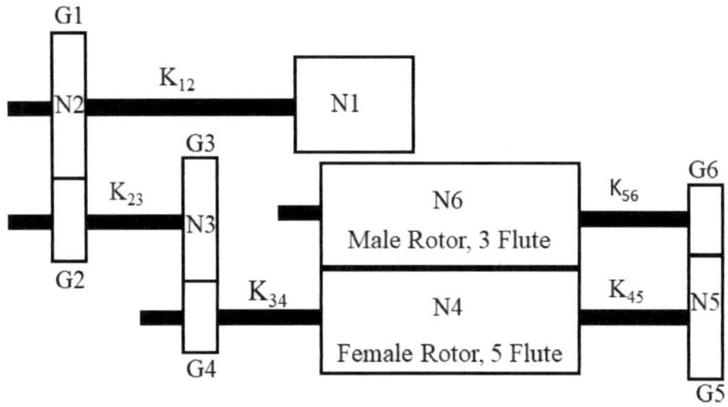

(a) Schematic, line diagram, un-reduced model

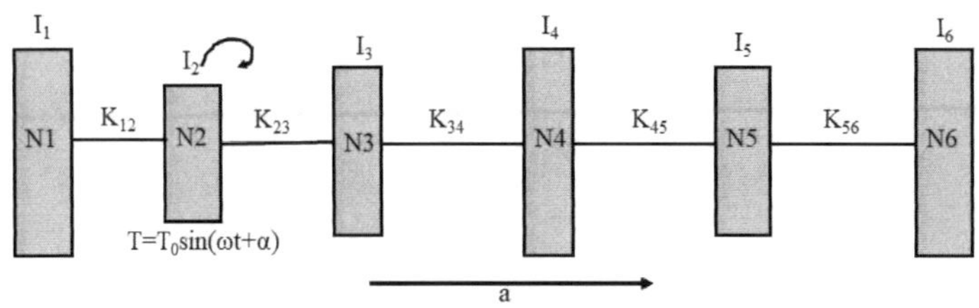

(b) Schematic, equivalent or reduced model

Figure 5. Line diagram of un-reduced & equivalent forced torsional model of a screw compressor

is unity, there is no damping in the system and the deflection or deformation will be infinitely large. This is not possible in practice because of the presence of damping in the system. The ratio $\frac{1}{1-(\frac{\omega}{\omega n})^2}$ gives an indication of the amplification in the shear stress or torque at resonance [8].

5. Results and discussions

5.1. Benchmarking the models

As it is with every model it is important to test the accuracy of the model using benchmark data. In this case, two test cases are presented and the results compared to the references. Test case 1 is taken from reference [6] and it represents the drive train of a steel rolling machine as shown in figures 6(a) and 6(b), with the former representing the un-reduced transmission system and the latter the reduced model with the stiffness and inertias reduced to the input shaft at the electric motor. This example is for the case when the forcing function, T(t)=0, which determines the natural torsional modes of the transmission system. The arrows 'a', 'b' and 'c' are used to determine the direction of the torque in order to determine the torque balance at the branched point or node 3. That is the sum of the torques from branches 'a' and 'b' must be equal to the torque entering branch 'c'. This boundary condition is useful for correctly predicting the deflection at the branched node 3.

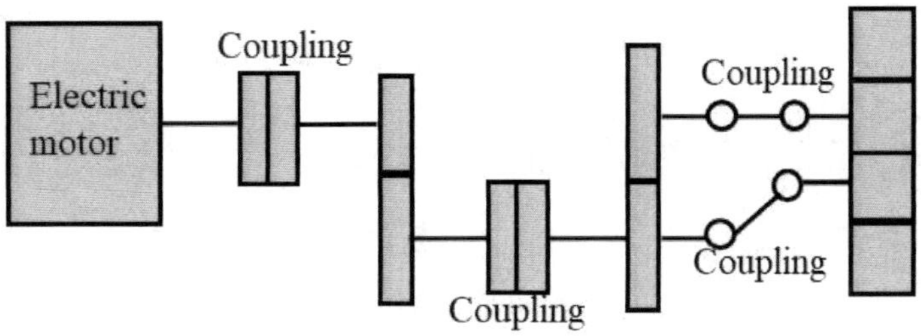

(a) Test case 1 unreduced transmission system

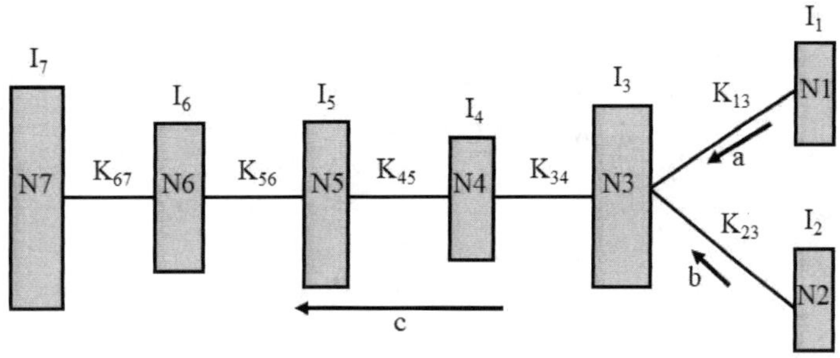

(b) Test case 1 reduced transmission system

Figure 6. The transmission system of a steel rolling machine taken from [6]

For shafts elements in series the effective stiffness is calculated using:

$$\frac{1}{K_{eff}} = \frac{1}{K_1} + \frac{1}{K_2} + \cdots + \frac{1}{K_N} \tag{3}$$

For this test case these values are computed to be: K_{13}=K_{23}=44192 Nm/rad; K_{34}=148740 Nm/rad; K_{45}=1.1632e6 Nm/rad; K_{56}=2.62e6 Nm/rad and K_{67}=7.59e6 Nm/rad.

For a given gear stage that consists of a gear wheel and a pinion or a wheel, an idler and a pinion, the effective or combined inertia can be determined. If I_w is the wheel moment of inertia, I_p the moment of inertia of the pinion and I_d the moment of inertia of the idler, then for the case without an idler we can write: $I_{eff} = I_w + i_{wp}^2 I_p$, where i_{wp} is the gear ratio between wheel and pinion (ratio of number of teeth on wheel to number of teeth on pinion=$\frac{G_w}{G_p}$. If the wheel is first connected to an idler and the idler to the pinion, then $I_{eff} = I_w + i_{wd}^2 I_d + i_{wp}^2 I_p$, where $I_{wd} = \frac{G_w}{G_d}$. Adding an idler does not change the overall step-up between the wheel and the pinion. That is $i_{wp} = i_{wd} \times i_{dp} = \frac{G_w}{G_d} \times \frac{G_d}{G_p} = \frac{G_w}{G_p}$. For this test case we have: I_1=I_2=0.07245 kgm^2; I_3=0.04196 kgm^2; I_4=0.20620 kgm^2; I_5=0.381 kgm^2 and I_6=1.1721 kgm^2. Note that the subscripts in the variables for the stiffness and the mass moment of inertia represent the node numbers.

Using the Holzer's method, assume that θ_1=1 and θ_2=1 and take branch "a" as the first branch and compute the relative angular deflection at node 3. After this, take branch "b" as the second and compute the relative angular deflection at node 3. Note that $\theta_{3a} = \theta_{3b}$. Proceed to branch "c" to compute the residual torque, $T_e(\omega)$, at node 7 and this must equal zero from which the angular frequencies (ω) can be determined. In this case 6 angular frequencies are computed, with the first being a rigid body mode with an angular frequency of zero. The model described above is implemented in MathCAD with the plotted $T_e(\omega)$ versus (ω) shown in figure 7(a). The zeros of this plot are as follows: ω_1=578.596 rad/s; ω_2=1540 rad/s; ω_3=2398 rad/s;

ω_4=2855 rad/s and ω_5=3829 rad/s. The corresponding torsional frequencies are determined using $\omega_N = 2\pi f_N$. This gives: f_1=92 Hz, f_2=245 Hz, f_3=382 Hz, f_4=454 Hz and f_5=609 Hz.

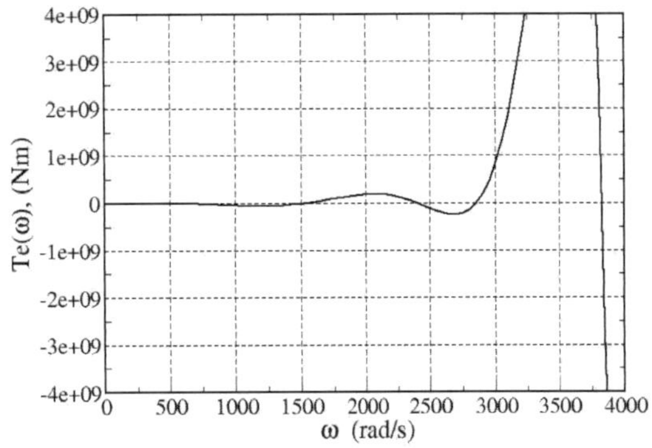

(a) Plot of residual torque

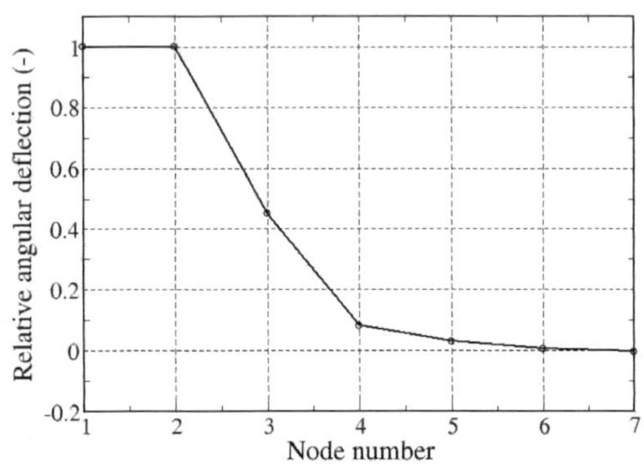

(b) Angular deflection, mode 1

Figure 7. Residual torque, Te(ω), versus angular freq., ω & the relative angular deflection

The mode shapes are also implemented in the MathCAD model with mode 1 shown in figure 7(b).

Test case 2 is for a forced torsional model in which $T(t) \neq 0$ and it is assumed that the external torque or excitation is acting on rotor 2 with $T_2 = 10000sin(5t)$ and $T_1 = T_3 = T_3 = T_4 = 0$ and taken from reference [7] and shown in figure 8(a). In this example, the forcing frequency is 5rad/s and so the angular frequency of the system is 5. To develop the forced lumped mass torsional model it is assumed that the first angular deflection is θ_1 and then the angular deflections at the other discs are expressed in terms of θ_1. The residual torque is then determined as a function of θ_1 and this must equal zero, from which θ_1 is determined. That is $T_e(\theta_1) = T_4(\theta_1) = \frac{8 \cdot \frac{73375 \cdot \theta_1}{256} + \left(29000 \cdot \theta_1 + 63125 - 8 \cdot \frac{1375 \cdot \theta_1}{16}\right)}{8} = 0$ fom which $\theta_1 = -2.063$. Working backwards, the angular deflections θ_2, θ_3 & θ_4 are determined as shown in figure 8(b). This is followed by determining the twist of the various shafts in which for shaft 1 we have $\theta_{M1} = \frac{T_1}{k_{t1}} = -0.516$. For shaft 2 we have $\theta_{M2} = \frac{T_2}{k_{t2}} = 0.291$ and for shaft 3 we have $\theta_{M3} = \frac{T_3}{k_{t3}} = -0.123$. The twisting moments in Nm in the various shafts are given by $T_1 = -5156$ Nm, $T_2 = 2911$ Nm & $T_3 = 2451$ Nm. Based on these numbers, the maximum stress in each section of the shaft can be determined, if the diameter of the shaft is known. The shear stress is given by $\tau = \frac{TR}{J}$, where τ is the shear stress, R is the shaft radius and J is the polar moment of inertia.

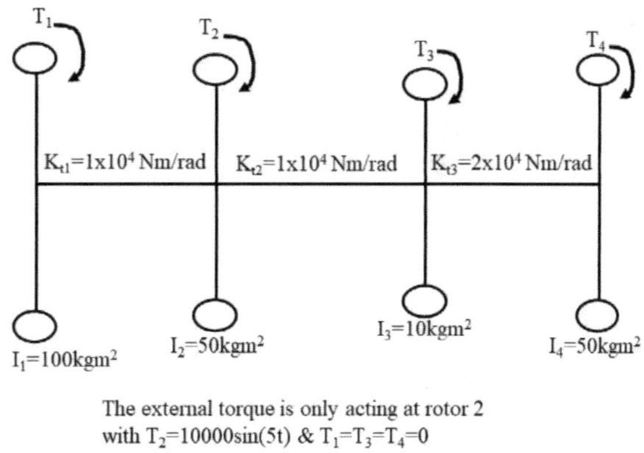

(a) Line diagram, reduced transmission system

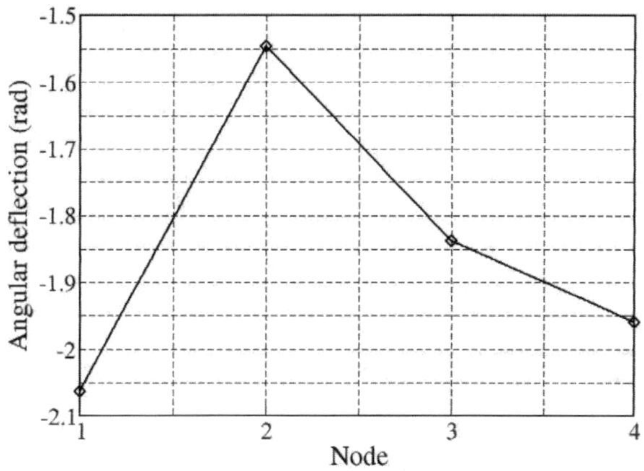

(b) Angular deflection

Figure 8. Forced torsional model example with line diagram and the angular deflection taken from [7]

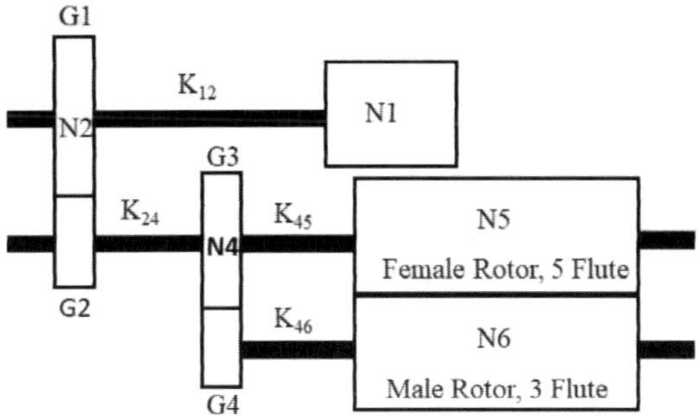

(a) Line diagram, screw compressor

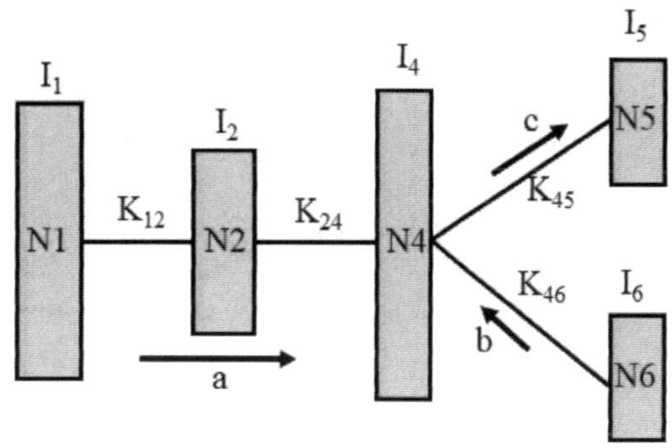

(b) Reduced or equivalent torsional model

Figure 9. Line diagram and the equivalent model of the screw compressor

IOP Conf. Series: Materials Science and Engineering **1322** (2024) 012001 doi:10.1088/1757-899X/1322/1/012001

5.2. Determination of the excitation function

To enable the actual quantification of the torque and hence the shear stress in the input shaft and other shaft elements in the drive train forced torsional model is developed for a screw compressor with the line diagram and the equivalent or reduced model depicted in figure 9. The input speed range is 1800 rpm to 3240 rpm. To develop the forced torsional model curve fitting is used to determine the forcing torque function that is generated by the electric motor. The torque, denoted by T(t) and discussed in section 2.1, is used in the torsional model as forcing function and the corresponding angular deflections in the shafts and the twisting moments are determined and compared with measurement data. The sampling rate of the data is 20 kHz and a total of 10000 data points are taken at a prop-shaft angle of 15° and at a critical input speed of 2340 rpm. The time step, $\Delta t = \frac{1}{Sampling\ rate} = 5x10^{-4}$ s. The torque is measured at the motor shaft using a torque transducer.

The time series data and the FFT showing the power spectra density (PSD) of the torque is shown in figure 10. Matlab was used for doing the FFT and no biasing of the data using windowing was done and so the default rectangular window was used. The PSD of the torque is obtained using $20log10(T_{abs})$, where T_{abs} is the absolute torque obtained from the FFT of the torque time series. This is done to determine the fundamental frequency and to curve fit the measured time series data in order to determine the analytical forcing torque function. The fundamental frequency is 78.2 Hz, which is 2 × input speed. The first sub-harmonic is 39.1 Hz or 2346 rpm, which is 1 × input speed. The implication here is that the 1 × input speed is not excited.

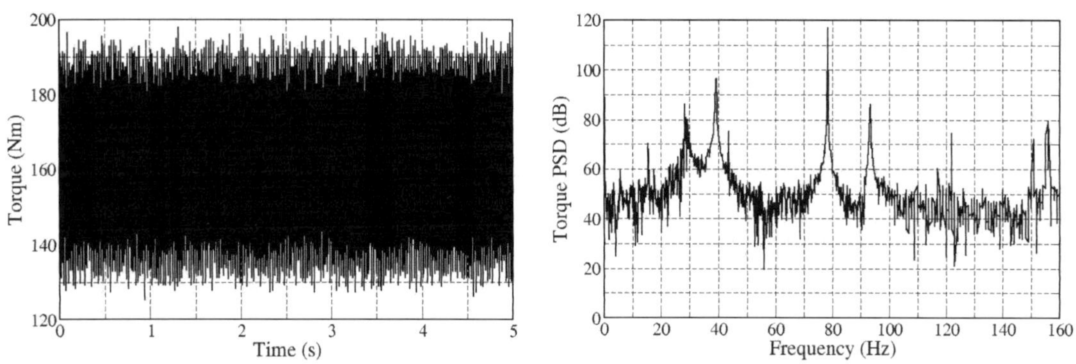

(a) Time series (b) PSD of torque

Figure 10. Time series and PSD of the measured torque of a screw compressor

The curve fitting tool in Matlab that uses the linear least-squares algorithm method was used for fitting the analytical function to the measured times series of the torque as shown in figure 11. The goodness of fit in this case is 0.8339 with 95% confidence interval. The times series data fitted was up to 5 seconds but for clarity it is plotted for only 0.25 seconds.

The torque forcing function obtained with a positive torque amplitude can be written as:

$$T(t) = 33sin\left(491.35t + \frac{\pi}{2}\right) + 161.1 \tag{4}$$

5.3. Forced torsional model implementation

The forced torsional model developed is implemented in MathCAD for the first torsional mode including the residual torque used to determine the angular deflections and hence the torque in the various shaft sections. The mode shape is determined as shown in figure 12 in which there

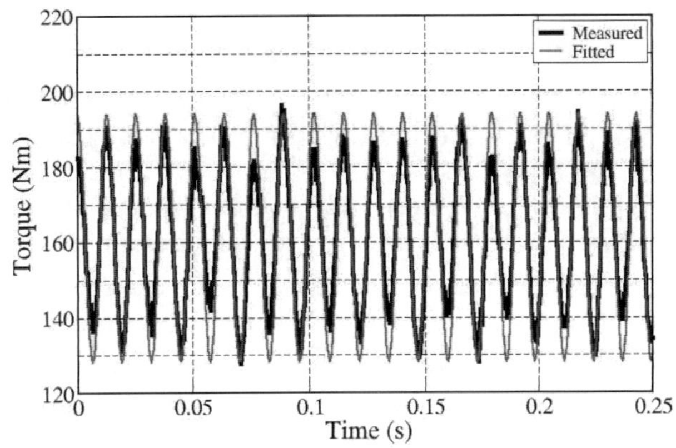

Figure 11. Curve fit of torque time series data taken at a critical speed of 2340rpm

is a 180° phase shift or anti-phase motion between the drive and the compressor. This must be avoided in practice because when it happens the compressor will try to drive the motor instead of the motor driving the compressor. For the example presented here the angular twisting in the shafts of the machine is not severe and test showed that there was no torsional resonance. The twisting moments obtained are T_1=4.379 Nm, T_{24}=33.365 Nm, T_6=-25.247 Nm and T_{45}=6.107 Nm. With this, the twist of the various shafts can be obtained using: For shaft 1: $\theta_{M1} = \frac{T_1}{k_{12}}$ and for shaft 2: $\theta_{M2} = \frac{T_{24}}{k_{24}}$; for shaft 3: $\theta_{M3} = \frac{T_6}{k_{46}}$ and for shaft 4: $\theta_{M4} = \frac{T_{45}}{k_{45}}$.

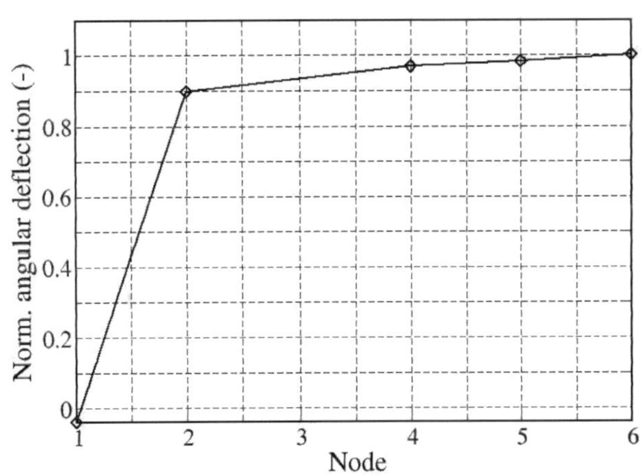

Figure 12. Predicted forced mode shape of an oil free screw compressor

6. Conclusions

Forced torsional analyses of dry screw compressors is presented. It begins with benchmarking the models based on existing data in the literature to check the veracity of the methodology. The forced models are based on times series data of torque measurements that are curve fitted to determine the forcing function used in the model development. The following conclusions can be drawn:

- Because the input shaft is connected to the prop-shaft via a coupling, the torque cycles twice per shaft revolution (2× input speed) and this explains why a torque peak is seen at 2× input speed on the machine, including the noise & vibration spectra.
- In the torque equation, it is observed that the coefficient of time, t, (that is, $\frac{4n\pi}{60}$) holds only when the input speed, n, is not critical. For example, the screw compressor simulated is running at 2340 rpm input speed and $\frac{4n\pi}{60}$ =490.08 rad/s, which is identical to the ω value in the forcing torque function and this is equal to 2 × input speed.
- The twisting moments determined at the input shaft was in line with measurements. Because the drive train was not set into torsional resonance there was no torque peak observed in the test as expected.
- With the developed models and methodology it will be possible to robustly design the drive train of screw compressors to prevent torsional vibration and torque peak in the shafts and thus reduce the noise produce by these machines.

As future work, strain gauge telemetry will be used to measure the torque time series in each shaft section for further validation of the models developed and presented in this work.

Nomenclature

K	Torsional stiffness	[Nm/rad]
I	Mass moment of inertia	[kgm^2]
ω	Angular frequency	[rad/s]
θ	Angular deflection	[rad]
T(t)	Torque function	[Nm]
T_0 or T_1	Forcing torque amplitude	[Nm]
α	Phase shift	[rad]
i	Gear ratio	[-]
PSD	Power spectra density	[dB]
n	Rotational speed	[rpm]
N	Node in torsional model setup	[-]
P_0	Rated motor power	[W]
f	Frequency	[Hz]
t	Time	[s]
τ	Shear stress	[N/m^2]
J	Polar moment of Inertia	[m^4]

References

[1] Smith E H *Mechanical Engineer's Reference Book* 12th Edition (Elsevier)
[2] Leader M E and Kelm R D 2004 *Practical implementation of torsional analysis and field measurement* (Vibration Inst. Annual Meeting)
[3] Feese T and Maxfield R 2008 *Torsional vibration problem with motor/ID fan system due to PWM variable frequency drive* (Proc. of the 37th Turbomachinery Symposium)
[4] Mobley R K 1999 *A model for the transient pulsation generation at the discharge of a screw compressor by a shock tube analogy* (IOP Conf. Ser.: Mater. Sci. Eng.) **425** 012022
[5] Glover G K 1977 *Mechanical Vibrations* 3rd Edition (NEM CHAND and BROS ROORKEE)
[6] Yuwen Y 2004 *Analysis of Multi-Branched Torsional Vibration for Design Optimization* PhD Thesis (West Virginia University College of Engineering and Mineral Resources)
[7] Den Hartog J P 1947 *Mechanical Vibrations* 3rd Edition (McGRAW-Hill Book Company, INC)
[8] Willie J, Asal W and Sachs R 2014 *Analysis of the Noise and Vibration of a Dry Screw Compressor* (VDI-Berichte 2228)
[9] Willie J and Sachs R 2015 *Structural and torsional vibration analysis of a dry screw compressor* (IOP Conf. Ser.: Mater. Sci. Eng. **90** 012005)
[10] Willie J and Sachs R 2016 *Structural and torsional vibration and noise analysis of a dry screw compressor.* (Process Mechanical Engineering **0(0)** 1–10)

IOP Conf. Series: Materials Science and Engineering **1322** (2024) 012002 doi:10.1088/1757-899X/1322/1/012002

Investigation of Pressure Pulsation and Vibration of the Internally Geared Screw Compressor

J Zhu[1,2], M Read[1], A Kovacevic[1] and V Sishtla[2]

[1]Department of Engineering, City, University of London, London, UK
[2]Carrier Corporation, Syracuse NY, USA

E-mail: jin.zhu@carrier.com

Abstract. The Internally Geared Screw Machine (IGSM) is a relatively novel and promising type of positive displacement compressor concept that takes inspiration from classic gerotor pumps. The IGSM uses two rotors rotating in the same direction about parallel axes while maintaining continuous contact between both rotors which isolates several working chambers. By using ported end plate, the IGSM can determine the timing of the fluid entering and exiting the working chamber. There are several key advantages to the IGSM over traditional twin-screw compressors including reduction of the rotor to casting leakage, elimination of the blow hole area, reduction in sliding velocity at the point of contact, and a uniform circumferential rotor temperature profile. Although there is some research regarding the geometric profiling and chamber modeling of this type of machine, there is no research currently available on the pressure pulsation and vibration characteristics of the IGSM. The purpose of this paper is to provide an initial look into the pressure pulsation and vibration characteristics of the IGSM and how they differ from that of a traditional twin-screw machine. By explicating the key differences between the IGSM and the industry standard twin-screw compressor, this paper aims to better understand an important yet underdeveloped area of IGSM design.

1. Introduction

The history of the screw compressor is a relatively old technology that can be dated back to 1878. Since its inception, there have been substantial improvement made to the design and operation of screw compressors including increased efficiency, lower noise and vibrations, as well as improvements to many other compressor properties. However, research into screw compressors usually adheres to either twin screw machines, or offshoots of twin screw machines. For example, Akei et al. proposed the use of a mini screw to achieve high flow so that it can use the low GWP refrigerant DR-12[1]. Little research have been conducted regarding new styles of rotary compressor that defy the conventional twin screw setup.

The Internally Geared Screw Machine (IGSM) is a novel compressor that takes inspiration from the gerotor pump. Similar to the gerotor pump, the IGSM has 2 meshing internal rotors that has continuous contact with each other to create separate working chambers. To achieve the desired compression, a stationary ported end plate is used so that the fluid can only leave once the desired volume ratio is achieved. Figure 1 shows a 3D model of the IGSM.

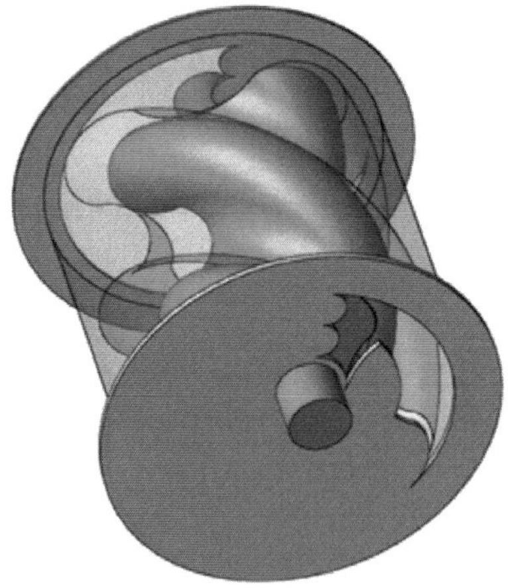

Figure 1. Model of the Internally Geared Screw Machine (IGSM)

Since this is a novel compressor that is still undergoing development, there is limited research available for this topic. The IGSM compressor design was proposed by Read et al.[2] where the author detailed the geometric analysis of the machine as well as several operational characteristics. In [3], the authors detailed a chamber modeling approach to achieve performance calculations, and in [4], the authors used CFD to investigate the performance of an oil injected model. Amongst the work currently being pursued, one area that has yet to be explored is the pressure pulsations and how that affect the vibration and structural integrity of the system.

The study of pressure pulsations and vibration for screw compressors is paramount in the successful application of such a machine to industrial applications. There are many benefits to having a compressor that have low vibration characteristics. For example, pressure pulsation causes cyclical loading which can cause fatigue stress, which in turn wears down the life of a mechanical component such as the rotors or the housing causing the need for the component to be prematurely replaced before its life cycle end. Another practical reason is that high vibrations may cause the overall structure to degrade and cause problems in the attached piping and sensor systems. The purpose of this paper is to preform an initial investigation of the pressure pulsation and vibration characteristics of the IGSM and compare these results to the traditional twin screw compressor.

2. Methodology

It is well established that the dominant source of fluid-induced vibrations for a screw machine is caused by the discharge pressure pulsations. This type of vibration analysis is a complicated process as it is in essence a fluid-structure interaction (FSI) problem. Therefore, purely mathematical models are either inaccurate or fails to provide the necessary range to perform such analysis. In order to accurately model this process, it is more advantageous to use a high fidelity numerical model that can accurately display the pressure profile. For this study, the method used to evaluate the system vibrations is based on a 1-way FSI analysis. A fully coupled FSI system is likely unnecessary as the structural vibrations will have minimal effect on the fluid since the scale of the vibrations that will occur in the compressor shell is too small that have a substantial effect on fluid flow. For this process, the discharge pressure pulsations are treated as forcing function acting upon the discharge port, which in turn causes structural deformations and vibrations.

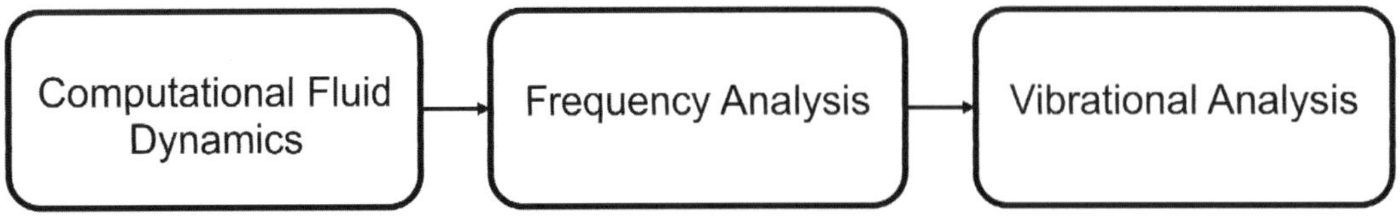

Figure 2. Vibration analysis procedure

Figure 2 shows the analysis procedure used to evaluate the surface deformation and velocity of the compressor. There are 3 main steps to this process: CFD analysis of the compressor, frequency analysis of the pulsating discharge flow, and vibration analysis of the compressor system. Further analysis can be extended to look at the associated acoustic pressure field due to the system vibration, but that aspect is omitted from this study. The purpose of the CFD is to generate the pulsating pressure distribution onto the compressor discharge port. After obtaining the discharge pressure pulsation from the CFD, these results are then transformed to the frequency domain via spacial spectrum analysis. Frequency domain is generally what is used for vibration analysis as frequency data will provide characteristics such as resonant frequencies and peak amplitudes. Finally, the frequency domain results are then used as a forcing function to calculate the surface velocity of the compressor via FEA analysis. Note that the resultant vibrations are dependent on the shape and mass of the structure. In reality, compressors in HVAC applications are mounted on heat exchanger a chiller system, thus the fixed support used in the harmonic response module is not actually fixed as it will mechanically interact with the chiller components.

3. Numerical Simulation

As mentioned previously, the fundamental fluid-solid interaction that encircles pressure pulsation induced vibrations is complicated in nature and thus require full fidelity CFD simulations to fully capture the flow characteristics. The CFD simulation was performed using ANSYS CFX, and the vibration analysis was performed using ANSYS harmonic response. To simulate the comprssor rotors, SCORG was used to generate the moving mesh. The static meshes used for the other fluid domains were generated using ANSYS meshing. For the purposes of this study, the effects of oil was not considered, but it will be explored in a future study.

Table 1: Cases considered for this study

Cases	Suction Pressure	Suction Temperature	Discharge Pressure	Speed
1	1 bar	25 ^{o}C	2 bar	12000 RPM
2	1 bar	25 ^{o}C	3 bar	12000 RPM
3	1 bar	25 ^{o}C	4 bar	12000 RPM

Three cases were considered for this study as shown in table 1. For each of the 3 cases, the suction pressure was set at 1 bar, the suction temperature was set at 25 ^{o}C, and the shaft speed was set at 12000 RPM. For the discharge pressure, a ramping function was used to linearly vary the pressure from 1 bar to 2 - 4 bar for the 1st rotation and kept constant at 2 - 4 bar for the remainder of the simulation.

Table 2. SCORG parameters for generating rotor mesh

	IGSM	Twin-Screw
Circumferential Division	120	40
Radial Division	10	6
Angular Division	72	40
Interlobe Division	N/A	40

Table 3. Mesh statistics for CFD simulation

	IGSM	Twin-Screw
Suction Elements	30960	1478046
Rotor Elements	1677600	1643520
Discharge Elements	341214	404741
Pipe Extension Elements	1423200	113685

Table 2 shows the parameters used to generate the rotor mesh and table 3 shows the mesh statistics of each of the different fluid domains. Note that the interlobe parameter that defines the mesh between male and female rotors does not exist for the IGSM. Looking at the mesh statistics, it is evident that

IOP Conf. Series: Materials Science and Engineering **1322** (2024) 012002 doi:10.1088/1757-899X/1322/1/012002

the suction elements of the IGSM is much lower than every other fluid domain. This is due to the fluid volume of the suction port being significantly smaller than either the rotor and pipe extension. Although the volume of the discharge port is even smaller than the suction port, the discharge port is purposely refined since that is the surface area of interest. The elements for the pipe extension is much larger for the IGSM than the twin screw because the it was attached to the IGSM discharge thus it was refined along with the discharge.

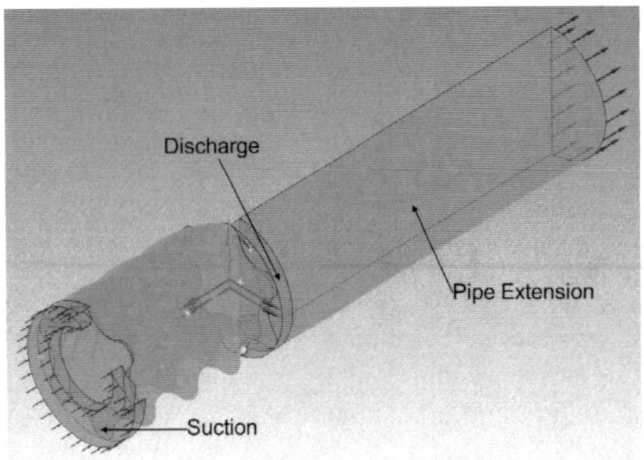

(a) IGSM Model

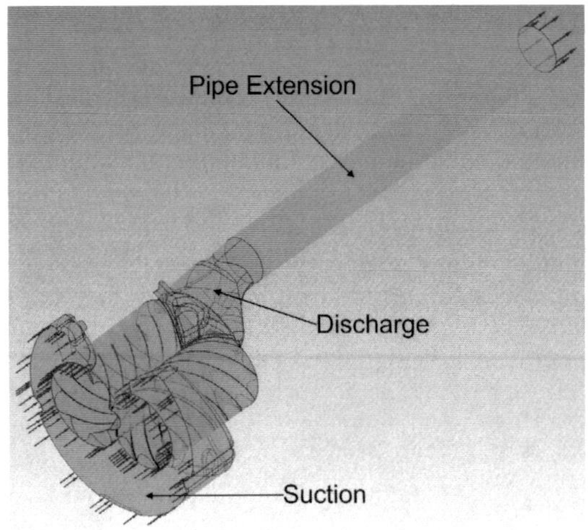

(b) Twin-Screw Model

Figure 3. Models used for CFD simulations of IGSM and Twin Screw compressors.

The fluid models that were used for the CFD simulations are shown in figure 3. Figures 3a and b shows the IGSM model and twin screw model respectively. The rotor configuration used for the IGSM is 5-6 for the inner and outer rotors respectively, and the rotor configuration for the twin screw is 7-8 for the male and female rotors respectively. The pipe extension domain is placed artificially to allow for better convergence of the fluid downstream of the discharge port.

(a) IGSM Model

(b) Twin Screw Model

Figure 4. Models used for mechanical simulations of IGSM and Twin Screw compressors.

IOP Conf. Series: Materials Science and Engineering **1322** (2024) 012002 doi:10.1088/1757-899X/1322/1/012002

To complete the vibration analysis of both compressors, a solid model is needed for the FEA analysis. Figures 4a and b shows the solid models for the IGSM and Twin Screw respectively. In terms of scale, it should be noted that the IGSM model is significantly smaller than that of the twin screw model. This is because the IGSM is a currently ongoing project and not all the parts of the IGSM model has been fully realized whereas the twin screw model is fully designed and operational. Further analysis will be conducted in the future once the IGSM has a more complete design.

4. Results and Discussion

As previously discussed, CFD simulations were performed for both the IGSM and twin screw compressors at 2 - 4 bar discharge pressure. The pressure pulsation results are compared against the twin-screw compressor to examine differences in the pulsation characteristics. The vibration cannot be directly compared between the two machines as the twin screw compressor is much larger as compared to the IGSM. Three sets of results will be discussed in this section: section 4.1 will explore the discharge pressure pulsation that is applied to the discharge port, section 4.2 will consider the frequency response of the system, and section 4.3 will look at the surface velocity of the system.

4.1. Discharge Pressure Pulsation

The purpose of this section is to analyze the pressure pulsations that is applied to the discharge port. Figure 5 shows the mean pressure pulsations that is acting on the discharge port for a discharge pressure of 2, 3, and 4 bar with a suction pressure of 1 bar. The left hand side of figure 5 show the IGSM pressure pulsations and the right hand side of figure 5 shows the twin screw pressure pulsations.

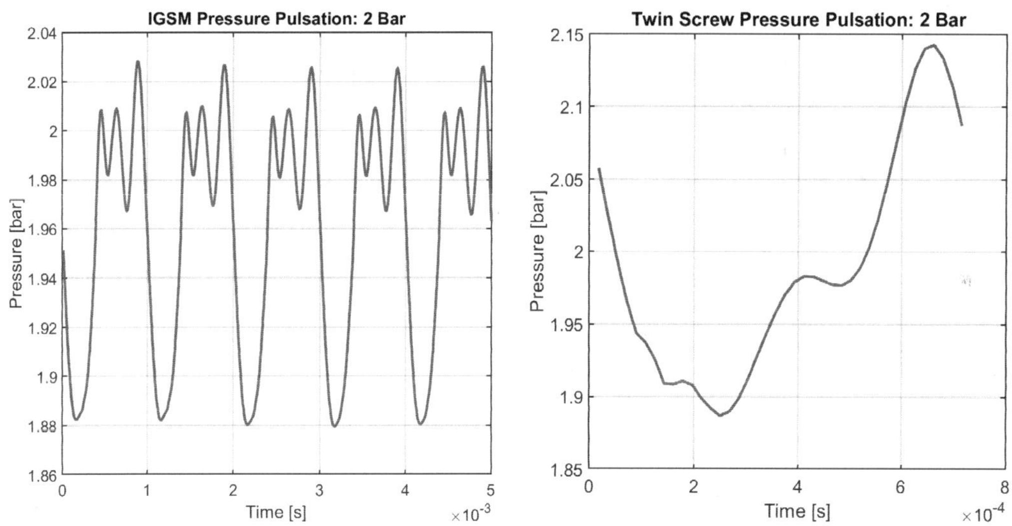

(a) IGSM: $P_{out} = 2$ bar (b) Twin Screw: $P_{out} = 2$ bar

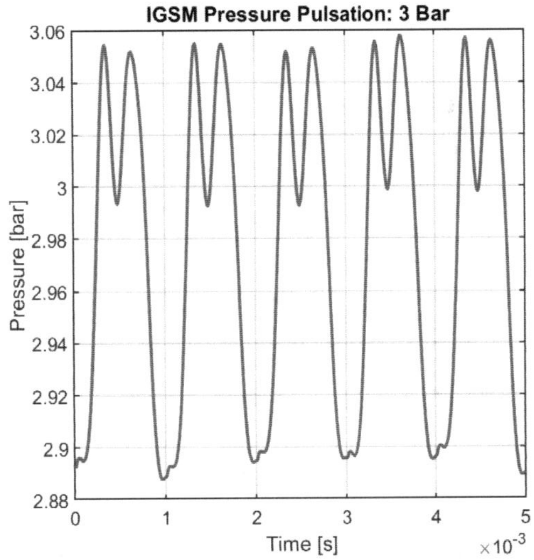

(c) IGSM: P_{out} = 3 bar

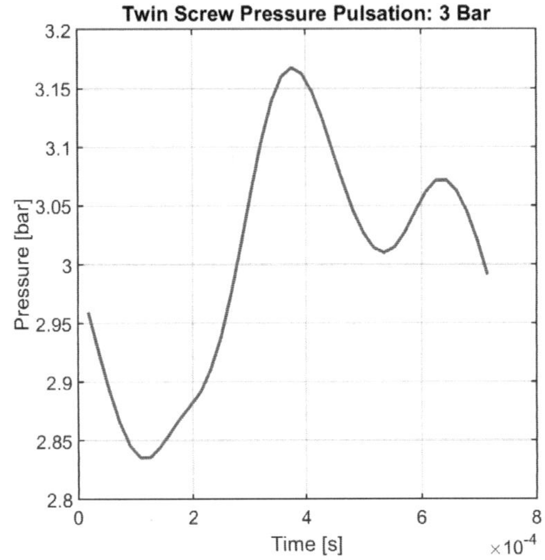

(d) Twin Screw: P_{out} = 3 bar

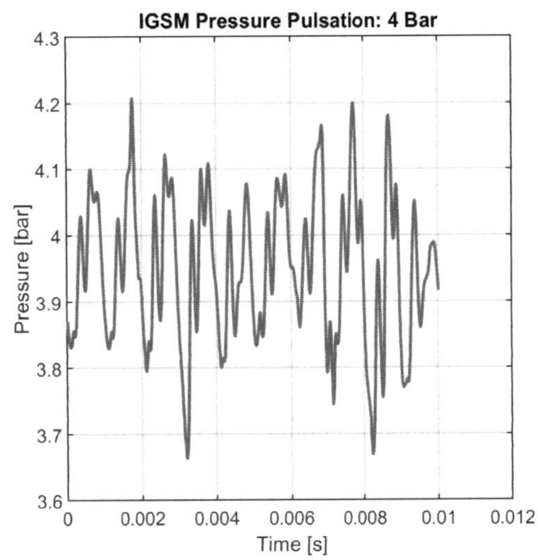

(e) IGSM: P_{out} = 4 bar

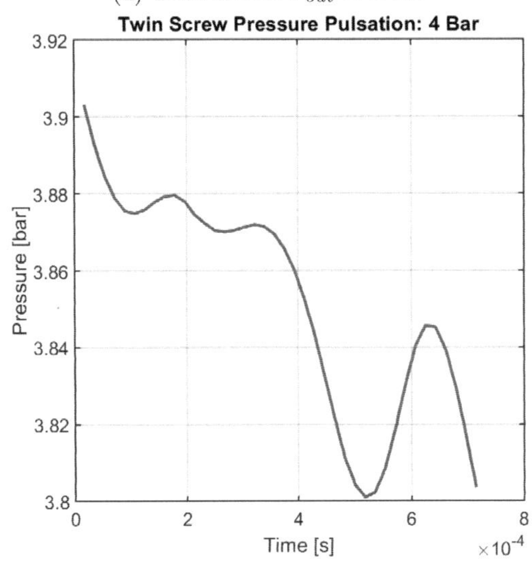

(f) Twin Screw: P_{out} = 4 bar

Figure 5. Discharge pressure pulsations for IGSM (left) and Twin Screw (right) at 2, 3 and 4 bar outlet pressure

One key property of the IGSM pressure pulsations as compared to the twin screw compressor is the existence of the micro oscillations that occurs at each cycle. These micro oscillations occur for both that 2 bar and 3 bar case, but does not exist for the 4 bar case. The reason that the oscillations no longer occur is due to the additional noise that is present in the 4 bar case. Since no oil is considered in this study, the sealing effect that oil provides also no longer exist. At a pressure ratio of 4, the inter-lobe leakage becomes too much to gather acceptable results. This effect can be seen in both the IGSM and twin screw cases, as although there is a clear oscillating presence, a distinct pattern no longer exist. This behaviour is especially present for the 2 bar case where 3 of these micro oscillations occur. This behaviour also seems to be damped down as the pressure ratio increases as the number of micro oscillations decreased at an outlet pressure of 3 bar as compared to the 2 bar case.

Table 4 shows the RMS pressures for each of the 3 cases for both the IGSM and twin screw cases. From this result, it is evident that the RMS pressure of the traditional twin-screw is closer to the set discharge pressure as compared to the IGSM. This is because the cyclical over and under compression is equally prevalent in the twin screw, i.e, there is just as much over compression as there is under

Table 4. RMS pressure of the pressure pulsations for the IGSM and twin screw compressors

Discharge Pressure	IGSM	ΔP_{IGSM}	Twin Screw	ΔP_{TS}
2 bar	1.9603 bar	0.0397	1.9891 bar	0.0109
3 bar	2.9769 bar	0.0231	3.0039 bar	0.0039
4 bar	3.9502 bar	0.0498	3.8530 bar	0.1470

Table 5. Peak-to-peak pressure of the pressure pulsations for the IGSM and twin screw compressors

Discharge Pressure	IGSM	Twin-Screw
2 bar	0.1487 bar	0.2560 bar
3 bar	0.1701 bar	0.3286 bar
4 bar	0.5446 bar	0.1032 bar

compression, whereas the IGSM exhibit much more under compression as it does over compression. Therefore, the RMS value for the IGSM is skewed towards a lower value. Another difference between the pressure pulsations for the IGSM and the twin screw is the peak-to-peak pressure value. Although the pressure pulsations for the IGSM skews towards under compression, the peak-to-peak pressure decreases significantly as compared to the twin screw compressor (as shown in table 5). This result does not hold for the 4 bar case due to the noise present in the simulations.

4.2. Frequency Response
The purpose of this section is to look at the frequency response of the IGSM and twin screw in response to the surface velocity of the structure. The main purpose of frequency response is to regard the behaviour of the structure while it is under harmonic loading, which allows one to ascertain the dominant frequency of the system.

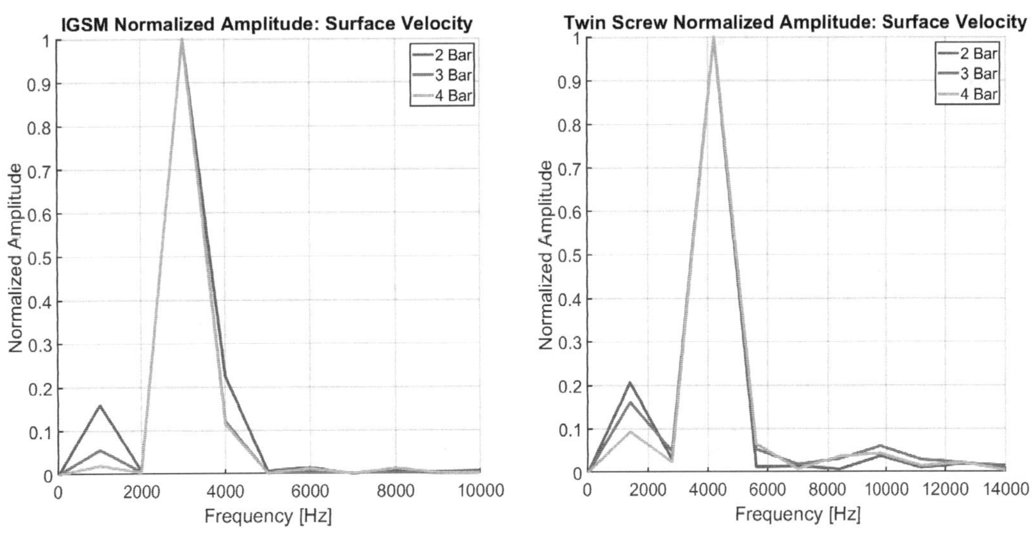

(a) IGSM frequency response due to surface

(b) Twin Screw frequency response due to surface

Figure 6. Frequency response due to surface velocity for IGSM and Twin Screw at the first 10 harmonics

Figure 6 shows the frequency response of the first 10 harmonics for both the IGSM and twin screw compressor due to surface velocity. Since both compressors are running at 200 Hz and the male lobe count for the IGSM and twin screw compressor are 5 and 7 respectively, the fundamental frequency is 1000 Hz and 1400 Hz respectively. Looking at the result of figure 6, it can be seen that the dominant frequency reside at the 3rd harmonic and the 2nd dominant frequency is the 1st harmonic for both the IGSM and twin screw compressors. It should also be noted that there is very little influence at the 2nd harmonic as well as every harmonic past the 3rd harmonic. As for the 1st and 4th harmonic, both systems show that the same trend that increasing the pressure ratio decreases the influence of the 1st harmonic. In regards to the harmonic response of the system, it can be said that the IGSM behaves very similar to the twin screw compressor.

4.3. Surface Velocity

The purpose of this section is to examine the surface velocity of both the IGSM and twin screw compressor in response to the pulsating discharge pressure. Normally, one would extend this analysis to the connecting pipe as well to examine weakness in the piping design, but since a full discharge design is not yet available for the IGSM, this analysis is omitted for this study. It should be mentioned that the surface velocity of a structure is heavily dependent on both the discharge port design and the structure of the compressor. Since the IGSM here is still in the initial phase of design while the twin screw is a fully designed compressor, it is therefore difficult to make a direct comparison between the two compressors. However, what can be examined is to show that the magnitude of the surface velocity is near to that of the twin screw so that there is confidence a full design will be similar enough to the twin screw and no key design features need to be implemented.

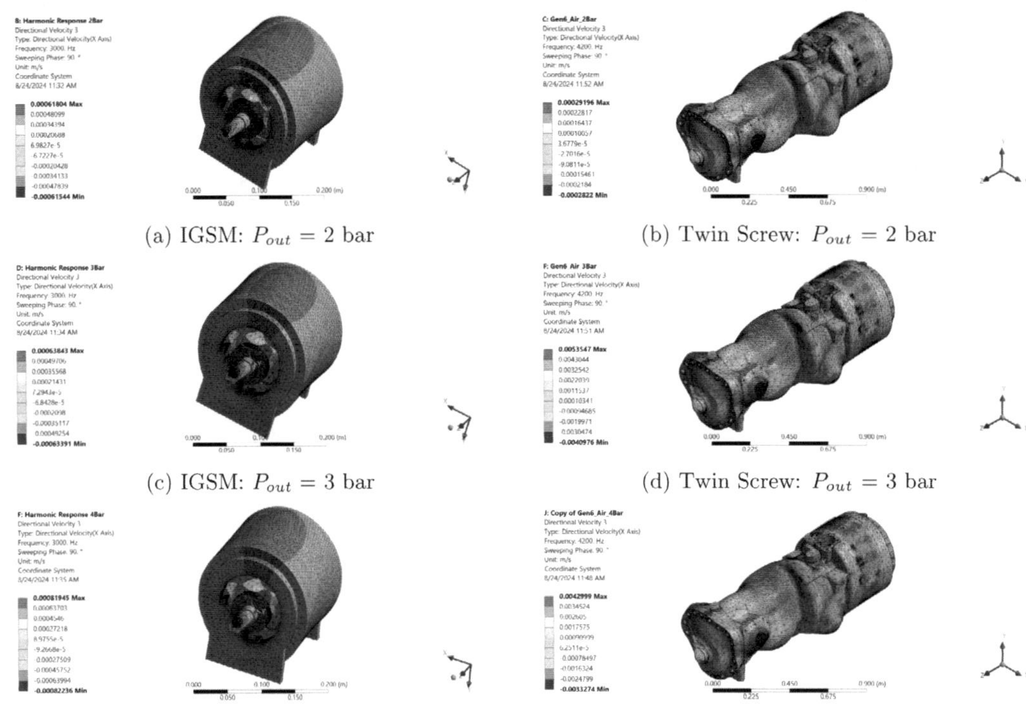

(a) IGSM: $P_{out} = 2$ bar (b) Twin Screw: $P_{out} = 2$ bar

(c) IGSM: $P_{out} = 3$ bar (d) Twin Screw: $P_{out} = 3$ bar

(e) IGSM: $P_{out} = 4$ bar (f) Twin Screw: $P_{out} = 4$ bar

Figure 7. Surface velocity for IGSM and twin screw compressor at 2, 3 and 4 bar outlet pressure.

Figure 7 show the directional surface velocity of the IGSM and twin screw at the 3rd harmonic. As previously shown, the 3rd harmonic is the dominant frequency and the direction of velocity is shown is radial as that direction is best indicator of surface vibration of the system. One interesting feature about the IGSM is that there seems to be a region of low surface velocity around the top of the compressor that travels down in regions increasing in surface velocity as it reaches the fixed support at the bottom. This effect may be due to the symmetrical nature of the under designed structure as this does not occur in the twin screw machine.
Table 6 and 7 shows the mean surface velocity at the first 5 harmonics for the IGSM and twin screw compressor respectively. Recall that it was shown the largest amplitude shown in figure 6 occurs at the

Table 6. IGSM mean surface velocity for 2, 3, and 4 outlet pressure at the first 5 harmonics (unit: m/s).

	F_1	F_2	F_3	F_4	F_5
$P_{out} = 2$ bar	1.26E-06	3.04E-07	2.51E-05	1.40E-05	3.56E-07
$P_{out} = 3$ bar	1.04E-06	1.66E-07	4.15E-05	2.27E-06	3.84E-08
$P_{out} = 4$ bar	3.05E-06	5.11E-08	5.40E-05	1.36E-05	2.35E-07

Table 7. Twin Screw mean surface velocity for 2, 3, and 4 outlet pressure at the first 5 harmonics (unit: m/s).

	F_1	F_2	F_3	F_4	F_5
$P_{out} = 2$ bar	1.68E-05	2.86E-06	6.11E-06	4.90E-06	5.76E-07
$P_{out} = 3$ bar	2.47E-05	2.68E-06	1.47E-04	1.05E-06	1.45E-06
$P_{out} = 4$ bar	1.16E-05	6.30E-07	1.14E-04	8.41E-06	1.84E-07

3rd harmonic for both the IGSM and twin screw compressor and that result is confirmed here as the largest velocity magnitudes fall at that harmonic. Another characteristics that can be seen is that the magnitude of the surface velocity is shown to be in the neighborhood of the twin screw across discharge pressures. Although not a perfect trend, it can also be seen on average that the surface velocity is lower for the IGSM. This result is to be expected as a fully designed discharge port will have more area affected by the pulsating flow thus resulting the more force applied to the structure.

5. Future Work

A study of the pressure pulsation and vibration characteristics were conducted for the novel IGSM. There are several key takeaways regarding this study including the existence of the micro oscillations that occur for the IGSM but not for the twin screw. There are also several similarities in that both machines is dominated by the 3rd harmonic of the system and have similar surface velocity values. As mentioned previously, a full port and case design has not been actualized yet, so one aspect of the future work is to reconfirm the results for a fully designed IGSM. Other future work include better understanding the micro oscillations that occur in the IGSM pressure pulsations and the effects that oil injection will play in the vibrations of the system.

References

[1] Akei M, Sishtla V and MacBain S 2018 *Mini Screw: the development of a high-CFM compact compressor for LGWP A1 low pressure refrigerant* International Conference on Screw Machines

[2] Read M, Smith I K and Stosic 2017 N *Internally geared screw machines with ported end plates* IOP Conference Series: Materials Science and Engineering Vol. 232

[3] Lacevic H, Kovacevic A and Read 2023 M *An Investigation of Internally Geared Screw Compressor Performance Using a Chamber Modelling Approach* 13th International Conference on Compressors and Their Systems

[4] Rane S, Kovačević A and Read M 2023 *Development of Numerical Grid and CFD Model for Analysis of Oil-Injected IGSM* 13th International Conference on Compressors and Their Systems pp. 503–514

[5] Stosic N, Smith I K and Kovacevic A 2003 *Optimisation of screw compressors* Applied Thermal Engineering 23 1177 - 1195

[6] Stosic N, Smith I and Kovacevic A 2005 *Screw compressors: mathematical modelling and performance calculation* Springer Science & Business Media

IOP Conf. Series: Materials Science and Engineering **1322** (2024) 012003 doi:10.1088/1757-899X/1322/1/012003

Influence of fluid properties and model parameters with regard to stratified gas-liquid gap flows

L Burchardt[1], J Vorspohl[2], F Sabozin[3], M Meinke[2], W Schröder[2] and A Brümmer[1]

[1]Chair of Fluidics, TU Dortmund University, Dortmund, Germany
[2]Institute of Aerodynamics, RWTH Aachen University, Aachen, Germany
[3]Chair of Thermodynamics, Ruhr University Bochum, Bochum, Germany

E-mail: lasse.burchardt@tu-dortmund.de

Abstract. In rotary positive displacement machines - like twin-screw compressors - clearance flows have a big impact on the quality of the energy conversion. To partially seal the gap connections and thereby reduce internal losses, an auxiliary fluid, e.g. oil, is often injected into these machines. If a chamber model method is used to simulate the thermodynamic operating behavior of such a wet-running twin-screw machine, two-phase mass flow rates through the gaps must be calculated as accurately as possible in order to achieve a high simulation accuracy. This paper presents an extended model for simulating such two-phase gap flows. The model allows the simulation of a gap flow, taking into account the possible outgassing of a gas (e.g. refrigerant) from a liquid phase (e.g. oil) along the gap. The model is used to analyse the influence of various fluid properties (e.g. solubility, density, viscosity) or boundary conditions (e.g. pressure ratio, quality at the inlet of the gap) in relation to the simulated gap mass flow rate.

1. Introduction

Due to several advantages, including their simple and robust design, and high efficiency across a broad spectrum of operating conditions, twin-screw machines find applications in various industrial sectors. They are used, for example, to generate compressed air, in process gas technology, or in HVAC (Heating, Ventilation and Air Conditioning) systems to compress refrigerants. Liquid-injected twin-screw machines in particular are widely used, as they offer several benefits compared to dry-running machines. When using oil as the auxiliary fluid, the good lubrication properties of the oil eliminate the need for a synchronisation gearbox between the rotors. The partial sealing of gaps between moving parts in the machine is another use of the injected liquid, which reduces internal losses [1].

The resulting two-phase gap flows must be taken into account in chamber model simulations calculating the thermodynamic performance of a wet-running twin-screw machine. In this method, the geometry of the machine is subdivided into individual chambers where mass and energy conservation must hold for each chamber and phase. Mass and energy exchange between the chambers through connections, which in reality occur through gaps, among other pathways, must be considered in simulations using models that calculate two-phase mass flow rates [2].

Models for calculating two-phase mass flow rates can be classified into different groups based on the effects they consider. Although most of them take into account the compressibility of the gaseous phase, some models also have the characteristic of an incompressible flow. An example of this can be found in the study by Bell et al., which is purely pressure-driven [3]. Another subdivision can be made based on the velocity ratio of the phases in the flow. In homogeneous two-phase flow models, the assumption of maximum friction between the phases is made. The velocities of the phases are equal, so the slip S, defined as the ratio of the mean velocity of the gas phase to the mean velocity of the liquid phase, is $S = 1$. A two-phase nature of the flow can be accounted for in various ways. For instance, Nikolov and Brümmer use smaller flow coefficients, which describe the ratio of the actual to the theoretical mass flow rate, compared to single-phase gas flows in dry-running machines [4]. A further compressible, homoge-neous flow model was developed by Gasche et al. specifically for an oil-R134a refrigerant mixture. This non-isothermal model additionally accounts for the degassing of the refrigerant due to the decreasing pressure across the gap [5].

In two-phase flow models that account for a velocity difference between the phases ($S \neq 1$), there are various approaches to calculating the slip. Morris [6], as well as Nikolov and Brümmer [7], derive models based on the assumption of annular flow with entrainment, employing Chisholm's method [8] to compute the slip. Wang et al. pursue a different approach, considering in their incompressible model of stratified flow both a moving gap boundary and shear stresses at the interface between the phases [9]. To calculate the slip, the model by Lin is used [10].

The listed approaches demonstrate that there are many different models for calculating two-phase gap mass flow rates, each possessing distinct characteristics and accounting for various effects of two-phase flows. As part of this study, a further model for calculating a two-phase gap mass flow is derived, which should take into account as many of the aforementioned properties as possible: compressible gas-phase flow, moving gap boundaries, separated flow, and degassing. Different fluid properties and boundary conditions will be assessed to estimate the factors most influencing the calculated two-phase gap mass flow rates.

2. Theoretical Approach

The flow model is to be derived for a stationary, laminar, pressure-driven two-phase stratified flow through a plane-parallel gap with moving a boundary (Couette-Poiseuille flow). This geometric characteristic (see figure 1) is approximately present in the screw machine, as described in [9], in the housing gap between the rotor and the housing. The starting point for the equations of the model is the integral momentum conservation per phase. This simplifies to the following when neglecting the convective terms and the specific weight force [11]:

$$\frac{\partial p}{\partial z} = \eta \cdot \left(\frac{\partial^2 c_z(y)}{\partial y^2} \right) \tag{1}$$

Additionally, the pressure is constant across a cross-section of the gap, so the following applies:

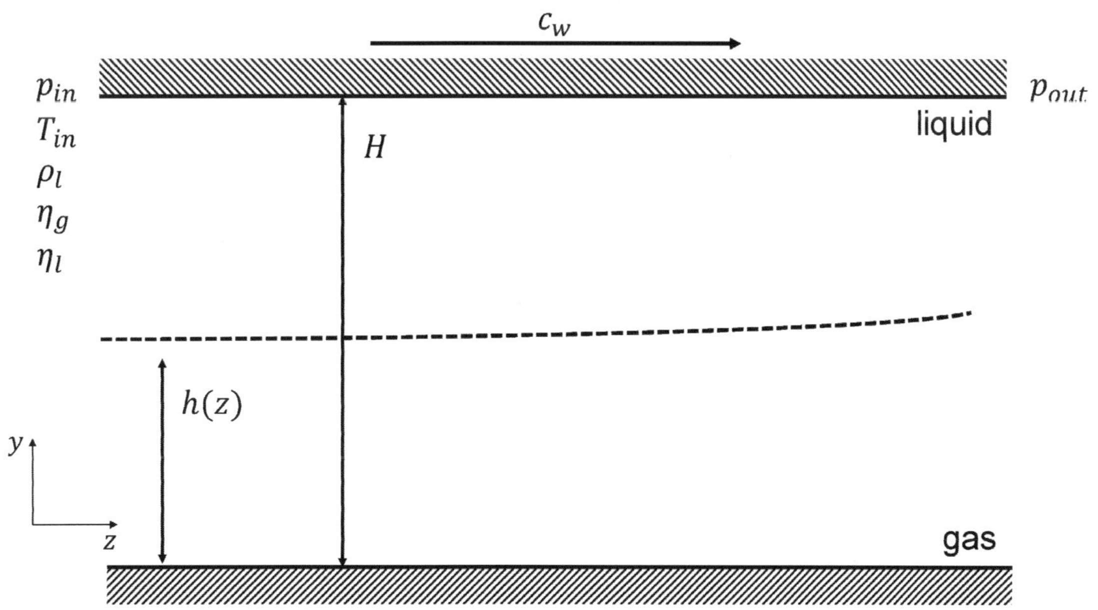

Figure 1. Stratified two-phase flow through plane-parallel gap with moving boundary

$$\frac{\partial p}{\partial y} = 0 \tag{2}$$

To obtain the velocity profile, (1) is integrated twice with respect to y and then solved for $c_z(y)$:

$$c_z(y) = \frac{\partial p}{\partial z} \cdot \frac{y^2}{2\eta} + K_1 \cdot y + K_2 \tag{3}$$

This equation applies to both phases of the stratified flow. The integration constants K_i must be determined using boundary and transition conditions for each phase. Due to the centrifugal forces occurring in a rotating system, it is assumed that the liquid is located in the upper part of the gap. Therefore, the condition for the velocity of the gas phase $c_{z,g}$ at the lower wall (rotor surface) applies:

$$c_{z,g}(y = 0) = 0 \tag{4}$$

For the velocity of the liquid $c_{z,l}$ at the moving gap boundary (housing in relative system of the rotor), the following holds:

$$c_{z,l}(y = H) = c_w \tag{5}$$

Here, c_w represents the relative wall velocity. At the interface between the faces, two transition conditions apply. Firstly, the velocities of the respective phases at this boundary must be equal:

$$c_{z,g}(y = h) = c_{z,l}(y = h) \tag{6}$$

Secondly, the shear stresses τ_{yz} calculated through the relationship

$$\tau_{yz}(y) = \eta \cdot \frac{\partial c_z(y)}{\partial y} \tag{7}$$

must also be equal at the phase boundary:

$$\tau_{yz,g}(y = h) = \tau_{yz,l}(y = h) \tag{8}$$

With these four conditions, the integration constants for calculating the velocity profiles of each phase can now be determined. They are given by:

$$K_{1,g} = \frac{\eta_l}{\eta_g} \cdot K_{1,l} \tag{9}$$

$$K_{2,g} = 0 \tag{10}$$

$$K_{1,l} = \frac{\frac{1}{2} \cdot \frac{\partial p}{\partial z} \cdot \left[\frac{1}{\eta_l} \cdot (h^2 - H^2) - \frac{1}{\eta_g} \cdot h^2\right] + c_w}{\frac{\eta_l}{\eta_g} \cdot h + H - h} \tag{11}$$

$$K_{2,l} = c_w - \frac{\partial p}{\partial z} \cdot \frac{H^2}{2\eta_l} - K_{1,l} \cdot H \tag{12}$$

The mass flow rates $\dot{m}_i$ can now be calculated by integrating the phase velocities over their respective height, and multiplying by their density ρ_i and the width of the gap w:

$$\dot{m}_g = \rho_g \cdot w \cdot \left[\frac{\partial p}{\partial z} \cdot \frac{1}{6\eta_g} \cdot h^3 + \frac{1}{2} \cdot K_{1,g} \cdot h^2\right] \tag{13}$$

$$\dot{m}_l = \rho_l \cdot w \cdot \left[\frac{\partial p}{\partial z} \cdot \frac{1}{6\eta_l} \cdot \left(H^3 - h^3\right) + \frac{1}{2} \cdot K_{1,l} \cdot \left(H^2 - h^2\right) + K_{2,l} \cdot (H - h)\right] \tag{14}$$

IOP Conf. Series: Materials Science and Engineering **1322** (2024) 012003 doi:10.1088/1757-899X/1322/1/012003

3. Calculation algorithm

The solution of this non-linear system of equations ((13) and (14)) is carried out using an algorithm, which will be described in detail below, with the possibility of degassing not yet being considered.

Fundamental to the solution is the discretisation of the gap in the z-direction. For a partial step, the assumption of constant fluid properties and conditions applies. At the inlet of the gap, the viscosities η_l, η_g and the densities ρ_l, ρ_g are initially set. Additional the pressure p_{in} and temperature T_{in} are predefined. In each cross-section, the same pressure and temperature are assumed for both phases. The density of the liquid is regarded as constant, while the density of the compressible gaseous phase can be calculated using the ideal gas equation:

$$p = \rho_g \cdot R_s \cdot T \tag{15}$$

At the outlet of the gap, the outlet pressure p_{out} is also specified. This serves as a control in the algorithm to ensure that the mass flow rates determined match the boundary conditions. Since degassing is not initially considered, the mass flow of each phase across the gap, and thus at every local step, must remain constant to satisfy conservation of mass. This condition is used to rearrange the equations (13) and (14) so that only zeros appear on the left side, allowing the resulting system of equations to be solved using a multidimensional Newton-Raphson method [12] at each local step. The unknowns are the pressure gradient $\partial p/\partial z$ and the height of the gas phase h, which are initially specified at the inlet. The state of both phases is re-determined at each local step. The procedure is repeated until all local steps are completed and the outlet of the gap is reached. A comparison of the calculated and specified pressure difference across the gap is used to verify that the boundary conditions are met. If they are not, the initial pressure gradient at the inlet of the gap is systematically adjusted using a bisection method [12]. This process continues until the calculated pressure difference across the gap deviates by less than 0.1% from the specified value, ensuring that the calculated mass flow rates match the specified boundary conditions.

Various assumptions can be made for the state changes of the phases resulting from the pressure drop along the gap. The relation between pressure and temperature change is determined by the polytropic correlation:

$$\frac{T_2}{T_1} = \left(\frac{p_2}{p_1}\right)^{\frac{n-1}{n}} \tag{16}$$

For assuming an isothermal change in state, the polytropic exponent is $n = 1$. In this case, the change in gas density arises solely from the pressure variation according to (15). With a polytropic exponent $n \neq 1$, changes in temperature of the gas and liquid can also be considered. This not only affects the density of the gas - the density of the liquid is still regarded as constant - but also the dynamic viscosity of each fluid. This transport property is much more dependent on temperature than on pressure, so the influence of the latter is often neglected. The dynamic viscosity of liquids can be estimated in the temperature range from 0 to 100°C using the empirical Andrade approach [13]:

$$\eta_l(T) \approx \eta_{l,0} \cdot \exp\left(\frac{T_A}{T + T_B} - \frac{T_A}{T_B + T_0}\right) \tag{17}$$

Here, T_A and T_B are fluid-dependent constants, and $\eta_{l,0}$ is the reference value of dynamic viscosity at the reference temperature $T_0 = 273.15$ K. The relationship indicates that the transport property of liquid decreases with increasing temperature. With gases, the correlation is reversed. Sutherland's semi-empirical formula [13]

$$\eta_g(T) \approx \eta_{g,0} \cdot \frac{T_0 + T_S}{T + T_S} \cdot \left(\frac{T}{T_0}\right)^{3/2} \tag{18}$$

which also neglects the influence of pressure, shows an increase in dynamic viscosity of the gas with increasing temperature. The Sutherland-constant T_S is a characteristic for the respective gas.

Another effect of two-phase flow that this model aims to incorporate is degassing from a liquid due to a pressure drop and the associated decrease in the solubility of a gas in the liquid. The objective is to calculate the mass flow rate of degassed gas that transitions to the gaseous mass flow rate within a local step. For this purpose, an approach based on the mole fraction of the dissolved gas in the liquid x_g, which depends on pressure and temperature, is utilized. This is the ratio of the molar amount of gas present in the liquid N_g to the total amount of substance of the pure liquid $N_{l,t}$:

$$x_g = \frac{N_g}{N_{l,t}} \tag{19}$$

The amount of gas that degasses within a local step ΔN_g can thus be determined by the difference in the mole fraction of the gas between two spatial points, due to their different states, and the total amount of substance of the liquid:

$$\Delta N_g = N_{l,t} \cdot (x_g(p_1, T_1) - x_g(p_2, T_2)) \tag{20}$$

The latter is determined by the volume increment of the local step dV, which is approximately the product of the length of the local step dz, the width of the gap w, and the height of the liquid phase at the starting point of the local step, and by the ratio of the liquid density ρ_l to the molar mass of the liquid M_l:

$$N_{l,t} = dV \cdot \frac{\rho_l}{M_l} \approx w \cdot dz \cdot (H_1 - h_1) \cdot \frac{\rho_l}{M_l} \tag{21}$$

To calculate the change in the gas mass flow rate due to the degassed amount of gas $\Delta\dot{m}_g$, the time required for the liquid flow to cover the local step is computed based on the velocity of the liquid $c_{liq,m}$, averaged over the height of the phase at the starting point of the local step, and the length dz. When multiplied by the molar mass of the gas M_g, this finally gives the change in the gas mass flow rate due to degassing in kg s^{-1}:

$$\Delta\dot{m}_g = \frac{\Delta N_g \cdot M_g \cdot c_{liq,m}}{dz} \tag{22}$$

4. Results

The calculations of the mass flow rates are performed under different boundary conditions and various assumptions, as well as for different fluid pairs, allowing the influences of individual effects and parameters to be assessed. The constant geometric and physical boundary conditions are listed in table 1. Table 2 and table 3 present the fluid properties of the liquids and gases used. The properties of the gases are aligned with dry air, nitrogen (N_2), or carbon dioxide (CO_2). For the liquids, water (H_2O), decane ($C_{10}H_{22}$), or squalane ($C_{30}H_{62}$) serve as references. The fluids used are oriented towards those whose combinations are planned to be experimentally investigated within the Research Unit "Archimedes" (FOR5595) in the future. To vary the boundary conditions, various dimensionless numbers are used. For instance, the pressure ratio

$$\Pi = \frac{p_{out}}{p_{in}} \tag{23}$$

describes the relationship between the inlet and outlet pressure of the gap. The gas volume fraction at the entry of the gap is characterised by the dimensionless number ϵ, which relates the height of the gas phase at the inlet h_{in} to the total height of the gap H:

$$\epsilon = \frac{h_{in}}{H} \ . \tag{24}$$

The circumferential Mach number Ma_U associates the wall velocity c_w with the speed of sound of the gas a_g:

$$Ma_U = \frac{c_w}{a_g} \ . \tag{25}$$

To represent and interpret the results, the Reynolds number Re is used, which relates the mass flow rate $\dot{m}$ per width w to the dynamic viscosity η:

$$Re = \frac{\dot{m}}{\eta \cdot w} \ . \tag{26}$$

Table 1: Constant boundary conditions

Designation	Symbol	Value
gap height	H	0.0001 m
gap length	l	0.02 m
number of local steps		100
temperature (inlet)	T_{in}	323.15 K
pressure (outlet)	p_{out}	$1 \cdot 10^5$ Pa

Table 2: Fluid properties (liquid)
(T = 323.15 K; p = $1 \cdot 10^5$ Pa)

Fluid	Property	Value
water	ρ_l	1000 kg m^{-3}
	η_l	1 mPa s
	$a_{l,in}$	1480 m s^{-1}
decane	ρ_l	710 kg m^{-3}
	η_l	0.6 mPa s
	$a_{l,in}$	1136 m s^{-1}
squalane	ρ_l	790 kg m^{-3}
	η_l	2 mPa s
	$a_{l,in}$	1300 m s^{-1}

Table 3: Fluid properties (gas)
(T = 323.15 K; p = $1 \cdot 10^5$ Pa)

Fluid	Property	Value
dry air (ideal gas)	η_g	0.002 mPa s
	R_s	287 J kg^{-1} K^{-1}
	$a_{g,in}$	360.6 m s^{-1}
nitrogen	η_g	0.0019 mPa s
	R_s	296.8 J kg^{-1} K^{-1}
	$a_{g,in}$	366.5 m s^{-1}
carbon dioxide	η_g	0.0016 mPa s
	R_s	188.92 J kg^{-1} K^{-1}
	$a_{g,in}$	278.1 m s^{-1}

4.1. Variation of fluid pairs

The calculation of gaseous mass flow rates per gap width for the different fluid pairs is initially conducted under the assumption of an isothermal flow ($n = 1$). The viscosity of the gas and the liquid remains constant, while the density of the compressible phase depends solely on the pressure.

To initially investigate the influence of the viscosities of the fluids on the flow behavior, the left diagram in figure 2 illustrates the volumetric flow rate of the gas $\dot{V}_g$ per gap width w as a function of the pressure ratio Π. The wall velocity $c_w = 20$ m s^{-1} is constant, as is the gas volume fraction $\epsilon = 0.3$ at the inlet of the gap. When comparing the trends, which rise parabolically with the pressure ratio for all fluid combinations, it becomes evident that the flow velocity and hence the volumetric flow rate increase with decreasing viscosity of the fluids. In good approximation, the greater the product of the viscosities of the gas and liquids, the smaller the volumetric flow rate of the compressible phase per gap width. Therefore, the combination of decane and carbon dioxide exhibits the highest value, while the combination of long-chain squalane with dry air shows the lowest rate.

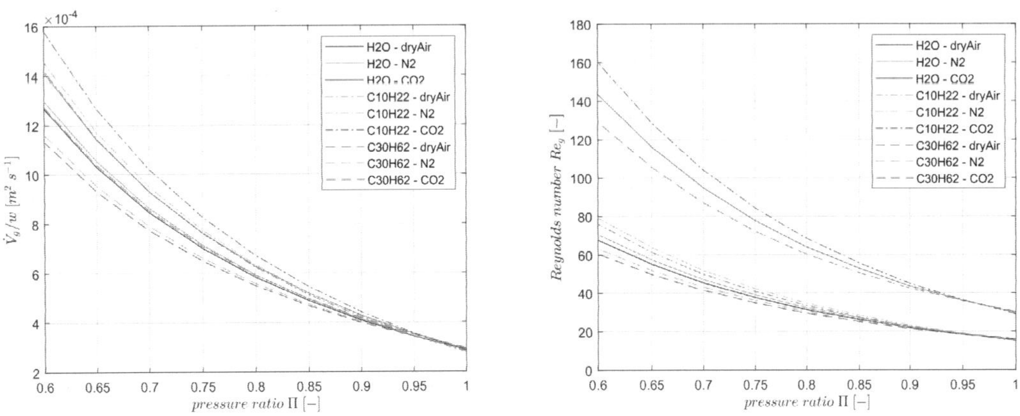

Figure 2. Volume flow of the gas $\dot{V}_g$ per gap width w (left) and Reynolds number of the gas Re_g (right) as a function of the pressure ratio Π ($c_w = 20$ m s^{-1}; $\epsilon = 0.3$; $n = 1$)

If the gaseous Reynolds numbers Re_g are considered, the density of the gas is another influencing parameter. In the right diagram in figure 2, these are also depicted as a function of the pressure ratio with the same boundary conditions ($c_w = 20$ m s^{-1} and $\epsilon = 0.3$) and assumptions ($n = 1$). Even in a pure Couette-flow on the right side of the diagram ($\Pi = 1$), the influence of the gas density is clearly

IOP Conf. Series: Materials Science and Engineering **1322** (2024) 012003 doi:10.1088/1757-899X/1322/1/012003

evident. Due to the smaller specific gas constant R_S, the density of carbon dioxide at the same pressure and temperature is over 50% higher than that of nitrogen and dry air, which have very similar properties (see table 3). Therefore, the gas mass flow rates of carbon dioxide, and consequently, the Reynolds numbers of the gas are greater than those of the combinations with the other gases, regardless of the liquid. For all fluid pairs, a parabolic trend is again observed as the pressure ratio decreases. The fact that the two-phase flows with squalane as the liquid display the lowest Reynolds numbers further emphasizes the damping effect of higher viscosity.

For the fluids under consideration, the Reynolds number of the gas is predominantly influenced by the gaseous mass flow rates. Consequently, in this work, the ratio of Reynolds numbers accurately reflects the ratio of mass flow rates. This is due to the relatively small differences in the dynamic viscosities of the gases. However, for other fluid combinations with greater differences in dynamic viscosity, this relationship may differ and should be considered in detail.

In addition to the prevailing pressure ratio across the gap, the following sections examine the effects of other boundary conditions on the gap mass flow rate. On the left side of figure 3, the gaseous Reynolds numbers are shown as a function of circumferential Mach number for a constant pressure ratio $\Pi = 0.7$ and a constant gas volume fraction $\epsilon = 0.3$ at the inlet of the gap for the various fluid combinations. In all trends, a linear increase in the Reynolds number with increasing circumferential Mach number can be observed. However, the rate of increase with circumferential Mach number clearly depends on the density of the gas. Thus, the Reynolds number of the heavier carbon dioxide rises more steeply than that of dry air or nitrogen. With respect to fluid viscosity, it is evident that a lower liquid viscosity leads to higher mass flow rates, which in turn result in greater Reynolds numbers for the compressible phase. A comparison of the curves of the fluid combinations with dry air and nitrogen shows a slightly steeper increase for the latter gas due to the lower assumed viscosity. However, this influence appears to be marginal.

Another variation was made regarding the gas volume fraction at the gap inlet and thus the quality, which is the mass ratio of the compressible to the incompressible phase. On the right side of figure 3, the gaseous Reynolds number for the different fluid combinations are shown as a function of the volume fraction. The pressure ratio $\Pi = 0.7$ and the wall velocity $c_w = 20$ m s^{-1} are kept constant. Fundamentally, all graphs show a parabolic increase with the rising gas volume fraction at the gap inlet. This increase is again steeper for the two-phase flows with carbon dioxide as the compressible phase compared to the other gases. Additionally, the results illustrate the conclusion that a higher viscosity of the liquid leads to smaller gaseous Reynolds numbers.

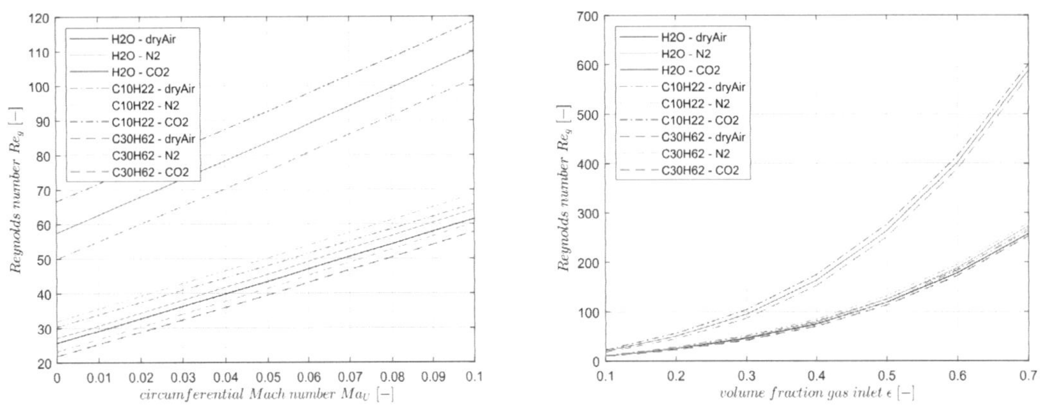

Figure 3. Reynolds number of the gas Re_g as a function of the circumferential Mach number Ma_U (left)($\Pi = 0.7$; $\epsilon = 0.3$; $n = 1$) and the volume fraction of gas at inlet ϵ (right) ($c_w = 20$ m s^{-1}; $\Pi = 0.7$; $n = 1$)

4.2. Non-isothermal flow

So far, the calculations of the gas mass flow rates in the gap have been based on the assumption of isothermal flow ($n = 1$). The following section examines the effects of considering a temperature change due to the pressure drop along the gap. For this purpose, according to equation (16), the temperature at each local step is recalculated as a function of pressure for different assumed polytropic exponents n. As mentioned above it is assumed, that the temperature of gas and liquid is the same in each cross section. Therefore a change of temperature along the gap affects both the density of the compressible phase (see equation (15)) and the viscosity of the gas and the liquid (see equations (13) and (14)). While the density of the gas increases with decreasing temperature, the viscosity of the gas decreases and the viscosity of the liquid increases. To evaluate the consideration of temperature changes, the dimensionless mass flow coefficient

$$\beta = \frac{\dot{m}_{g,n\neq 1}}{\dot{m}_{g,n=1}} \tag{27}$$

is used. This coefficient relates the gas mass flow rate at an assumed polytropic exponent $n \neq 1$ to that of an isothermally assumed flow ($n = 1$). Figure 4 shows the mass flow coefficient for flows of water (H_2O) and nitrogen (N_2) and water and carbon dioxide (CO_2) for different polytropic exponents as a function of the pressure ratio. The constants used for the temperature-dependent calculation of the viscosity of the fluids are also listed in table 4 [13]. The trends indicate an initial increase to above one for all fluid combinations and polytropic exponents in the range of high pressure ratios, meaning small pressure differences across the gap. In this area, the increase in the density of the compressible phase and the decrease in the viscosity of the gas with decreasing temperature outweigh the increase in the viscosity of the liquid, resulting in a larger mass flow rate compared to the isothermal case. However, starting from a pressure ratio of approximately 0.9, the weighting changes. As the pressure ratio increases, the temperature drop across the gap becomes greater for the same inlet temperature. The associated increase in the viscosity of the liquid outweighs the increase in the gas density and the decrease in the gas viscosity. As a result, the mass flow coefficient becomes $\beta < 1$ and continue to decrease. This effect is amplified by assuming a higher polytropic exponent due to a lower temperature at the outlet. Thus, the gas mass flow rate decreases by more than 10% for both fluid combinations at a pressure ratio $\Pi = 0.6$ with a polytropic exponent $n = 1.3$.

Table 4: Constants for calculating viscosity

Fluid	Property	Value
water	$\eta_{l,0} \cdot 10^6$	1792 Pa s
	T_A	506 K
	T_B	-150 K
nitrogen	$\eta_{g,0} \cdot 10^6$	16.65 Pa s
	T_S	105 K
carbon dioxide	$\eta_{g,0} \cdot 10^6$	13.74 Pa s
	T_S	253 K

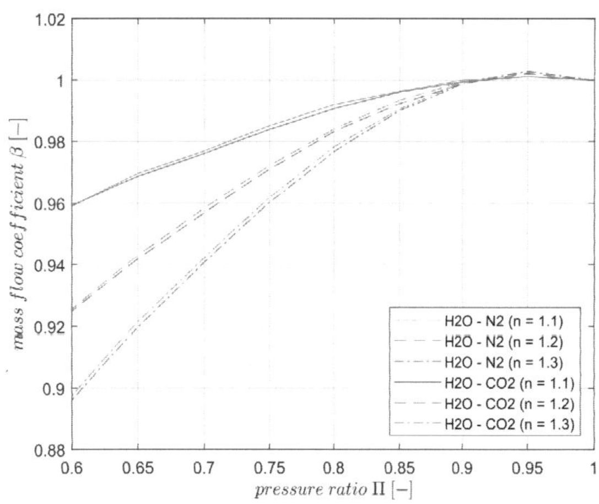

Figure 4. Mass flow coefficient β as a function of the pressure ratio Π for different polytropic exponents n ($c_w = 20$ m s^{-1}; $\epsilon = 0.3$)

4.3. Consideration of degassing

Another physical effect that should be considered for its impact on the calculated mass flow rates of the stratified two-phase flow is the degassing of gas from the liquid due to the decreasing pressure along the gap and the associated decrease in the solubility of the gas in the liquid. According to the calculation procedure outlined in section 3, it is necessary to calculate the molar fraction of the gas in the liquid at different local points, depending on the state at these points. Since again an isothermal flow is assumed in the following, the molar fraction is therefore only dependent on the pressure. The calculation is performed using the software tool TREND [14], developed at the Ruhr University Bochum. In TREND, it is assumed that the gas phase and the liquid phase are in equilibrium. The molar masses of the fluids used for the two-phase flows with degassing are listed in table 5.

To evaluate the impact of considering degassing, the mass flow coefficient δ is introduced. This sets the gas mass flow $\dot{m}_{dg}$, which is calculated on the assumption that degassing takes place, in relation to the gaseous portion of the mass flow $\dot{m}_g$ that results without taking degassing into account:

$$\delta = \frac{\dot{m}_{dg}}{\dot{m}_g} \tag{28}$$

Figure 5 shows the mass flow coefficient δ for a two-phase flow of water (H_2O) and nitrogen (N_2) as a function of the pressure ratio for different gas volume fractions at the inlet of the gap. The equations of state used in TREND for this mixture are listed in [15]. For a pressure ratio $\Pi = 1$, $\delta = 1$ for any gas volume fraction at the inlet. This is logical because, in a pure Couette-flow, there is no pressure drop along the gap and thus no change in solubility. For smaller pressure ratios, i.e. larger pressure differences across the gap, the gaseous mass flow rate increases when degassing is taken into account for all considered boundary conditions. This applies in particular the smaller the gas volume fraction of the two-phase flow at the gap inlet. For example, at $\epsilon = 0.1$ and a pressure ratio $\Pi = 0.6$, the calculated gas mass flow rate with degassing is more than twice as large as that without degassing. This is, due to the small absolute gas mass flow rate at low gas volume fractions at the inlet (see figure 3) and the consequently larger relative impact of the degassed mass flow rate.

Another significant factor influencing how much the consideration of degassing affects the calculated gas mass flow rate is the general solubility of the gas in the liquid. An example of this is the combination of carbon dioxide and water compared to nitrogen and water, whereby the former gas has a much higher solubility in water. This, in turn, results in a higher degassed mass flow rate for the same pressure drop. For instance, at a pressure ratio $\Pi = 0.9$ and $\epsilon = 0.7$, in a two-phase flow of water and carbon dioxide, the gas mass flow rate increases by nearly 29%, due to the consideration of degassing, whereas the mass flow rate of nitrogen, as a compressible phase, only increases by 0.7%.

Table 5: Molar masses

Fluid	**Property**	**Value**
water	M_l	18.015 g mol^{-1}
nitrogen	M_g	28.014 g mol^{-1}
carbon dioxide	M_g	44.01 g mol^{-1}

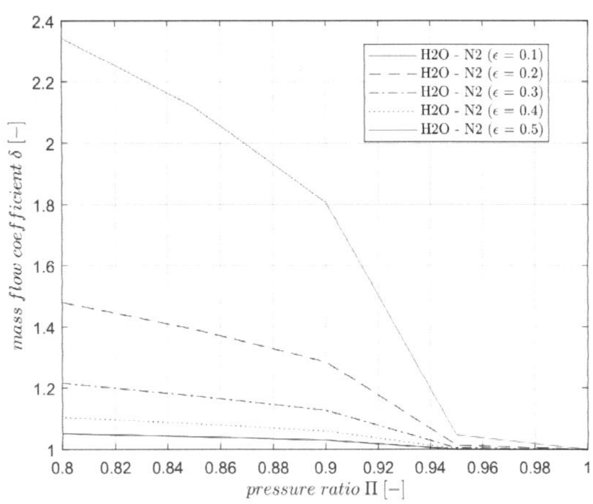

Figure 5. Mass flow coefficient δ as a function of the pressure ratio Π for different gas volume fractions ϵ at the inlet of the gap ($c_w = 20$ m s^{-1}; $n = 1$)

4.4. Evaluation of the assumptions

After assessing the influence of various boundary conditions and fluid properties on the calculated gaseous Reynolds numbers and the two-phase mass flow rates for the developed model, a conclusive evaluation of the assumptions made is conducted based on the results.

One parameter for evaluation is the Reynolds number Re, which can be used to classify single-phase flows into laminar and turbulent flows. To evaluate the assumption of two-phase, laminar stratified flow, the Reynolds number for the liquid and gaseous phase is calculated separately in the computations.

Under the given boundary conditions, the Reynolds numbers for the gas flows of all fluid pairs are below 2300, placing them within the laminar flow regime. It is observed that the values increase with higher wall velocity and a larger volume fraction of gas at the gap inlet. However, the maximum value reaches approximately 1000 in a flow of decane and carbon dioxide ($c_w = 40$ m s^{-1}, $\Pi = 0.6$, $\epsilon = 0.7$). In this combination, the gas mass flow rate is the highest and the dynamic viscosity of the compressible fluid is the lowest. In contrast, the situation is different for incompressible liquid flow. Although the Reynolds numbers for the majority of fluid combinations and boundary conditions also fall within the laminar range, they transition to the turbulent flow regime at low gas volume fractions at the inlet of the gap and high wall velocities. This is particularly true for flows with decane as the liquid phase due to its low dynamic viscosity. In such cases, the Reynolds numbers sometimes range between 4000 and 5000. Additionally, it should be noted that in recent studies of liquid annular seals, the transition region from laminar to turbulent flow occurs at Reynolds numbers ranging from 1000 to 3000. [16]

The assumption of neglecting convective terms in the momentum conservation can also be evaluated using the Reynolds number. This is valid if the product of the Reynolds number and the flow inclination angle is much less than 1. This condition is not met under certain boundary conditions at high Reynolds numbers, particularly for the liquid phase of the flow, but also for the gaseous phase. Alternative modeling approaches that do not neglect convective terms can be found in, among others, [17] and [18].

A further evaluation of the model and the assumptions can be conducted using the Mach number Ma, which relates the flow velocity c of a fluid to its speed of sound a:

$$Ma = \frac{c}{a} \tag{29}$$

To determine the Mach number of each phase in the two-phase flow, the phase velocities are averaged over the respective cross-sectional areas. Due to the higher speed of sound in the considered liquids (see table 2), the Mach numbers of the liquid phase are well below 0.3 for all boundary conditions and fluid pairs. Conversely, due to the lower speed of sound and higher flow velocities, the compressible gas flows show different results. Especially with large pressure differences across the gap, high wall velocities and large gas volume fractions at the inlet of the gap, the Mach numbers are above 0.3. The highest Mach number among the examined fluid pairs and boundary conditions is observed in a two-phase flow of decane and carbon dioxide, with $Ma = 0.53$. Under the considered boundary conditions and fluid combinations, the Mach number does not reach 1. However, under different conditions, such as lower pressure ratios, this could occur. Since this model does not yet account for choked flow, there is still potential for improvement in this regard.

In the study, further simplifications were made in the consideration of degassing. It is assumed that the fluid properties of the gas and liquid phases correspond to those of the lower and higher boiling component in the mixture, respectively, as these substances constitute the majority of each phase. However, this is a considerable simplification and must be investigated in more detail in future works.

5. Conclusion

In the present work, a generic stratified flow model is developed to calculate two-phase mass flow rates through a planar gap with moving boundaries. This model aims to estimate the impact of various fluid properties, boundary conditions, and physical effects considered in the calculations on the gap mass flow rates.

A variation of the considered fluid combinations reveals that the gas mass flow rates are primarily influenced by the density of the gas, while the viscosity of the liquid phase is the second most significant fluid property. Regarding the boundary conditions at the gap, the pressure ratio and the gas volume fraction at the inlet exhibit a parabolic influence, whereas the wall velocity affects the mass flow rate of the compressible phase linearly. Considering the temperature change due to the pressure drop along the gap leads to a reduction in the gas mass flow rate at pressure ratios $\Pi < 0.9$ due to the increase in the viscosity of the liquid. The relative impact of degassing due to the pressure drop on the calculated gas-phase mass flow rate is, in addition to the gas volume fraction at the inlet of the gap, strongly dependent on the overall solubility of the gas in the liquid.

The assessment of the assumptions underlying the flow model is carried out through the calculated Reynolds and Mach numbers. It is found that the assumptions are generally reasonable, although they may need to be re-evaluated for certain fluid combinations and boundary conditions. Therefore, this work should serve as a foundation for the further development of two-phase flow models that also account for different flow patterns, turbulent or choked flows. Information on flow patterns, which can be obtained through both experimental methods and high-resolution numerical simulations, will be realised for this purpose. The simplified representation of degassing in this model also presents potential for further improved approaches.

Acknowledgments
The authors gratefully acknowledge the Deutsche Forschungsgemeinschaft (DFG, German Research Foundation) for financially supporting this work within the Research Unit FOR 5595 Archimedes ("Oil-refrigerant multiphase flows in gaps with moving boundaries - Novel approaches for experiment, modeling and simulation") - Project Number 510921053.

References

[1] Rinder L 1979 *Schraubenverdichter* (Wien: Springer-Verlag) DOI 10.1007/978-3-7091-8554-4
[2] Grieb M and Brümmer A 2022 Thermodynamic simulation and experimental validation of an oil-free twin-screw expander *IOP Conference Series: Materials Science and Engineering* **1267** 012015 DOI 10.1088/1757-899x/1267/1/012015
[3] Bell I H, Lemort V, Groll E A, Braun J E, King G B and Horton W T 2012 Liquid-flooded compression and expansion in scroll machines–part i: Model development *International Journal of Refrigeration* **35** 1878–1889 DOI 10.1016/j.ijrefrig.2012.07.010
[4] Nikolov A and Brümmer A 2017 Investigating a small oil-flooded twin-screw expander for waste-heat utilisation in organic rankine cycle systems *Energies* **10** 869 DOI 10.3390/en10070869
[5] Gasche J L, Andreotti T and Maia C R M 2012 A model to predict r134a refrigerant leakage through the radial clearance of rolling piston compressors *International Journal of Refrigeration* **35** 2223–2232 DOI 10.1016/j.ijrefrig.2012.07.015
[6] Morris S D 1991 Compressible gas-liquid flow through pipeline restrictions *Chemical Engineering and Processing: Process Intensification* **30** 39–44 DOI 10.1016/0255-2701(91)80007-C
[7] Nikolov A and Brümmer A 2023 Two-phase mass flow rate through restrictions in liquid-flooded twin-screw compressors or expanders *International Journal of Refrigeration* **148** 152–167 DOI 10.1016/j.ijrefrig.2023.01.017
[8] Chisholm D 1985 Two-phase flow in heat exchangers and pipelines *Heat transfer engineering* **6** 48–57 DOI 10.1080/01457638508939624
[9] Wang C, Xing Z, Chen W, Sun S and He Z 2019 Analysis of the leakage in a water-lubricated twin-screw air compressor *Applied Thermal Engineering* **155** 217–225 DOI 10.1016/j.applthermaleng.2019.04.001
[10] Lin Z H 1982 Two-phase flow measurements with sharp-edged orifices *International Journal of Multiphase Flow* **8** 683–693 DOI 10.1016/0301-9322(82)90071-4
[11] Spurk J and Aksel N 2019 *Strömungslehre: Einführung in die Theorie der Strömungen.* (Springer Vieweg) DOI 10.1007/978-3-662-58764-5
[12] Schwarz H R and Köckler N 2013 *Numerische Mathematik* (Springer-Verlag) DOI 10.1007/978-3-8348-8166-3
[13] Truckenbrodt E 2013 *Fluidmechanik: Band 1: Grundlagen und elementare Strömungsvorgänge dichtebeständiger Fluide* (Springer-Verlag) DOI doi.org/10.1007/978-3-540-79018-1
[14] Span R, Beckmüller R, Hielscher S, Jäger A, Mickoleit E, Neumann T, Pohl S, Semrau B and Thol M 2020 *TREND. Thermodynamic Reference and Engineering Data 5.0* (Lehrstuhl für Thermodynamik, Ruhr-Universität Bochum, Bochum)
[15] Gernert J and Span R 2016 Eos–cg: A helmholtz energy mixture model for humid gases and ccs mixtures *The Journal of Chemical Thermodynamics* **93** 274–293 DOI 10.1016/j.jct.2015.05.015
[16] Hyun S H and Delgado A 2024 Dynamic performance of liquid smooth annular seal operating in the transition regime *Journal of Engineering for Gas Turbines and Power* **146** 091019 DOI 10.1115/1.4064805
[17] San Andres L A 1991 Analysis of variable fluid properties, turbulent annular seals *Journal of Tribology* **113** 694–702 DOI 10.1115/1.2920681
[18] Kuhr M M, Lang S R and Pelz P F 2022 Static force characteristic of annular gaps–experimental and simulation results *Journal of Tribology* **144** 111804 DOI 10.1115/1.4054792

High-resolution simulations of gas-liquid Couette-type sealing gap flows

J Vorspohl[1], L Burchardt[2], M Meinke[1], A Brümmer[2] and W Schröder[1]

[1]RWTH Aachen University, Institute of Aerodynamics and Chair of Fluid Mechanics, Aachen, Germany
[2]TU Dortmund University, Chair of Fluidics, Dortmund, Germany

E-mail: j.vorspohl@aia.rwth-aachen.de

Abstract. The two-phase flow in sealing gaps of twin-screw compressors is investigated using highly resolved coupled multiphase simulations. Oil-injected rotary-type positive displacement compressors (RPDC) such as twin-screw machines are used for the compression of refrigerants in HVAC (Heating, Ventilation, and Air Conditioning) systems. The efficiency of such machines is largely determined by the inevitable two-phase surge and gap flows. The simulation of flows in such compressors is still a challenge due to the complexity of the two-phase flow in the narrow gaps (<0.3mm) with organic working fluids and moving boundaries. Due to the rotation of the screws relative to the housing, gaps are present between the rotors and the end plates of the housing and between the tips of the rotors and the cylindrical housing walls. Oil injection can increase the efficiency of screw compressors or expanders by more than 10%, since oil enhances the sealing of the gaps where leakage occurs. On the other hand, additional momentum losses are caused by the two-phase gap flows. The oil injection, however, increases the hydraulic losses which depend on the geometry of the gap, the liquid, and the kinematics of the established multiphase flow in the gap. In this work, a highly efficient, high-resolution numerical method is used for the simulation of the gas-liquid multiphase flow in the screw compressor gap between the rotor tip and the stationary housing. The Lattice-Boltzmann method (LBM) defines the basis of the method used in this work, forming an in-house developed framework in which each fluid phase is solved by a separate solution algorithm. To capture the motion of the liquid-gas phase boundary, a level-set method is used. In the present work, a generic gap flow is considered. A Couette-type flow configuration, where the sealant flow in the gap is driven by the moving inner wall, is investigated. The Reynolds number based on the height of the channel is Re = 20000. Finally, we discuss the impact of bubble-laden sealing gaps on the momentum losses of the system.

1. Introduction

Oil-injected twin-screw compressors have wide-spread application in the industry. The overall performance heavily depends on the quality of the different clearances of the machine. The injected oil significantly improves the sealing quality, i.e., reduces the leakage flow of the working gas. On the other hand, the momentum loss in the clearances is increased, and thus the overall power torque.

The state-of-the-art method for the design of screw machines is the analysis by means of a chamber model. Here, the machine is modeled as a combination of different chambers which undergo a forced change in volume, defined by the geometry and rotational speed. In this approach, the leakage flow through the different clearances is modeled by various empirically based methods. Utri et al. investigated the mass flow rate through male rotor clearances of dry-running screw machines by means of numerical simulations [1]. Reynolds-averaged Navier-Stokes (RANS) simulations with shear stress transport (SST) turbulence model are performed to determine the gap flow for different pressure ratios and rotation speeds, including the effect of moving walls. For small pressure ratios and small Reynolds numbers, the deviation from the theoretical mass flows for an isentropic nozzle flow is maximal. For liquid flooded machines, Nikolov et al. compared the theoretical mass flow rate based on orifice or nozzle flow models for varying inlet geometries with experimental data [2]. The discrepancy between the theoretical and experimental mass flow rate is modeled by a novel flow coefficient for different pressure ratios Π and mass dryness fractions x. Vasuthevan et al. performed 2D volume of fluid (VOF) simulations of a rectangular profile within a cylinder to investigate the different mechanisms of momentum loss in wet-running screw compressors [3]. Different circumferential Reynolds numbers and liquid volume ratios are considered, while the pressure ratio is kept at $\Pi = 1$. The momentum loss in the rotor housing clearance is found to be small compared to the losses associated with the surge flow in front of the rotor tip. Although the latter heavily depends on the liquid volume ratio of the machine, that is, the amount of liquid that is circulated by the rotor movement, the momentum loss in the clearance is not affected, indicating constant flow conditions in the gap region. The two-phase flow in oil-injected screw machines is investigated by Basha et al. [4] using different numerical methods for multiphase flows. Although the mixture model used can best predict machine performance, the oil distribution within the machine is best met by an Eulerian-Eulerian approach.

To better understand the interaction of the two-phase flow in rotor housing clearances for oil injected screw machines, further investigations are needed. In this work, the two-phase gap flow of such machines is modelled by a generic Couette-type flow. First, the single-phase case, i.e. a liquid filled gap without any working gas leakage, is considered. In a second step, initially spherical bubbles are added to the gap. For both cases, the overall flow structure and the momentum losses are discussed.

2. Method

The simulations including grid generation and post processing are carried out using m-AIA, an open source solver framework with a focus on fluid dynamics [5].

In this work, a fully coupled approach for the simulation of liquid-gas two-phase flows is used. The flow field in both phases is simulated using a separate lattice Boltzmann (LB) method. The interface between both phases is captured using a level-set method. Continuity, the effect of surface tension and the jump condition in the stress tensor across the interface are enforced by means of a modified boundary condition for both flow solvers. All solution algorithms operate on a joint hierarchical Cartesian grid, which enables the efficient exchange of coupling data between all solvers without any I/O needed. In the following, all numerical methods are briefly discussed.

2.1. Lattice Boltzmann method

Based on the Boltzmann equation from kinetic gas theory, the lattice Boltzmann (LB) method was derived by discretizing the phase space and splitting the equation into a "streaming" and a "collision" step. The general form reads as

$$f_i(\mathbf{x} + \mathbf{c}_i \Delta t, t + \Delta t) = f_i^c(\mathbf{x}, t) = f_i(\mathbf{x}, t) + \Omega_i(\mathbf{f}) + F_i. \tag{1}$$

with f being the particle probability density function (PPDF), $\mathbf{c}_i$ the discrete particle velocity for direction i, f^c the post-collision state, Ω_i the collision operator, and F_i the forcing term. From here on, the PPDF $f_i(\mathbf{x}, t)$ is simply referred to as f_i. In the present work, a three-dimensional discretization with a lattice consisting of 27 discrete velocities (D3Q27) is used. By calculating the moments of the PPDF, the density ρ and velocity $\mathbf{u}$ are obtained. In discrete form, this reads as

$$\rho = \sum_i f_i(\mathbf{x}, t) \quad \text{and} \quad \mathbf{u} = \frac{1}{\rho} \sum_i \mathbf{c}_i f_i(\mathbf{x}, t). \tag{2}$$

For the collision operator, different options are available from the literature. Here, the Bhatnagar-Gross-Krook (BGK) operator [6] and the cumulant collision operator [7] are briefly discussed. The BGK collision operator, which is suitable for low Reynolds number flows, reads

$$\Omega_i(\mathbf{f}) = \omega_{BGK}(f_i^{eq} - f_i), \tag{3}$$

with f_i^{eq} being the Maxwell equilibrium distribution function given by

$$f_i^{eq} = w_i\rho \left[1 + \frac{\mathbf{c}_i \cdot \mathbf{u}}{c_s^2} + \frac{(\mathbf{c}_i \cdot \mathbf{u})^2}{2c_s^4} - \frac{\mathbf{u} \cdot \mathbf{u}}{2c_s^2}\right], \tag{4}$$

where w_i are the weighting factors, which are $8/27$, $2/27$, $1/54$, and $1/216$ for the center, the Cartesian, the edge diagonal, and space-diagonal directions. The speed of sound is denoted as c_s. The collision frequency ω_{BGK} is given by

$$\omega_{BGK} = \frac{\Delta t c_s^2}{\nu + \frac{1}{2}\Delta t c_s^2}, \tag{5}$$

with ν denoting the viscosity of the fluid. In the present work, the forcing term is set to

$$F_i = -\frac{\rho g}{c_s^2} w_i \mathbf{c}_i \cdot \mathbf{e}_z, \tag{6}$$

to account for the effect of gravity g. For high Reynolds number flows, the cumulant collision operator has been developed [7]. As suggested by the name, the PPDF is transformed into countable cumulants c_α, in order to obtain Galilean invariant and statistically independent quantities. The collision is then performed for the cumulants, which reads

$$c_\alpha^c = c_\alpha + \omega_\alpha (c_\alpha^{eq} - c_\alpha), \tag{7}$$

where c_α^{eq} denotes the Maxwell equilibrium distribution (4) in cumulant space. In the present study, the relaxation parameters in cumulant space are chosed as

$$\omega_1 = \omega_{BGK}, \quad \omega_\alpha = 1 \quad \text{for} \quad \alpha \neq 1. \tag{8}$$

In order to support local grid refinement, the method of Dupuis and Chopard is used [8]. Since $\Delta t \approx \Delta x$, the collision frequency for different grid spacings is adapted to ensure that the viscosity ν is constant (5). For solid walls, the interpolated bounce back method, described by Bouzidi et al. [9] is used. The missing PPDFs from the boundary are constructed as

$$f_{\bar{i}}(\mathbf{x}, t+1) = \begin{cases} 2q f_i^c(\mathbf{x}, t) + (1-2q) f_i^c(\mathbf{x} - \mathbf{c}_i, t) + \Delta f_i & \text{if } q < \frac{1}{2} \\ \frac{1}{2q} f_i^c(\mathbf{x}, t) + \frac{2q-1}{2q} f_i^c(\mathbf{x} - \mathbf{c}_i, t) + \frac{1}{2q}\Delta f & \text{if } q \geq \frac{1}{2}, \end{cases} \tag{9}$$

where $\bar{i}$ denotes the opposite direction of i and q the normalized distance $\frac{d}{\Delta x}$ from the cell center to the wall. For static walls, the momentum source term Δf_i is zero, for walls moving with the velocity $\mathbf{u}_B$ it is given by

$$\Delta f_i = \frac{2w_i}{c_s^2}(\mathbf{c}_i \cdot \mathbf{u}_B). \tag{10}$$

2.2. Level-set method

To capture the interface between the liquid and gas fluid phases, a level-set method is used [10]. The scalar variable φ is used to identify the computational domain for each phase, as well as the interface between them, such that

$$\varphi(\mathbf{x}, t) \begin{cases} < 0 & \Omega_l \quad \text{liquid phase} \\ = 0 & \Gamma_i \quad \text{interface} \\ > 0 & \Omega_g \quad \text{gas phase.} \end{cases} \tag{11}$$

Additionally, the level-set is initialized and subsequently treated such that φ describes the signed distance to the interface. The temporal evolution of the interface is predicted by advecting the scalar field φ by the local fluid velocity, which reads

$$\frac{\partial \varphi}{\partial t} + \tilde{\mathbf{u}} \cdot \nabla \varphi = 0. \tag{12}$$

To solve this transport equation, a fifth-order upwind-central scheme for the spatial discretization and a third-order total variation diminishing Runge-Kutta scheme for the time integration is employed [11]. Here, the transport equation is only solved in a narrow tube around the interface Γ, as described in [10]. In general, the velocity field $\tilde{\mathbf{u}}$ used for the transport of the interface is only known at the interface itself. By using so-called hyperbolic extension, the velocity $\mathbf{u}_I$ is artificially extended in the interface normal direction [12]. During the advection of the interface, the signed distance property $\|\nabla\varphi\| = 1$ is not guaranteed to be preserved. To remedy this, the level-set has to be reinitialized regularly. In the present work, a high order constrained reinitialization is used, which ensures the invariance of the zero set $\varphi = 0$, i.e., the position of the interface, as described by Hartmann et al. [13]. Conveniently, the geometric properties at the interface, such as the normal vector and the curvature, can be calculated as

$$\mathbf{n} = \frac{\nabla\varphi}{\|\nabla\varphi\|} \qquad \kappa = \nabla \cdot \mathbf{n}. \tag{13}$$

2.3. Coupling conditions

In the present work, a two-fluid approach for the simulation of the two-phase flow in the clearance geometry is employed, i.e., the solution algorithms are coupled by means of boundary conditions at the two-phase interface. Similarly to the boundary treatment at solid walls (9), a bounce back approach is chosen. The momentum source term following [14] reads

$$\Delta f_i = \frac{2w_i}{c_s^2}(\mathbf{c}_i \cdot \mathbf{u}_I) - \frac{2w_i}{c_s^2}\mathbf{\Lambda}_i q(1-q)[\mathbf{S}]. \tag{14}$$

The first term of (14) ensures the continuity at the phase interface, i.e., $[u_I] = 0$, where $[\cdot]$ denotes the jump of a given quantity across the interface. The second term accounts for the stress tensor jump at the interface, with $\mathbf{\Lambda}_i$ being the second order velocity tensor. The stress jump can be decomposed into normal and tangential components

$$[S] \colon \vec{n} \otimes \vec{n} = \frac{1}{2\bar{\eta}}([p] + 2\sigma\kappa) - \frac{[\eta]}{\bar{\eta}} \colon \vec{n} \otimes \vec{n}$$

$$[S] \colon \vec{n} \otimes \vec{t}_j = -\frac{[\eta]}{\bar{\eta}} \colon \vec{n} \otimes \vec{t}_j.$$

Here, $\bar{\eta}$ denotes the average of the dynamic viscosity $\eta = \rho\nu$ over both phases and σ the surface tension of the chosen material pairing.

3. Results

In the present work, a generic two-phase clearance flow setup is considered. The general setup is shown in Fig. 1. Here, we consider the clearance volume moving with the male rotor such that the upper wall, i.e., the housing, is moving with u_W. To save computational resources, the setup is considered periodic in the x- and the z-direction, i.e., in the streamwise and spanwise direction. In the present work, the Reynolds number is chosen as $Re = \frac{\rho_l u_W h}{\eta_l} = 20000$, where h is the channel height. Since the rotor tip radius is large compared to the gap height, i.e., $R_i \gg h$, the channel curvature is neglected. In accordance with [15], the computational domain is chosen as $[8 \times 1 \times 8]h$. For the grid resolution, a wall resolution of $y^+ \approx 1$ is chosen, resulting in $\approx 0.5 \cdot 10^9$ cells.

3.1. Single-phase flow

In a first step, the single phase flow through the clearance is considered. Here, the gap is entirely filled with sealant, i.e., no working gas is present. The simulation is initialized with the laminar solution, i.e., the linear velocity profile with additional random perturbations in all velocity components. After a transition phase of around $10^3\Delta t$, a fully turbulent flow field is present. In Fig. 2, the instantaneous flow field at $t = 400 \cdot 10^3\Delta t$ is shown. The iso-surfaces for $Q = 0.5$ are colored with $\|\mathbf{u}\|/u_W$. The Q-Criterion is defined as

$$Q = \frac{1}{2}[(\mathrm{tr}(\nabla\mathbf{u}))^2 - \mathrm{tr}(\nabla\mathbf{u} \cdot \nabla\mathbf{u})^2]. \tag{15}$$

In contrast to the laminar flow field, the shear rates are drastically increased at the upper and lower wall. This leads to concentrated vortex production at the walls, as seen in Fig. 2. In Fig. 3a the time averaged streamwise velocity $\bar{\mathbf{u}}/u_W$ in the center plane at $y = h/2$ is shown. There are some significant deviations from the theoretical velocity $\bar{\mathbf{u}}(y = h/2)/u_W = 0.5$. These deviations are due to large-scale structures which are present in turbulent Couette-like flows. The intensity of the large-scale deviations also varies over the channel height, as seen in Fig. 3b, where $\bar{\mathbf{u}}/u_W$ is shown at $y = 3/4h$. These structures have been reported in the literature [16, 15]. They are the primary cause for the large correlation length in

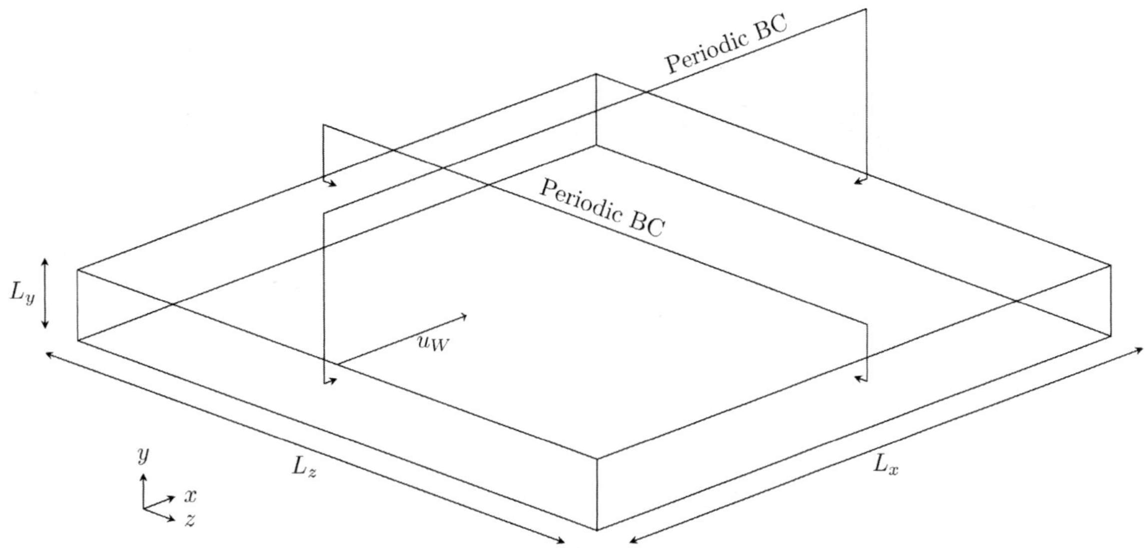

Figure 1. Simulation setup for the clearance flow. The domain size is $[8 \times 1 \times 8]h$. The flow in stream-and spanwise direction is considered periodic. The upper wall is moving with u_W.

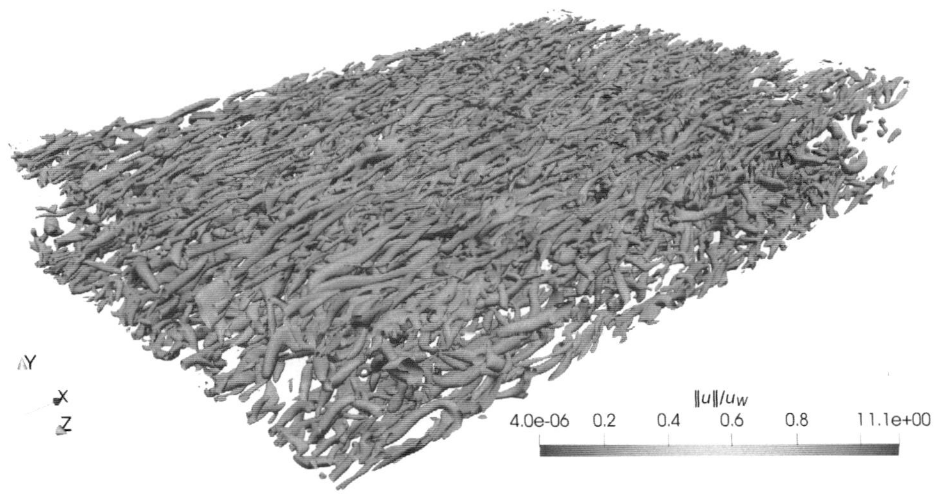

Figure 2. Instantaneous flow field at $t = 400 \cdot 10^3 \Delta t$. Iso-surfaces of the Q-criterion with $Q = 0.5$ are shown. The surface color is defined by the velocity magnitude $\|\mathbf{u}\|/u_W$.

such flows. This is the main reason for the relatively large computational domain in the streamwise and spanwise direction. If the spanwise domain size is not large enough, confinement effects become larger [15]. In the context of clearance flows in twin screw machines, this leads to non-uniform mass flow rates and momentum losses along the rotor tip. To the authors' knowledge, there has been no evidence of such spanwise fluctuations in rotor-housing clearances, neither experimentally nor numerically. In Fig. 4a, the averaged velocity profile is shown. Besides the temporal averaging over $T_{avg} = 500 \cdot 10^3 \Delta t$, the streamwise velocity is averaged in the stream- and spanwise direction to account for the aforementioned large scale

IOP Conf. Series: Materials Science and Engineering **1322** (2024) 012004 doi:10.1088/1757-899X/1322/1/012004

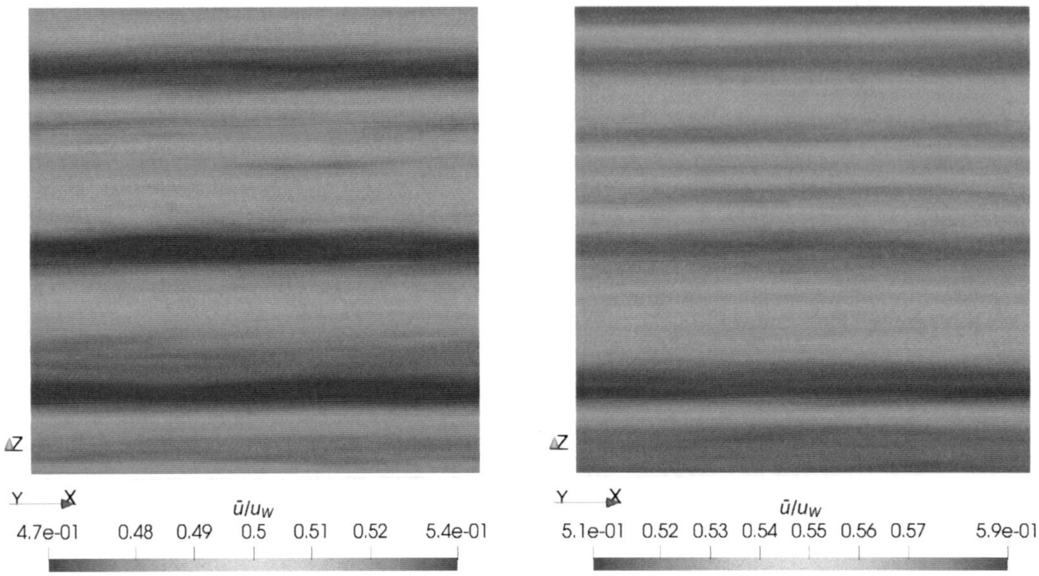

(a) (b)

Figure 3. Time averaged streamwise velocity component on $y/h = 0.5$ (a) and $y/h = 0.75$ (b).

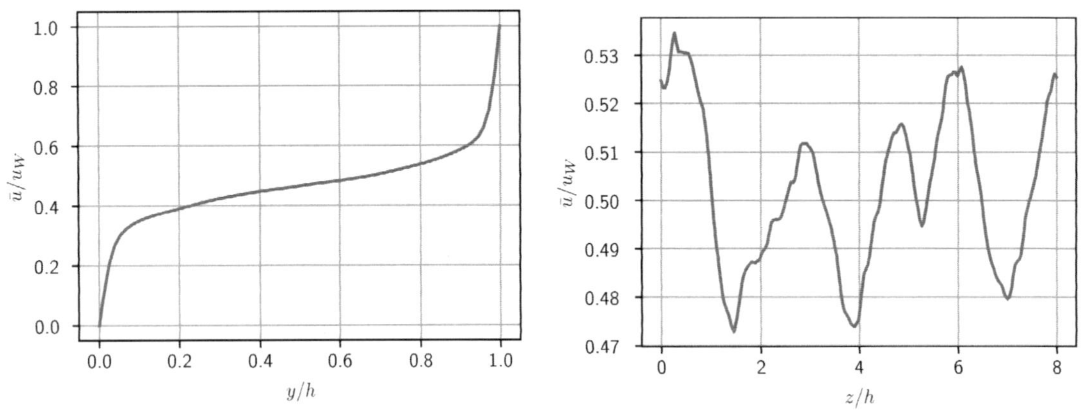

(a) Velocity over channel height. Additionally averaged in span- and streamwise direction.

(b) Velocity over spanwise direction. Additionally averaged in streamwise and wall-normal direction.

Figure 4. Time average streamwise velocity component.

structures. For the chosen T_{avg} the observed variation in the averaged quantities is below 1%. Compared to the laminar flow field, the skin friction is drastically increased. For the shown flow field the increase is $\tau_W / \tau_{W,lam} = 6.23$.

The spanwise fluctuations are shown in Fig. 4b, where the streamwise velocity is averaged in streamwise and wall-normal direction, i.e., indicating the variation of the volume flux through the clearance. Here, the deviation of the volume flux caused by the meandering motion of the large scale structures is up to 3.5%.

3.2. *Two-phase flow*

The single-phase simulations are extended to account for the effect of a second phase in the clearance flow field. We consider a gas phase with density and viscosity ratios of $\rho_l/\rho_g = 1000$ and $\eta_l/\eta_g = 100$. Initially, we assume spherical bubbles with a diameter of $d/h = 0.1$. The effect of surface tension is described by the Eötvös number

$$Eo = \frac{\rho_l g d^2}{\sigma}, \tag{16}$$

which is chosen as $Eo = 10^{-3}$. To initialize the two-phase simulation, 1024 bubbles are placed equidistantly in the computational domain. The flow field is initialized using the resulting flow field from the single phase simulation. To avoid destructive oscillation due to unbalanced pressure jumps at $t = 0$, the surface tension σ is ramped up over $T_\sigma = 10^4 \Delta t$. In Fig. 5a, the instantaneous flow field in the center

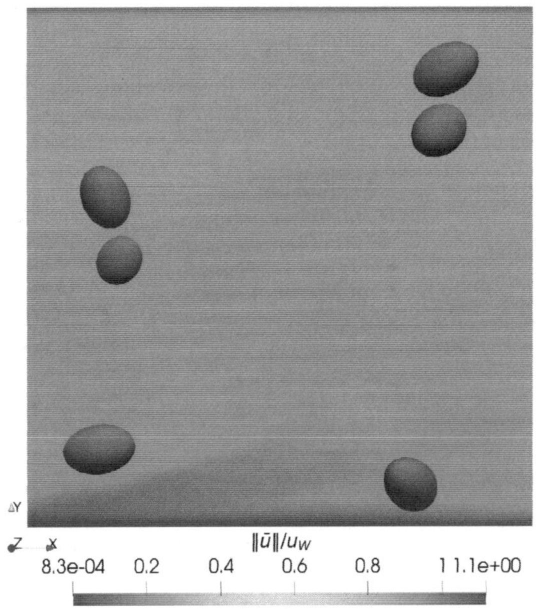

(a) Instantaneous flow field in the center plane for $3 < x < 4$, including bubble contours at $t = 400 \cdot 10^3 \Delta t$.

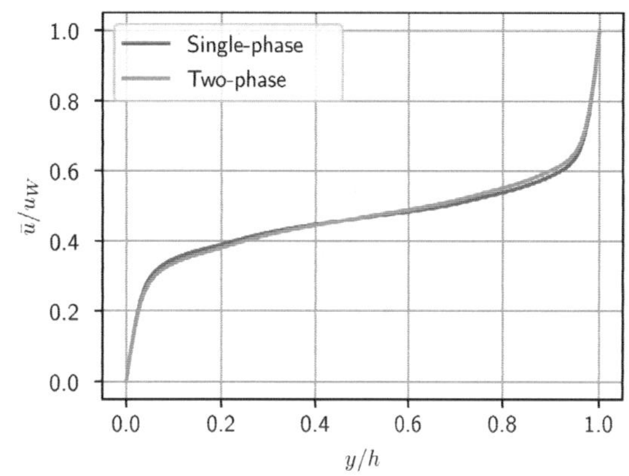

(b) Time averaged streamwise velocity component over spanwise direction. Additionally averaged in streamwise and wall-normal direction.

Figure 5. Results for the two-phase bubble laden clearance flow configuration.

plane at $t = 400 \cdot 10^3 \Delta t$ is shown. Due to the high surface tension, the bubble shape is nearly spherical. In regions with higher shear rates, the bubbles are visibly deformed towards an ellipsoidal shape. The averaged streamwise velocity component for the single- and the two-phase system is shown in Fig. 5b. The skin friction compared to the laminar single-phase system for the two-phase case is $\tau_W/\tau_{W,lam} = 5.9$. For the present volume fraction of $V_{gas}/V = 0.84\%$, the observed reduction in momentum loss is 5.3%.

4. Conclusion

A high-resolution simulation for generic sealing gap flows in twin-screw machines is presented. For the single phase flow, the occurring flow patterns due to large scale flow structures are discussed. It remains an open question if the observed patterns and the resulting spanwise variation in mass flow rate and wall shear stress are a) also present for the complex geometry of real-life machines and b) relevant for the overall performance of such.

For the two-phase system, the influence of small, initially spherical bubbles in the gap flow is analyzed. Depending on the current position of the bubbles, ellipsoidal shapes can be observed. Even for relatively low volume loadings, a significant decrease in the wall shear stress is found.

This work is a starting point for future research endeavors. Next, the effect of pressure gradients across the clearance will be investigated. Additionally, the effect of rotation will be included, leading to a Taylor-Couette-Poisseuille-type flows. The radial acceleration for high circumferential velocities is suspected to lead to phase separation in the gap, leading to extensive bubble-bubble and bubble-wall interactions. This requires novel numerical methods and algorithms, which have to be suitable for the application on high performance computing hardware. Further simulations for more complex, real-life geometries are needed to investigate if the observed spanwise fluctuations are present and relevant in actual twin-screw machines.

Acknowledgement
This work was funded by the Deutsche Forschungsgemeinschaft (DFG, German Research Foundation) within the Research Unit FOR 5595 Archimedes (Oilrefrigerant multiphase flows in gaps with moving boundaries – Novel microscopic and macroscopic approaches for experiment, modeling, and simulation) – Project number 510921053.

References

[1] Utri M, Höckenkamp S and Brümmer A 2018 *Fluid flow through male rotor housing clearances of dry running screw machines using dimensionless numbers* IOP Conference Series: Materials Science and Engineering 425 doi:10.1088/1757-899X/425/1/012033

[2] Nikolov A and Brümmer A 2023 *Two-phase mass flow rate through restrictions in liquid-flooded twin-screw compressors or expanders* International Journal of Refrigeration 148 pp 152–167 doi: 10.1016/j.ijrefrig.2023.01.017

[3] Vasuthevan H and Brümmer A 2018 *Multiphase-flow simulation of a rotating rectangular profile within a cylinder in terms of hydraulic loss mechanisms* IOP Conference Series: Materials Science and Engineering 425 doi:10.1088/1757-899X/425/1/012002

[4] Basha N, Kovacevic A and Rane S 2019 *Analysis of oil-injected twin-screw compressor with multiphase flow models* Designs 3 pp 1–19 doi:10.3390/designs3040054

[5] Institute of Aerodynamics 2024 m-aia doi:10.5281/zenodo.13350586

[6] Bhatnagar P, Gross E and Krook M 1954 *A model for collison processes in gases. i. small amplitude processes in charged and neutral one-component systems* Physical Review pp 511–525 doi:10. 1109/ ITCC.2003.1197559

[7] Geier M, Schönherr M, Pasquali A and Krafczyk M 2015 *The cumulant lattice boltzmann equation in three dimensions: Theory and validation* Computers and Mathematics with Applications 70 pp 507–547 doi:10.1016/j.camwa.2015.05.001

[8] Dupuis A and Chopard B 2003 *Theory and applications of an alternative lattice boltzmann grid refinement algorithm* Physical Review E - Statistical Physics, Plasmas, Fluids, and Related Interdisciplinary Topics 67 7 doi:10.1103/PhysRevE.67.066707

[9] Bouzidi M, Firdaouss M and Lallemand P 2001 *Momentum transfer of a boltzmann-lattice fluid with boundaries* Physics of Fluids 13 pp 3452–3459 doi:10.1063/1.1399290

[10] Sethian J A and Osher S 1987 *Fronts propagating with curvature-dependent speed: Algorithms based on hamilton-jacobi formulations* Journal of Computational Physics 79 pp 12–49

[11] Shu C-W and Osher S 1988 *Efficient implementation of essentially non-oscillatory shock-capturing schemes* Journal of Computational Physics 77 pp 439–471 doi:10.1007/978-94-007-5404-1_5

[12] Chen S, Merriman B, Osher S and Smereka P 1997 *A simple level set method for solving stefan problems* Journal of Computational Physics 135 pp 8–29. doi:10.1006/jcph.1997.5721

[13] Hartmann D, Meinke M and Schröder W 2010 *The constrained reinitialization equation for level set methods* Journal of Computational Physics 229 pp 1514–1535 doi:10.1016/j.jcp.2009.10.042

[14] Thömmes et al 2009 *A lattice boltzmann method for immiscible multiphase flow simulations using the level set method* Journal of Computational Physics 228 pp 1139–1156 doi:10.1016/j.jcp.2008.10.032

[15] Pirozzoli S, Bernardini M and Orlandi P 2014 *Turbulence statistics in couette flow at high reynolds number* Journal of Fluid Mechanics 758 pp 327–343 doi:10.1017/jfm.2014.529

[16] Illingworth S J 2020 *Streamwise-constant large-scale structures in couette and poiseuille flows* Journal of Fluid Mechanics 889 A13 doi:10.1017/jfm.2020.86

Astigmatism Quantification for Depth Localization of Bubbles and Tracers across Curved Surfaces

H Lange[1], **C Sax**[1], **A Brümmer**[2] and **J Kriegseis**[1*]

[1] Institute of Fluids Mechanics, Karlsruhe Institute of Technology, Karlsruhe, Germany

[2] Chair of Fluidics, TU Dortmund University, Dortmund, Germany

E-mail: `kriegseis@kit.edu`

Abstract. The present combined theoretical/experimental study addresses the impact of astigmatism on the two-phase flow diagnostics across the curved surfaces of liquid test-rig containments. In the present context, the target application is the two phase leakage-flow diagnostics across the annular housing gaps of oil-injected rotary positive displacement compressors (RPDC). Earlier studies by the authors identified the Defocusing Particle Tracking Velocimetry (DPTV) and Interferometric Particle Imaging (IPI) as particularly promising combination of flow measurement techniques to investigate the liquid and disperse gas phases inside the annular housing gap of RPDCs. The test-rig-specific influence of astigmatism on the resulting optical transfer function for a quantitative evaluation of the recorded defocused particle images (PI) is first compared to the theoretically derived circular PI diameter upon pure defocussing and subsequently tested for both classes of PIs, i.e DPTV and IPI. To mimic the optical configuration of optically accessible lateral surfaces of typical RPDC test rigs, a circular beaker glass (CBG) of comparable diameter is chosen for the experimental campaign. The results are discussed and future efforts for advanced PI-evaluation strategies are outlined on the grounds of the drawn conclusions.

1. Introduction

In order to increase the efficiency of rotary positive displacement compressors (RPDCs), an in-depth knowledge of the two-phase surge and gap flow between two adjacent compressor chambers is crucial as recently elaborated by e.g. Nikolov & Brümmer [1]. In addition to the measurement of torque the accordingly required experimental investigations moreover rely on the non-trivial flow measurement of both liquid and gas phases in the annular housing gap of the RPDCs. Such gap-flow investigations largely center around (laser-)optical imaging techniques, which either address the qualitative characterization of the spatio-temporal two-phase flow behavior or furthermore aim at a quantitative evaluation of velocity information inside the RPDC.

Xin *et al.* [2] and Vasuthevan & Brümmer [3], for instance, addressed the interrelationship of influential factors for the performance of oil-injected RPDC by means of combined qualitative visualizations and integral torque measurements. Patel *et al.* [4] measured the temperature field inside an oil-free RPDC by means of laser-induced fluorescence (LIF), which was later complemented with planar – i.e. two dimensional & two component (2D2C) – particle image velocitmetry (PIV) in the annular housing-gap plane of the RPDC by the same group [5]. Since additional knowledge of the lateral distribution of flow field across the annular housing gap is of utmost importance to improve both prediction and optimization strategies towards advanced RPDC efficiency, volumetric (3D3C) flow-diagnostics approaches seem mandatory for the given two-phase flow scenario.

Tomographic PIV [6] and more recently also Lagrangian particle tracking (LPT) methods [7] enjoy increasing attention for quantitative 3D3C flow diagnostics. However, these volumetric standard approaches might have limited applicability in cases, where microscopic characteristic length scales such as e.g. the cross-section of micro channels or - in the present context - a thin housing gap are accompanied by macroscopic characteristic lengths scales along the channel-flow direction or the lateral surface, respectively. For such wide aspect-ratio flow scenarios defocused imaging approaches [8] provide a promising alternative, where size and/or shape of the defocused particle image (PI) rather than the multi-camera tomographic reconstruction provides information on the depth location of the PI along the optical axis imaging setup inside the measurement volume. In context of planar thin gap flows, Leister *et al.* [9, 10] recently analyzed the single-phase flow scenario inside 400-540 µm wide rotor-stator gaps of an open wet clutch experimentally by means of phase-resolved defocusing particle tracking velocimetry (DPTV). This study furthermore took advantage of the known no-slip boundary conditions at rotor and stator to apply an in-situ calibration approach [11, 12].

In addition to a DPTV-based evaluation of the liquid phase, in case of RPDCs, additional complementary diagnostics of the two-phase character of investigated flow is mandatory. Standard (high-speed) imaging approaches by means of reflected light processes are suitable for macroscopic sizing and tracking efforts of liquid-gas interfaces, where small-scale processes such as bubble onset and initial growth require additional diagnostics approaches. For such disperse cases e.g. the interferometric particle imaging (IPI) is a promising approach for image-based particle-size analysis [13], which has been repeatedly applied for both spray-characterization [14] and bubble-sizing studies [15, 16].

The general applicability of both DPTV and IPI for two-phase flow diagnostics in thin annular gaps has recently been successfully outlined by Leister *et al.* [17], where the measurement system was operated in side-scatter configuration - i.e. optical axes of laser and camera were oriented approximately perpendicular. To prepare for the limited optical access of e.g. RPDC-related investigations, Leister *et al.* [10] successfully conducted back-scatter DPTV measurements through a transparent disc. Likewise, Sax *et al.* [18] evaluated and testified the applicability of IPI in back-scatter orientation in a combined theoretical/experimental investigation. Both studies [10, 18] only considered planar surfaces for the optical access, which accordingly resulted in circular particle images (PI).

The curved boundaries of the annular housing gap of RPDCs introduce additional astigmatism to the optical transfer function (OTF) of the experimental approach. As concluded by Leister *et al.* [17], the corresponding shape deformation of the recorded defocused PIs is hypothesized to provide additional information for the depth localization, which might take additional advantage of established astigmatism particle image velocimetry (APTV) approaches [19, 20].

In continuation of earlier studies by Leister *et al.* [17] the purpose of the present study is the adaption of the OTF towards a combined APTV/DPTV evaluation of RPDC-related gap-flow diagnostics. Since the measurement volume inside the above-mentioned boundary curvature accordingly acts as a cylinder lens itself, no immediate application of standard APTV approaches [19, 20] is straight-forwardly possible. Consequently, a quantitative characterization of the experimental optical system including the surface curvature will be elaborated for the PI localization in a first step. Subsequently, the developed OTF is testified with a first set of manually labeled raw data originating from both DTPV and IPI PIs, where the vertical and horizontal axes of the PIs (respectively corresponding to the vertical and azimuthal cylinder axes) are compared. Also the resulting particle displacements are determined to comparatively

IOP Conf. Series: Materials Science and Engineering **1322** (2024) 012005 doi:10.1088/1757-899X/1322/1/012005

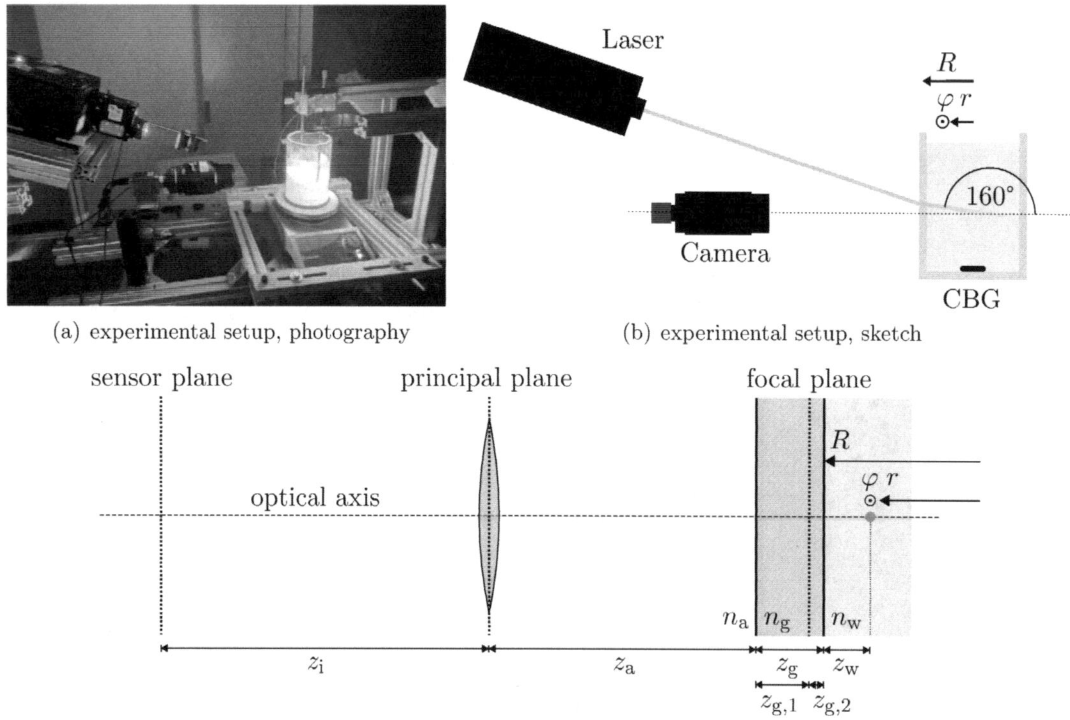

(a) experimental setup, photography

(b) experimental setup, sketch

(c) optical configuration and geometric distances; focal plane within glass for sufficiently defocused particle images

Figure 1. Experimental setup, photography (a) and sketch (b), optical configuration (c) of the experimental and processing procedure for the IPI and DPTV evaluations.

evaluate the theoretical (present OTF) and the retroactively determined [11, 12] no-slip PI diameter at the wall of the utilized circular beaker glass (CBG).

2. Experimental and Processing Procedure

2.1. Experimental setup

The experimental setup is comprised of a CBG (inner diameter $2R = 88.9\,\mathrm{mm}$, wall thickness $z_g = 5.8\,\mathrm{mm}$, acrylic glass refractive index $n_g = 1.492$), which sits on a magnetic stirrer to maintain a circumferential velocity of the water (density $\varrho_w = 993\,\mathrm{kg/m^3}$, dyn. viscosity $\mu_w = 6.5 \times 10^{-4}\,\mathrm{kg/ms}$ at $40°C$, refractive index $n_w = 1.333$); see Figure 1(a). The water is seeded with 20 µm diameter Polyamid particles (density 1150 $\mathrm{kg/m^3}$) for the DPTV evaluation, which results in a particle response time of $\tau_p \approx 34\,\mathrm{\mu s}$ and a correspondingly small Stokes number $St < 10^{-2}$ for the tracer-based velocimetry [21]. Note, however, that upon sufficient displacement from the focal plane the particle diameter itself no longer contributes to the diameter of the PI; see e.g. the treatise of Olsen & Adrien [22]. Furthermore, the electrolysis apparatus as already used by Sax *et al.* [18] is applied to produce a constant rate of bubbles for the IPI approach. A *PCO Pixelfly* camera (14 bit, maximum resolution 1392 × 1040 pixels) is equipped with an $f = 105\,\mathrm{mm}$ *Nikon Nikkor* lens to capture both classes of PIs. The CBG is illuminated with the raw beam of a *Quantel EverGreen* Nd:YAG ($\lambda = 532\,\mathrm{nm}$, 200 mJ) dual-cavity laser, which is oriented vertically in 160° back-scatter orientation to ensure reasonable IPI fringe-pattern visibility in back-scatter mode [18]. Double frame images are recorded at 15 Hz repetition rate with a pulse distance between the laser cavities of $\Delta t = 300\,\mathrm{\mu s}$.

Table 1. Experimental and optical parameters

Quantity	Value
focal length f	105 mm
f-number $f_{\#}$	2.8
aperture diameter $D_a = f/f_{\#}$	37.5 mm
magnification M	1.06
reproduction scale	6.1 µm/px
camera sensor pixel size	6.45 µm
refractive index air n_a	1
refractive index acrylic glass n_g	1.492
refractive index water n_w	1.333
thickness beaker glass wall z_g	5.8 mm
time interval Δt between frame A and B	300 µs

2.2. Optical considerations

Sufficiently displaced from the focal plane, the defocused particle image diameter a can be approximated as proportional to the optical distance z^* to the focal plane as e.g. elaborated in the treatise of Olsen & Adrian [22]. This assumption has been shown to hold true for the present class of defocus applications [9–12]. Accordingly, the combination of defocus sensitivity $\partial z/\partial a$ (change of particle depth position with particle image size, see also Equation 11) and the PI diameter at the glass-water interface $a(R)$ are the desired quantities for the OTF of the setup. The optical configuration of the imaging system is sketched in the vertical plane in Figure 1(c). Note that this plane serves as basis for the OTF, since no additional astigmatism is introduced in this plane, which in turn lead to robust estimates of the expected vertical PI axis a_v (minor axis of the elliptical PI). The horizontal axis a_h of the defocused PI (major axis of the elliptical PI), in contrast, will be influenced by the astigmatism caused by the curved beaker glass surface.

Focal length ($f = 105$ mm), f-number ($f_{\#} = 2.8$), camera sensor pixel size (6.45 µm), thickness of the beaker glass wall ($z_g = 5.8$ mm) and the refractive indices of air, acrylic glass and water are known optical parameters. A summary of these and further parameters can be found in Table 1.

As a first step a calibration target is recorded at the glass-water interface to determine the magnification $M = 1.06$ of the optical setup and the reproduction scale of 6.1 µm/px. Considering the definition of magnification [21]

$$M = \frac{z_i}{z_o} \tag{1}$$

and focus criterion [21]

$$\frac{1}{f} = \frac{1}{z_i} + \frac{1}{z_o}, \tag{2}$$

the image distance $z_i = 216.3$ mm and the optical object distance $z_o = 204.1$ mm can be calculated straight forwardly from the determined and subsequently known values for M and f [21] . Since the object distance z_o is comprised of both traveled distance in air and the known contribution inside the glass, i.e.

$$z_o = \frac{z_a}{n_a} + \frac{z_g}{n_g}, \tag{3}$$

the distance between principal plane and air-glass interface can be determined as $z_a = 200.2$ mm, cp. also Figure 1(c).

In order to obtain sufficiently defocused particle images at the glass-water interface and beyond, the camera is moved backwards by 600 µm after imaging the calibration target. This results in a modified value for $z_\mathrm{a} = 200.8\,\mathrm{mm}$, with the focal plane now situated within the glass; see Figure 1(c). The geometric location of the focal plane inside the glass is

$$z_{\mathrm{g},1} = \left(z_\mathrm{o} - \frac{z_\mathrm{a}}{n_\mathrm{a}}\right) \cdot n_\mathrm{g} = 4.9\,\mathrm{mm} \tag{4}$$

for the distance between the air-glass interface to the focal plane and likewise the distance to the glass-water interface $z_{\mathrm{g},2} = z_\mathrm{g} - z_{\mathrm{g},1} = 0.9\,\mathrm{mm}$. The optical distance for a particle at the glass-water interface can then be determined as

$$\frac{z_\mathrm{a}}{n_\mathrm{a}} + \frac{z_\mathrm{g}}{n_\mathrm{g}} = 204.7\,\mathrm{mm}. \tag{5}$$

According to Olsen & Adrian [22] the vertical axis a_v of a PI can be calculated by

$$a_\mathrm{v} = \sqrt{\underbrace{M^2 d_\mathrm{p}^2}_{\text{geometric image}} + \underbrace{5,95(M+1)^2\lambda^2 f_\#^2}_{\text{diffraction}} + \underbrace{\frac{M^2 z^{*2} D_\mathrm{a}^2}{(z_\mathrm{o}+z^*)^2}}_{\text{defocusing}}}, \tag{6}$$

with the particle size d_p, the laser wavelength λ and the aperture diameter $D_\mathrm{a} = f/f_\# = 37.5\,\mathrm{mm}$. For sufficient defocusing (sufficiently large z^*) and a large optical distance between focal plane and principal plane z_o in comparison to the particle's optical distance to the focal plane z^* the equation simplifies to

$$a_\mathrm{v} = \frac{M D_\mathrm{a}}{z_\mathrm{o}} z^*, \tag{7}$$

because the geometric image term and the diffraction term become negligible compared to the defocusing term [10]. Note that for sufficiently defocused particle images as generally applies for DPTV, particle size accordingly has negligible influence on the resulting PI size.

For the optical distance

$$z^* = \frac{z_{\mathrm{g},2}}{n_\mathrm{g}} = 0.6\,\mathrm{mm}, \tag{8}$$

the vertical axis of PIs at the glass-water interface can now be determined via Equation 7, which results in $a_\mathrm{v} = 116.85\,\mathrm{\mu m}$. For any particle beyond the glass-water interface, z^* expands to

$$z^* = \frac{z_{\mathrm{g},2}}{n_\mathrm{g}} + \frac{z_\mathrm{w}}{n_\mathrm{w}}, \tag{9}$$

which likewise also modifies the vertical PI diameter inside the CBG to

$$a_\mathrm{v}(z_\mathrm{w}) = \frac{M D_\mathrm{a}}{z_\mathrm{o}} \left(\frac{z_{\mathrm{g},2}}{n_\mathrm{g}} + \frac{z_\mathrm{w}}{n_\mathrm{w}}\right). \tag{10}$$

Note that the optical distance z_w in the water can be also translated into a CBG-fixed frame of reference, where the radial particle location r and its distance to the outer wall z_w can be converted via $z_\mathrm{w} = R - r$; see Figure 1(c).

The determination of the defocus sensitivity can be approximated via the slope of Equation 10, i.e.

$$\frac{\partial z}{\partial a} = \frac{\Delta z_\mathrm{w}}{\Delta a_\mathrm{v}} = \frac{z_\mathrm{o} n_\mathrm{w}}{M D_\mathrm{a}}, \tag{11}$$

where Δz_w is the displacement of a particle along the optical axis and Δa_v the corresponding change in the PI's vertical axis. For the given experimental configuration the defocusing sensitivity and vertical axis for particles at the glass-water interface are finally determined to $44\,\mathrm{\mu m/px}$ and $a_\mathrm{v}(z_\mathrm{w} = 0) = 18.12\,\mathrm{px}$.

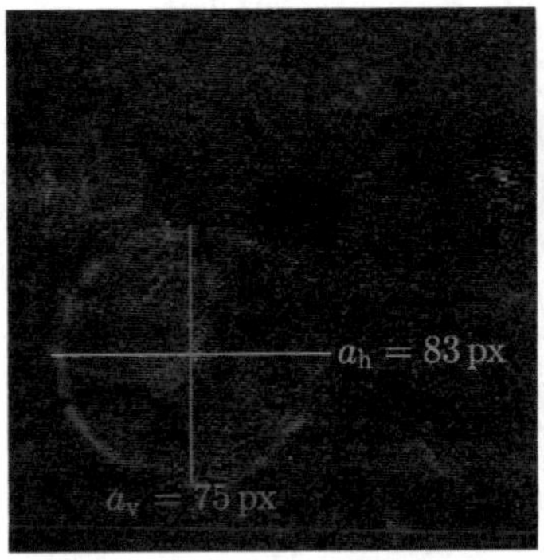

(a) preprocessed DPTV result

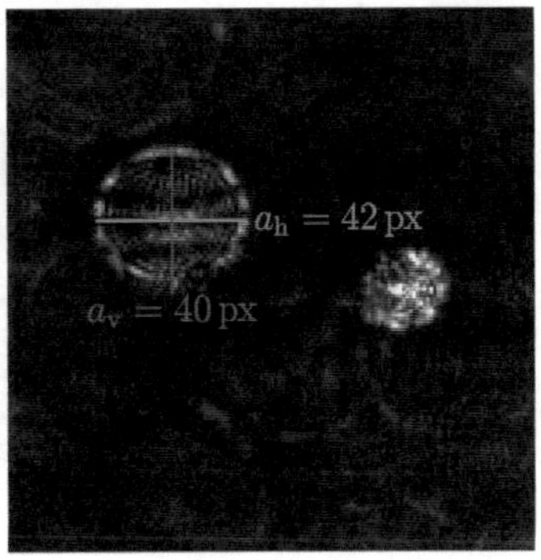

(b) preprocessed IPI result

Figure 2. Random PIs of the recorded tracer particles (a) and bubbles (b); manually labeled vertical and horizontal axes a_v and a_h are added to either PI in red and blue to the pre-processed images, respectively.

3. Results

Random results of both classes of recorded PIs are shown in Figure 2, which already comprise the manually-labeled vertical (red) and horizontal (blue) axes, a_v and a_h, of the ellipsoidal PIs. Note that the different PI light intensities between Figure 2(a) and 2(b) indicate the general competing effect of PI diameter and resulting signal-to-noise (SNR) ratio for a given scatter-light intensity (cp. also Leister *et al.* [10] in context of circular PIs).

To evaluate the impact of astigmatism originating from the CBG curvature on the resulting PI shape, both axes of the 342 manually labeled PIs of tracer particles (DPTV, green) and bubbles (IPI, red) are plotted against one another in Figure 3. Furthermore, a bisecting dashed line is added to the diagram to indicate the case of circular PIs as reference slope for the evaluation of the additional effect of astigmatism. As expected from the theoretical considerations in Section 2.2 no PI diameters below $a_v = 18.12\,\text{px}$ are registered, which retroactively confirms the applicability of the above approach. Moreover, the slope of the distribution of registered PI diameters falls considerably below the dashed reference. The resulting slope for the PIs is determined from a linear regression and also added to the diagram to have immediate comparability to the circular reference. This diminished slope is observed to likewise hold true for both classes of PIs, i.e. tracer particles (DPTV) and bubbles (IPI), which further confirms the independence of the combined defocus/astigmatism effect on the particle class.

A combined depth-codification and velocity estimation is displayed in Figure 4, where the horizontal PI displacement between two consecutive frames A and B of a double-frame image is shown against the vertical PI diameter a_v (both in pixels). A linear regression of the scattered data is added to the diagram with fixed no-slip boundary condition at the wall, i.e. zero displacement for $a_v = 18.12\,\text{px}$ as indicator for the expected quasi-linear near-wall velocity profile (cp. Leister *et al.* [17]). The scatter margin ($\approx 3\,\text{px}$) of the data points indicates the limited accuracy of the manually conduced labeling approach of the registered PIs. Nonetheless, the scattered results clearly reveal the expected slope of particle displacements. In order to convert the diagram also into physical coordinates, a second abscissa is added at the top of the diagram, which shows the wall distance $z_w = R - r = \frac{\Delta z_w}{\Delta a_v} a_v$. Similarly, the determined PI displacements are converted to circumferential velocities $u_\varphi(z_w)$ via reproduction scale and laser pulse distance, which reasonably resemble the expected velocity range for the stirred water in the CBG.

IOP Conf. Series: Materials Science and Engineering **1322** (2024) 012005 doi:10.1088/1757-899X/1322/1/012005

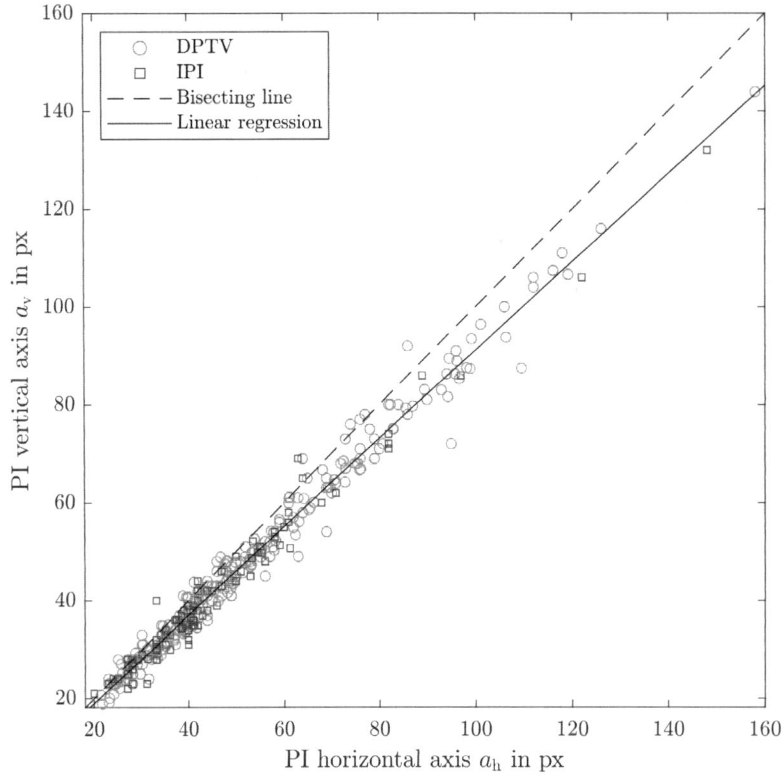

Figure 3. Evaluation of PI size and shape upon the combined influence of defocussing and astigmatism effects for 342 manually labeled particle images: The ordinate shows the vertical axis a_v, which only contains the imprint of defocussing, the abscissa shows the horizontal axis a_h, which furthermore comprises the stretching effect of astigmatism. A bisecting dashed line indicates the slope of circular PIs as reference for the linear regression of the registered PIs (solid line) the latter revealing a diminished slope.

4. Concluding Remarks

In continuation of earlier efforts towards quantitative laser-optical back-scatter diagnostics of the annular two-phase flow in RPDCs [10, 17, 18] the presented combined theoretical and experimental study addresses additional optical considerations as inevitably occurs for imaging measurements across curved surfaces. To mimic the impact of astigmatism on the OTF of the experimental approach, a CBG of comparable diameter as apply for typical RPDCs is chosen for the measurements, which reveals a mild yet significant impact of the curved surface on the resulting shape deformation of the recorded PIs. The correspondingly distinctively different defocus-sensitivity slopes for horizontal and vertical axes are, therefore, hypothesized to provide additional quantitative depth localization measures.

The derivation of the underlying OTF for the given problem comprises the three separate contributions of the air between the imaging equipment and the glass wall, the thickness of the wall, and the depth location inside the flow domain. Consequently, this OTF is straightforwardly applicable for future experimental efforts of similar geometrical/optical configurations. Moreover, the defocus PI-diameter determination for the no-slip condition at

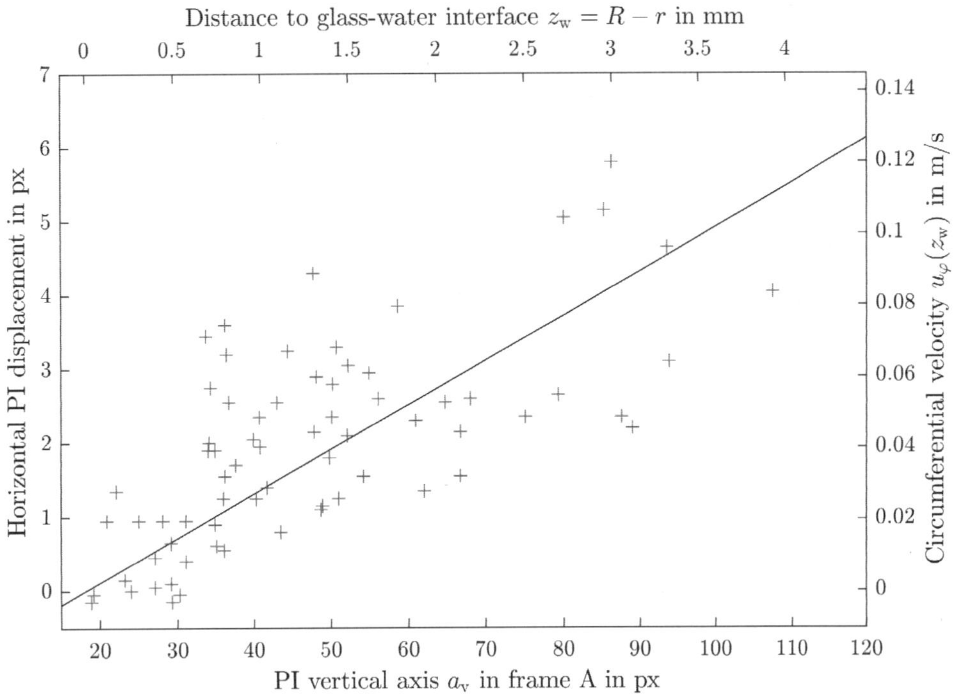

Figure 4. Particle-displacement evaluation for PI pairs. The lower and upper abscissae respectively show the vertical PI diameter a_v and correspondingly determined physical depth localization $z_w = R - r$. The latter can be calculated from Equation 10 as explained in Section 2.2; the left ordinate reveal the horizontal PI displacement in pixels between two consecutive frames A and B of a double-frame image; the right ordinate reveals the circumferential velocity $u_\varphi(z_w)$ of the respective particles as function of wall distance $z_w = R - r$. The two ordinates are equivalent and can be transformed into one another through the application of reproduction scale and time interval Δt. A linear regression of measured data is added to the diagram with fixed no-slip boundary condition at the wall.

the wall-liquid interfaces is shown to provide valuable a-priori information for the calibration of particle displacement curves. To improve the calibration procedure further, it seems promising to further merge the present OTF-based a-priori approach with both in-situ calibration [11, 12] but also a purpose-developed calibration image stack for the so-called general defocusing particle tracking (GDPT) [23, 24]. The latter is foreseen to also take immediate advantage of the development of the above-mentioned different defocus sensitivity slopes for horizontal and vertical axes.

As a final remark, the present study demonstrates the combined and simultaneous applicability of DPTV and IPI in backscatter configuration for the investigation of RPDC-related two-phase gap flows. However, to advance beyond the present proof-of-concept level, more sophisticated conventional and also machine-learning (ML) assisted pattern-identification approaches seem mandatory, which is indicated through the considerably scattered data margin of the presented results. Similar to ring detection in DPTV [25, 26], in this context current efforts at ISTM also center around an ML-based detection and classification of DPTV and IPI result, where the experimental results of the present study represent a preliminary subset of labeled ground-truth information for the supervised learning process.

Acknowledgments
This work was funded by the Deutsche Forschungsgemeinschaft (DFG, German Research Foundation) via Project Grant KR4775/4-1 within the Research Unit FOR 5595 Archimedes (Oil-refrigerant multiphase flows in gaps with moving boundaries — Novel microscopic and macroscopic approaches for experiment, modeling, and simulation) — Project Number 510921053.

References
[1] Nikolov A and Brümmer A 2023 Two-phase mass flow rate through restrictions in liquid-flooded twin-screw compressors or expanders *International Journal of Refrigeration* **148** 152–167 URL `https://doi.org/10.1016/j.ijrefrig.2023.01.017`
[2] Xin D, Feng J, Jia X and Peng X 2010 An investigation into oil—gas two-phase leakage flow through micro gaps in oil-injected compressors *Proceedings of the Institution of Mechanical Engineers, Part C: Journal of Mechanical Engineering Science* **224** 925–933 URL `https://doi.org/10.1243/09544062JMES1704`
[3] Vasuthevan H and Brümmer A 2019 Generic experimental investigation of hydraulic losses within twin-screw machines *The 9th International Conference on Compressor and Refrigeration.* **Xi'an (China)** 10pp
[4] Patel B, Kovacevic A, Charogiannis A, Alam M N and Schütte M 2021 The use of laser-induced fluorescence to measure temperature in the leakage gaps of oil-free positive displacement rotary machines *Measurement* **185** 110057 URL `https://doi.org/10.1016/j.measurement.2021.110057`
[5] Patel B, Kovacevic A, Plantegenet T and Tam T 2023 Study of leakage flow in oil-free positive displacement rotary machines using particle image velocimetry *Experimental Thermal and Fluid Science* **145** 110886 URL `https://doi.org/10.1016/j.expthermflusci.2023.110886`
[6] Scarano F 2013 Tomographic piv: principles and practice *Measurement Science and Technology* **24** 012001 URL `https://dx.doi.org/10.1088/0957-0233/24/1/012001`
[7] Schröder A and Schanz D 2023 3D lagrangian particle tracking in fluid mechanics *Annual Review of Fluid Mechanics* **55** 511–540 URL `https://dx.doi.org/10.1146/annurev-fluid-031822-041721`
[8] Barnkob R, Kähler C J and Rossi M 2015 General defocusing particle tracking *Lab on a Chip* **15** 3556–3560 URL `https://doi.org/10.1039/C5LC00562K`
[9] Leister R, Fuchs T, Mattern P and Kriegseis J 2021 Flow-structure identification in a radially grooved open wet clutch by means of defocusing particle tracking velocimetry *Experiments in Fluids* **62** 29 URL `https://doi.org/10.1007/s00348-020-03116-0`
[10] Leister R, Fuchs T and Kriegseis J 2023 Defocusing ptv applied to an open wet clutch: From macro to micro *Experiments in Fluids* **64** 94 URL `https://doi.org/10.1007/s00348-023-03623-w`
[11] Fuchs T, Hain R and Kähler C J 2016 In situ calibrated defocusing ptv for wall-bounded measurement volumes *Measurement Science and Technology* **27** 084005 URL `https://doi.org/10.1088/0957-0233/27/8/084005`
[12] Fuchs T, Hain R and Kähler C J 2016 Uncertainty quantification of three-dimensional velocimetry techniques for small measurement depths *Experiments in Fluids* **57** 73 URL `https://doi.org/10.1007/s00348-016-2161-5`
[13] Tropea C 2011 Optical particle characterization in flows *Annual Review of Fluid Mechanics* **43** 399–426 URL `https://doi.org/10.1146/annurev-fluid-122109-160721`
[14] Fansler T D and Parrish S E 2014 Spray measurement technology: a review *Measurement Science and Technology* **26** 012002 URL `https://doi.org/10.1088/0957-0233/26/1/012002`
[15] Dehaeck S and van Beeck J 2008 Multifrequency interferometric particle imaging for gas bubble sizing *Experiments in Fluids* **45** 1432–1114 URL `https://doi.org/10.1007/s00348-008-0502-8`
[16] Kawaguchi T, Akasaka Y and Maeda M 2002 Size measurements of droplets and bubbles by advanced interferometric laser imaging technique *Measurement Science and Technology* **13** 308–316 URL `https://doi.org/10.1088/0957-0233/13/3/312`
[17] Leister R, Brümmer A and Kriegseis J 2022 Laser-optical shear-flow analysis across the annular gap of a simplified displacement compressor model *IOP Conference Series: Materials Science and Engineering* **1267** 012003 URL `https://doi.org/10.1088/1757-899X/1267/1/012003`
[18] Sax C, Dreisbach M and Kriegseis J 2023 Interferometric particle imaging for particle sizing in the front-, side-, and back-scatter region *arXiv preprint arXiv:2303.16013* URL `https://doi.org/10.48550/arXiv.2303.16013`
[19] Cierpka C, Rossi M, Segura R and Kähler C J 2011 On the calibration of astigmatism particle tracking velocimetry for microflows *Measurement Science and Technology* **22** 015401 URL `https://doi.org/10.1088/0957-0233/22/1/015401`

[20] Rossi M and Kähler C J 2014 Optimization of astigmatic particle tracking velocimeters *Experiments in Fluids* **55** URL https://doi.org/10.1007/s00348-014-1809-2

[21] Raffel M, Willert C E, Scarano F, Kähler C J, Wereley S T and Kompenhans J 2018 *Particle Image Velocimetry* (Springer-Verlag GmbH) URL https://doi.org/10.1007/978-3-319-68852-7

[22] Olsen M G and Adrian R J 2000 Out-of-focus effects on particle image visibility and correlation in microscopic particle image velocimetry *Experiments in Fluids* **29** S166–S174 URL https://doi.org/10.1007/s003480070018

[23] Barnkob R and Rossi M 2020 General defocusing particle tracking: fundamentals and uncertainty assessment *Experiments in Fluids* **61** 110 URL http://dx.doi.org/10.1007/s00348-020-2937-5

[24] Barnkob R and Rossi M 2021 Defocustracker: A modular toolbox for defocusing-based, single-camera, 3d particle tracking *Journal of Open Research Software* **9** 22 URL https://doi.org/10.5334/jors.351

[25] Dreisbach M, Leister R, Probst M, Friederich P, Stroh A and Kriegseis J 2022 Particle detection by means of neural networks and synthetic training data refinement in defocusing particle tracking velocimetry *Measurement Science and Technology* **33** 124001 URL http://dx.doi.org/10.1088/1361-6501/ac8a09

[26] Sax C, Dreisbach M, Leister R and Kriegseis J 2023 Deep learning and hybrid approach for particle detection in defocusing particle tracking velocimetry *Measurement Science and Technology* **34** 095909 URL http://dx.doi.org/10.1088/1361-6501/acd4b4

IOP Conf. Series: Materials Science and Engineering **1322** (2024) 012006 doi:10.1088/1757-899X/1322/1/012006

On Rotor Profiling of Internally Geared Screw Machines

H Lacevic, A Kovacevic, N Stosic and M Read

Department of Engineering, City, University of London, London, UK

E-mail: halil.lacevic@city.ac.uk

Abstract. The internally geared screw machine represents a novel type of positive displacement compressor which consists of an inner and outer rotor. Both rotors rotate in the same direction but are each centered on offset parallel axes. The rotor profiles are designed to create multiple continuous contact points between the rotors, forming several separate working chambers whose volumes vary from minimum to maximum and back to minimum during a single rotation of the outer rotor. For a gas or two-phase working fluid, adjusting the discharge port geometry allows internal compression to occur before discharge. Previous research has focused on using the well-known rotor profiling method, which employs a circular pin to generate the inner and outer rotor profiles. Although the rack method for rotor profile generation has been described and investigated for conventional screw machine rotor profiles, it has never been applied to internally geared screw machine profile generation. This paper provides an initial description and application of the rack method for generating internally geared screw machine profiles. Potential benefits of using the rack method compared to conventional methods for rotor profile generation in internally geared machines are discussed. Additionally, the limitations of using the rack method for internal gearing are presented and illustrated through various examples and applications.

1. Introduction

Conventional twin screw compressors are a type of positive displacement machines used for gas compression and are widely used in industrial and refrigeration applications. Twin screw machines consist of two inter-meshing helical rotors (main and gate rotor) in a fixed casing. A number of separate working chambers with varying volume are formed between rotor lobes and casing as the rotors rotate. On a suction side, when the working chamber is first formed, gas is drawn into the working chamber, trapped, compressed and released through the discharge port on the other side of the machine. These type of compressor machines have presented several advantages, including high efficiency and the ability to handle varying loads and pressures. Geometry of a twin screw machine is very well known and various methods for generating rotor profiles are defined. Apart from the geometry, performance prediction tools including quasi one-dimensional chamber modelling [1] and computational fluid dynamics (CFD)[2] are widely used for twin screw compressors.

Most recent studies on screw compressors have focused on understanding the influence of the various leakages in conventional twin screw machines and improving the geometry of these machines to produce even more efficient configurations. There are not many studies investigating alternatives to the conventional twin screw machines, however a novel design, internally geared screw machine, providing slightly different configuration of rotors has been introduced by Read [3]. Internally geared screw machines are based on a design of gerotor pump which is commonly used in oil pumping applications. Gerotor pumps and internally geared screw machines consist of two internally geared rotors (outer and inner rotor) which mesh together. To produce separate working chambers, continuous contact between inner and outer rotor is required for both machines, however to reduce power transfer between the rotors in internally geared screw machines, a helical twist is added to both rotors, while gerotor pumps implement straight cut rotors. Other significant difference between gerotor pumps and internally geared screw machines is initial volume of working chamber. To produce efficient compression, a zero initial volume when the working chamber is first formed is required for internally geared screw machines.

Working chambers in internally geared screw machine are formed between outer and inner rotors and gas is drawn in these chambers, carried and released on the other side of the machine. Control of the periods at which the gas is allowed to enter or exit the working chamber allows the compression to occur. The control of these periods is done by providing specifically designed end ports for both suction and discharge side which were initially introduced by Read[3]. An example of end port plates on a model of internally geared screw machine is presented in Figure 1.

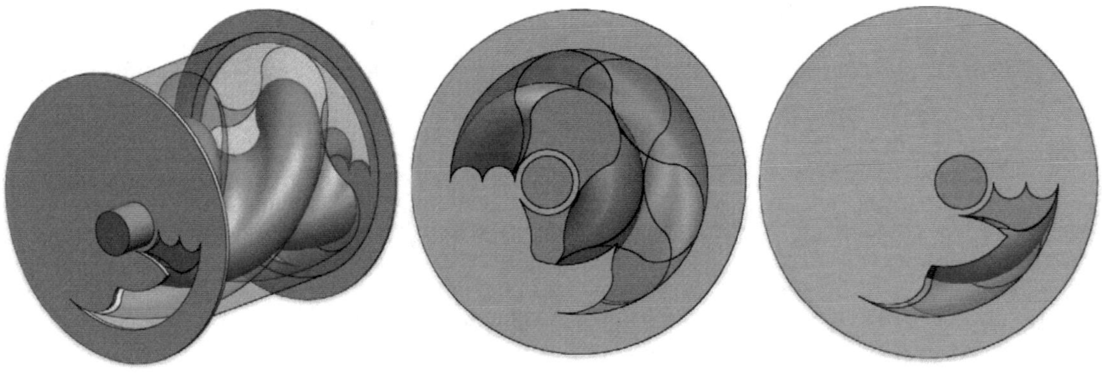

Figure 1. Internally geared screw machine model showing suction and discharge end port plates[3].

Internally geared screw machine has several potential advantages compared to the conventional twin screw machines as mentioned by Read[3]:

- Elimination of rotor-to-casing tip clearances present in conventional screw machines, as the outer rotor encloses the working chamber. The two rotors of the internally geared machine effectively share the same tip leakage areas.
- Elimination of the blow hole area characteristic of conventional machines.
- Reduced viscous drag due to the much lower relative speed at the contact lines between rotors, as compared to the rotor-casing leakage lines in a conventional machine.
- Reduced distortion due to thermal expansion, as the rotation of both rotors will lead to a uniform circumferential temperature in each during operation, with temperature variation limited to the axial direction.

The geometry of internally geared screw machine has been extensively investigated in previous studies [3, 4], which have implemented a variety of rotor profiling methods. In conventional twin screw machines, rotor profiles are commonly generated using the rack method, which offers several advantages compared to other methods, particularly in terms of manufacturing [5]. However, the rack method has not been previously applied to internally geared screw machines. This study focuses on investigating whether the rack method can be utilized for internally geared profiles, which require a zero initial working chamber volume and continuous contact between the rotors. This paper provides initial guidance for rotor profiling in internally geared screw machines and presents work in progress on exploring the potential benefits and limitations of applying the rack method.

2. Internally Geared Screw Machine Rotor Profiles

The design of a compressor machine begins with defining the shapes of rotor profiles, which significantly influence factors such as working chamber volume, leakage paths, and overall machine performance. An internally geared screw machine is consisted of two internally geared rotors (outer and inner) which rotate in the same direction about non-coincident parallel axes. The outer rotor profile forms a conjugate pair with the inner rotor, ensuring continuous points of contact between the rotors during rotation. This continuous contact enables the separation of the volume between the rotors into multiple working chambers. The volume of each working chamber varies with rotor position, depending on the chosen rotor geometry. Another crucial aspect affecting compressor machine performance is the minimum volume of a working chamber, which also depends on the selected geometry. To achieve efficient compression, a zero-minimum volume and continuous contact between the rotors are necessary. Various methods for generating rotor geometry for gerotor machines are described by Colbourne [6], Beard et al. [7], Vecchiato et al. [8], and Hsieh et al. [9].

Previous research concentrated on basic geometries as outlined by Moineau [10], where the internally geared rotor geometry is based on epi & hypocycloid curves. While these profiles offer a convenient method for generating rotor profiles, recent studies have shifted focus towards circular pin-generated rotor profiles, following the method detailed by Vecchiato et al[8]. Circular pin-generated profiles offer a broader range of profile shapes and a relatively straightforward approach to profile generation for internally geared screw machines.

The rack method, as described by Stosic [11], is commonly used for rotor profile generation in conventional twin screw machines. While the rack method has been widely used in conventional twin screw machines, it has not been applied to internally geared screw machines. Further investigation is required to determine whether the rack method can be adapted for internally geared profiles while maintaining a zero-minimum volume of the working chamber and ensuring continuous contact between the rotors. Additionally, it is essential to define the advantages and limitations of the rack method compared to other rotor profiling techniques.

2.1. *Circular Pin-Generated Rotor Profiles*

Internal pin-generated gearing can be used for generating meshing rotor profiles. The inner rotor must have one more or one less lobe than the outer rotor to ensure continuous contact[4]. In the case of internally geared screw machines, an extended epicycloid profile is formed on the gear with the smaller centrode when the profile of the gear with the larger centrode is partially defined by the circular pin. Circular pins on the outer rotor, consisting of N_1 lobes, are defined in the coordinate system S_1, fixed to the initial position of the outer rotor and centered on the outer rotor pitch circle[4]. Coordinate system S_2 is fixed to the initial position of the inner rotor and centered on the inner rotor pitch circle. Both coordinate systems S_1 and S_2 rotate with the rotors, while coordinate system S_f is fixed to the casing and defines a global coordinate system centered on the outer rotor pitch circle. The angle between coordinate systems S_1 and S_f defines the outer rotor rotational position Φ_1 relative to the global coordinate system.
The gearing ratio between the outer and inner rotors can be calculated as $m_{21} = \frac{N_2}{N_1}$, where $N_2 = N_1 - 1$ is the number of lobes on the inner rotor. In this case, the angle between coordinate systems S_2 and S_f defines the rotational position of the inner rotor $\Phi_2 = \frac{1}{m_{21}}\Phi_1$ relative to the global coordinate system S_f. The centers of coordinate systems S_1 and S_2 are separated by the axis distance between the rotors (E). The transformation of points from coordinate system S_1 to coordinate systems S_2 and S_f can be accomplished using transformation matrices as presented in Equations 1 and 2.

$$M_{12} = \begin{bmatrix} \cos(\Phi_2 - \Phi_1) & \sin(\Phi_2 - \Phi_1) & 0 & -E\sin(\Phi_1) \\ -\sin(\Phi_2 - \Phi_1) & \cos(\Phi_2 - \Phi_1) & 0 & -E\cos(\Phi_1) \\ 0 & 0 & 1 & 0 \\ 0 & 0 & 0 & 1 \end{bmatrix} \tag{1}$$

$$M_{1f} = \begin{bmatrix} \cos(\Phi_1) & -\sin(\Phi_1) & 0 & 0 \\ \sin(\Phi_1) & \cos(\Phi_1) & 0 & 0 \\ 0 & 0 & 1 & 0 \\ 0 & 0 & 0 & 1 \end{bmatrix} \tag{2}$$

The shapes of rotor profiles are entirely determined and defined by the radius of the circular pin (ρ) and its center distance (a) from the center of the global coordinate system S_f, as illustrated in Figure 2.

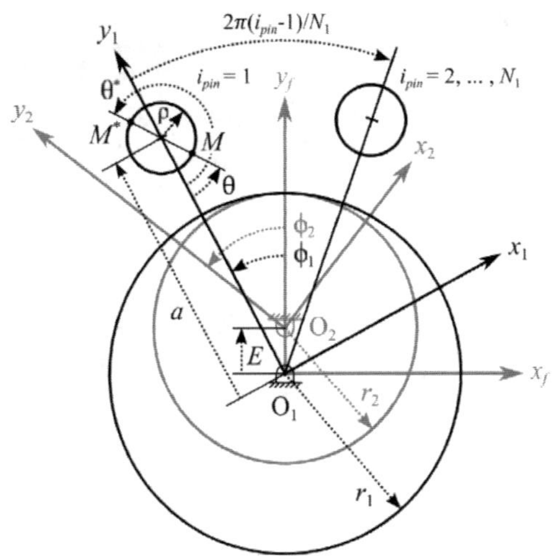

Figure 2. Circular pins and coordinate systems used for internally geared screw machine rotor generation in circular pin-generated profiles[4].

The points of a single circular pin are defined in coordinate system S_1 as a function of θ, where θ ranges from 0 to 2π, and can be expressed in matrix form, as depicted in Equation 3. The points of the inner rotor profile can be determined by rotating a single circular pin in coordinate system S_1 and then transforming it to the second coordinate system S_2, while applying the first equation of meshing as described by Vecchiato[8]. In this configuration, the inner rotor profile is entirely defined by the circular pin and can subsequently be used to generate points for the outer rotor profile. To find the outer rotor profile, the points of the inner rotor profile are transformed from coordinate system S_2 to S_1, while applying the second equation of meshing detailed by Vecchiato[8]. A visual representation of the circular pin-generation method is provided through examples of the inner and outer rotor profile enveloping, as depicted in Figure 3.

$$r_1 = [\rho\cos(\theta) \quad \rho\sin(\theta) \quad 0 \quad 1]^T \tag{3}$$

Figure 4 illustrates examples of profiles achievable through the circular pin-generation method. These profiles are characterized by the lobe number on the outer rotor, N_1, along with the non-dimensional parameters $\lambda = a/r_1$ and $\sigma = \rho/r_1$, and the axis spacing distance $E = r_1/N_1$. However, certain limitations exist on these values to prevent the occurrence of undercuts, which is extensively discussed by Read[4].

IOP Conf. Series: Materials Science and Engineering **1322** (2024) 012006 doi:10.1088/1757-899X/1322/1/012006

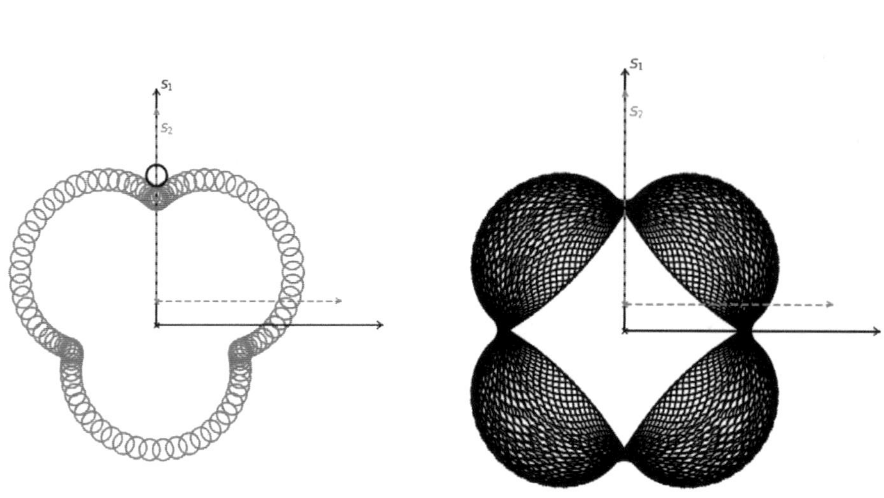

(a) Example of circular pin defined in S_1 coordinate system (black) enveloping the inner rotor profile in S_2 (red).

(b) Example of generated inner rotor profile enveloping the outer rotor profile.

Figure 3. Visual representation of circular pin-generation method.

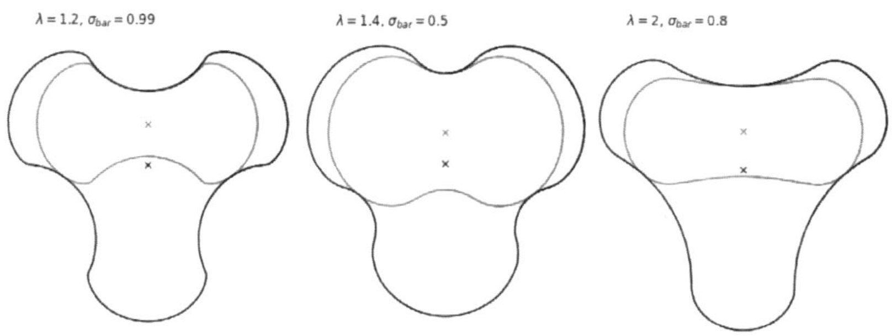

Figure 4: Example of profiles generated using circular pin-generation method with $N_1 = 3$.

2.2. Rack Generated Rotor Profiles

Conventional twin-screw machine rotor profiles are typically generated using the rack method, which leverages the principles of the Envelope Theory formulated by Euler [12] in the field of differential geometry. This theory allows for the mathematical calculation of the equation of a curve that envelopes a family of curves, making it applicable for generating equations of conjugate shapes on paired rotors. Initially introduced by Sakun [13], this method in rotor profiling involves defining a mathematical equation, known as the meshing condition, that must be satisfied by the generated meshing curve on a pair of rotors. The early approach focused on deriving analytical meshing conditions, which provided analytically representable conjugates of selected analytical curves such as circles, lines, and points. An improved approach, described by Stosic [11], significantly simplified the design procedure. This improvement led to much wider set of possible profile shapes and allowed profiles to be defined by various sections on the rack providing the ability to design various type of contact points at specific locations. Using the enveloping method, for any given rotor profile, a corresponding conjugate rotor profile depending on it's pitch circle can be found by applying the meshing condition. The rack rotor represents a special case when the pitch circle has "infinite radius". In other words, rack rotor represents straight rotor that is used to envelope main and gate rotor. The shape of the rack can be defined by segments which makes this method very powerful. An example of externally geared rotor profiles produced using the rack method is shown in Figure 5. In Figure 5 sine wave function is used for generating the rack curve. Rack method for conventional twin screw machines has the following procedure:

- Defining shape of the rack curve.
- Defining pitch circles for both main and gate rotors.
- Solving meshing condition between the rack and main rotor profile.
- Solving meshing condition between the rack and gate rotor profile.

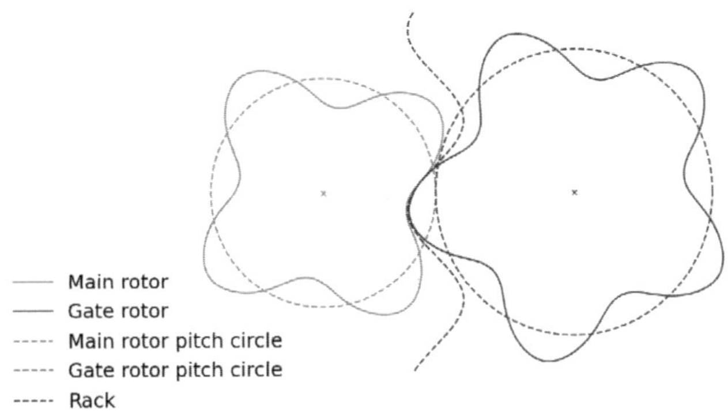

Figure 5. Profiles with external meshing produced using sine wave as a rack.

Although the rack method has been widely used in conventional twin-screw machines, it has never been applied to internally geared screw machines. However, the rack method is commonly used in internal gearing, particularly for generating involute profiles for gears. It is known that the rack method can produce internal gears, but the question remains whether it can produce internal gears that satisfy the necessary conditions for continuous contact between the rotors and a zero-minimum working chamber volume, which is essential for efficient compression in internally geared screw machines.

The rack method for internally geared screw machines begins by defining the rack curve in the global coordinate system. For any given rack curve (x_r, y_r), the coordinates of the inner and outer rotor profiles can be found using Equations 4 and 5, respectively. The pitch circle radii of the inner and outer rotors are denoted as r_{2p} and r_{1p}, while Ψ_i, $i = [1, 2]$ represents the solutions of the meshing condition between the rack and each rotor profile. The solutions to the meshing conditions, as described by Stosic [11], can be determined using Equations 6 and 7.

$$\begin{bmatrix} x_1 \\ y_1 \\ 0 \\ 1 \end{bmatrix} = \begin{bmatrix} \cos(\Psi_1) & -\sin(\Psi_1) & 0 & r_{1p}\Psi_1 \sin(\Psi_1) \\ \sin(\Psi_1) & \cos(\Psi_1) & 0 & -r_{1p}\Psi_1 \cos(\Psi_1) \\ 0 & 0 & 1 & 0 \\ 0 & 0 & 0 & 1 \end{bmatrix} \begin{bmatrix} x_r + E \\ y_r \\ 0 \\ 1 \end{bmatrix} \tag{4}$$

$$\begin{bmatrix} x_2 \\ y_2 \\ 0 \\ 1 \end{bmatrix} = \begin{bmatrix} \cos(\Psi_2) & -\sin(\Psi_2) & 0 & r_{2p}\Psi_2 \sin(\Psi_2) \\ \sin(\Psi_2) & \cos(\Psi_2) & 0 & -r_{2p}\Psi_2 \cos(\Psi_2) \\ 0 & 0 & 1 & 0 \\ 0 & 0 & 0 & 1 \end{bmatrix} \begin{bmatrix} x_r \\ y_r \\ 0 \\ 1 \end{bmatrix} \tag{5}$$

$$\Psi_1 = (y_r - \frac{\partial x_r}{\partial y_r}(r_{1p} - x_r - E))/r_{1p} \tag{6}$$

$$\Psi_2 = (y_r - \frac{\partial x_r}{\partial y_r}(r_{2p} - x_r))/r_{2p} \tag{7}$$

An example of internally geared profiles generated using the previously described rack method, with a sine wave used as the rack, is shown in Figure 6. From Figure 6, it is evident that the rack method can generate internally geared profiles that mesh around the pitch point, a typical characteristic in gears. However, it is also apparent that using the selected sine wave for the rack curve does not achieve continuous contact between the rotors.

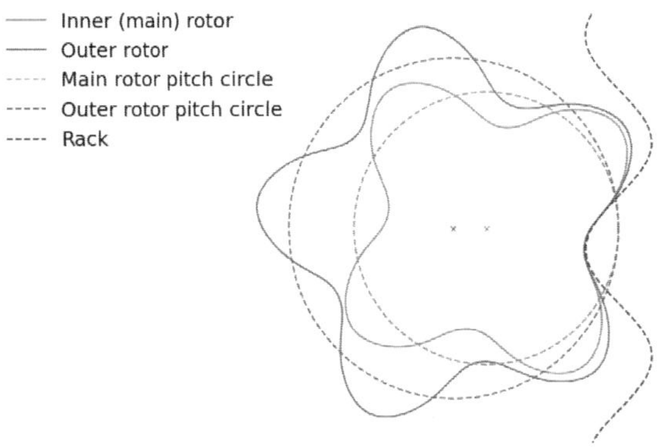

Figure 6. Internally geared profiles generated using the rack method and a sine wave as a rack curve.

As previously mentioned and shown in Figure 6, the rack method can produce profiles with internal meshing. However, whether the rack method can generate profiles that ensure continuous contact between the rotors and a zero-minimum volume of a working chamber has not been investigated. Main focus of this paper is to presents an initial investigation into the use of this method in internally geared screw machines, focusing on the cycloid curve for the rack profile. A cycloid is the curve traced by a point on the circumference of a circular wheel as it rolls along a straight line without slipping, characterized by its repetitive, arch-like shape. The equations for the coordinates of a rack curve based on the cycloid curve, as a function of the angle θ_c, are given in Equations 8 and 9, where $2R_c^{(a)}$ and $2R_c^{(d)}$ define the profile addendum and dedendum, respectively. To ensure that the rack curve defined by the parametric Equations 8 and 9 is continuous, the condition $R_c^{(a)} + R_c^{(d)} = E$ must be met. Examples of rack curves generated using these equations are shown in Figure 7.

$$x_r = \begin{cases} R_c^{(a)}(\theta_c - \sin(\theta_c)) & \text{if } (\theta_c \mod 4\pi) < 2\pi \\ R_c^{(d)}(\theta_c - \sin(\theta_c)) & \text{if } 2\pi \leq (\theta_c \mod 4\pi) < 4\pi \end{cases} \tag{8}$$

$$y_r = \begin{cases} R_c^{(a)}(1 - \cos(\theta_c)) & \text{if } (\theta_c \mod 4\pi) < 2\pi \\ R_c^{(d)}(-1 + \cos(\theta_c)) & \text{if } 2\pi \leq (\theta_c \mod 4\pi) < 4\pi \end{cases} \tag{9}$$

An example of valid rotor profiles with $N_1 = 5$ and $N_2 = 4$ generated using a cycloid-based rack curve with $R_c^{(a)} = [0.5, 0.25]$ is shown in Figure 8. This figure demonstrates that conjugate profiles have been successfully produced using the rack method and that the necessary conditions for achieving continuous contact between the rotors and maintaining a zero-minimum working chamber volume are satisfied.

Figure 7. Example of rack curves based on parametric equations given in Equation 8 and 9

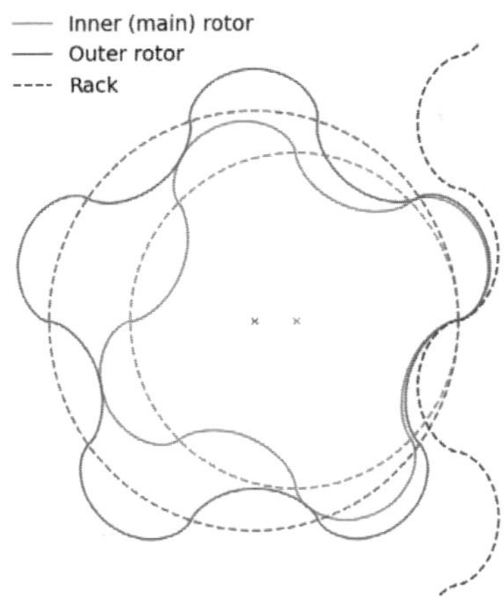

(a) Cycloid based rack curve with $R_c^{(a)} = 0.5$.

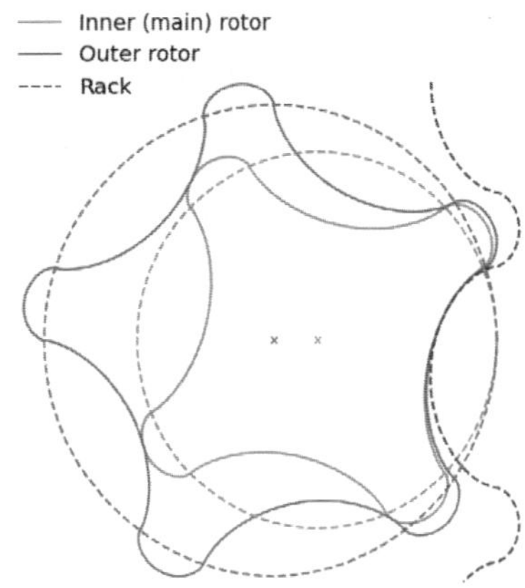

(b) Cycloid based rack curve with $R_c^{(a)} = 0.25$.

Figure 8. Example of valid rotor profiles generated using cycloid based rack curve.

3. Conclusion

Internally geared screw machines represent a novel type of compressor whose efficiency has not been fully investigated. Previous studies on these machines have primarily used circular pin-generated profiles. In contrast, conventional twin screw compressors typically employ the rack method for generating rotor profiles. The rack method offers several advantages, including the ability to produce a wider range of rotor profile shapes and easier manufacturing. Although the rack method has been used to produce internal gears where continuous contact is not required, its potential to generate rotor profiles with continuous contact and a zero-minimum working chamber volume has not been investigated until now.

This study applied the rack method for the first time to generate profiles for internally geared screw machines. The initial investigation introduced a rack curve based on cycloids, demonstrating that the resulting rotor profiles are valid for use in internally geared screw machines. The findings indicate that the rack method can produce rotor profiles that ensure continuous contact and maintain a zero-minimum working chamber volume. However, the limitations of this approach remain unknown.

Future research should focus on determining whether the cycloid-based rack curve is a special case and on generalizing the conditions needed to produce valid rotor profiles using the rack method for internally geared screw machines. Additionally, the limitations and potential benefits of this method compared to other available methods should be thoroughly examined and defined.

IOP Conf. Series: Materials Science and Engineering **1322** (2024) 012006 doi:10.1088/1757-899X/1322/1/012006

References

[1] Hanjalic K and Stosic N 1997 Development and optimization of screw machines with a simulation model—PART II: Thermodynamic performance simulation and design optimization *Journal of Fluids Engineering* 119(3) 664–670 doi:10.1115/1.2819296

[2] Kovacevic A, Stosic A and Smith I 2007 *Screw Compressors: Three Dimensional Computational Fluid Dynamics and Solid Fluid Interaction* (Berlin: Springer)

[3] Read M G, Smith I K, and Stosic N 2017 Internally geared screw machines with ported end plates *IOP Conference Series: Materials Science and Engineering* 232 012058 doi:10.1088/1757-899x/232/1/012058

[4] Read M, Smith I, and Stosic N 2020 Influence of rotor geometry on tip leakage and port flow areas in Gerotor-type twin screw compressors *Proceedings of the Institution of Mechanical Engineers, Part E: Journal of Process Mechanical Engineering* 236(1) 94–102 doi:10.1177/0954408920962412

[5] Stosic N, Smith I K, Kovacevic A and Mujic E 2011 Geometry of screw compressor rotors and their tools *Journal of Zhejiang University-SCIENCE A* 12(4) 310–326 doi:10.1631/jzus.a1000393

[6] Colbourne J 1974 The geometry of trochoid envelopes and their application in rotary pumps *Mechanism and Machine Theory* 9(3–4) 421–435 doi:10.1016/0094-114x(74)90025-1

[7] Beard J, Yannitell D and Pennock G 1992 The effects of the generating pin size and placement on the curvature and displacement of epitrochoidal gerotors *Mechanism and Machine Theory* 27(4) 373–389 doi:10.1016/0094-114x(92)90030-l

[8] Vecchiato D, Demenego A, Argyris J and Litvin F L 2001 Geometry of a cycloidal pump *Computer Methods in Applied Mechanics and Engineering* 190(18–19) 2309–2330 doi:10.1016/s0045-7825(00)00236-x

[9] Hsieh C-F and Hwang J-W 2006 Geometric design for a gerotor pump with high area efficiency *Journal of Mechanical Design* 129(12) 1269–1277 doi:10.1115/1.2779887

[10] Moineau R J L 1932 Gear mechanism

[11] Stošić N 1998 On gearing of helical screw compressor rotors *Proceedings of the Institution of Mechanical Engineers, Part C: Journal of Mechanical Engineering Science* 212(7) 587–594 doi:10.1243/0954406981521556

[12] Guggenheimer H W 1977 *Differential Geometry* (New York: Dover Publications)

[13] Sakun I A 1960 *Screw Compressors* (Moscow: Mashgiz) (in Russian)

Acknowledgement
Funding for this research was received from Carrier Inc., USA and PDM Analysis Ltd., UK.

IOP Conf. Series: Materials Science and Engineering **1322** (2024) 012007 doi:10.1088/1757-899X/1322/1/012007

Design and Improvement of Curved Envelope Meshing Pair Profile of Single Screw Compressor

W Lei, W Wu*, Z Zhang and C Peng

School of Energy and Power Engineering, Xi'an Jiaotong University, Xi'an, China

*E-mail: weifengwu@mail.xjtu.edu.cn

Abstract. In recent years, the wear problem of the wheel tooth in the single screw compressor (SSC had been solved due to the development and application of various envelope profiles. Among these profiles, the curved envelope profile shows performance advantages, which displays great application potential in the fields of high-temperature heat pump, steam compression, etc. However, SSC with curved envelop profile still faces the problem of incomplete design method, large leakage clearance, difficult processing and meshing pair interference under extreme conditions. Therefore, aiming at solving these problems, a coupling profile design method for the curved envelope meshing pair of the SSC was presented and studied. The profile equation of curved envelop meshing pair of a SSC with an ellipse profile as the characteristic line was established. The method of using standard tooth width was proposed for the quick geometric design of the SSC. The coupling profile design of the star wheel tooth angle and the groove bottom angle with variable diameter circular arc was invented. The equation of the transition arc surface of the bottom angle of the groove and the angle of the star wheel tooth was derived. It provided a theoretical basis for the design and improvement of meshing pair with curved envelop profile.

1. Introduction

The star wheel-screw meshing pair is the core component of the single screw compressor (SSC). The profile of the meshing pair is closely related to the working process and meshing characteristics of the compressor. Therefore, the theoretical research of meshing pair profile has attracted wide attention for a long time.

At present, the straight line envelope profile [1] has been used in most of the SSC products. The process of meshing pair with straight line envelope is simple and costs low. However, there is a serious wear problem in the straight line envelope profile. In order to avoid the problem, Zimmern proposed a column envelope profile [2]. The flank of the star wheel tooth is column, and the surface of the screw groove is the envelope surface of the column. However, the leakage of this type of profile is still quite large, so it is rarely used in engineering practice.

Significantly, based on the idea of 'dispersing abrasion', Feng et al. proposed multi-line envelope and multi-column envelope profiles. Feng et al. designed a dedicated milling machine for the screw processing of multi-column envelope, and the water-lubricated SSC processed in this way has stable performance and has realized industrial application [3].

Despite the above optimization, the utilization rate of the flank of the star wheel tooth is still low due to the discontinuity of the meshing area, and there are some unnecessary leakage channels. In order to improve these problems, Li et al.[4] developed a curved envelope profile, whose structural diagram is shown in Figure 1. It has a continuous and smooth meshing surface, reducing the size of the leakage channel and improving the sealing performance. Screws with curved envelope profile can be processed either by special machine tools or on CNC machine tools.

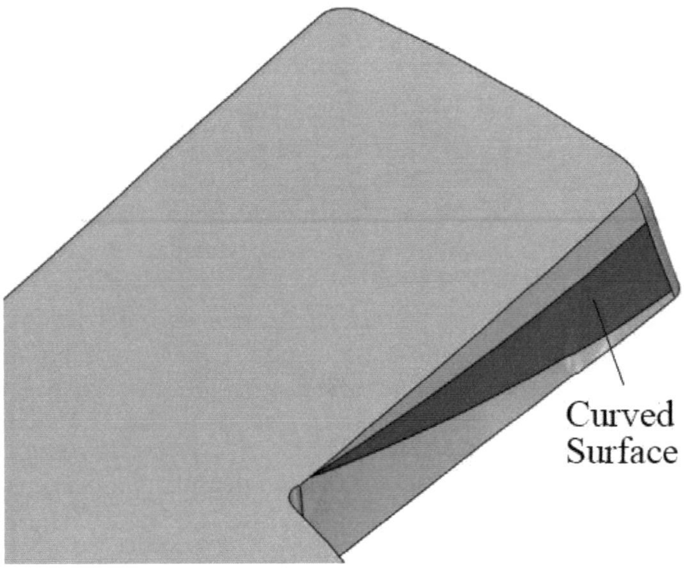

Figure 1. Diagram of the curved envelop profile

In addition to dispersing abrasion, suspension meshing technology can also effectively reduce the wear of meshing pairs. Post proposed a 'double slider' model and analyzed the hydrodynamic lubrication characteristics of the star wheel tooth of SSC [5, 6]. Jensen, Shen and other scholars focused on the geometric characteristics of the meshing pair to explain the influence of hydrodynamic lubrication on the performance of the meshing pair [7, 8]. Huang et al. proposed the theory of suspension meshing, pointing out that reasonable optimization and arrangement of the profile of the star wheel tooth flank could form effective dynamic oil film [9, 10]. The researches have proved that it is a technical route to fundamentally solve the wear problem of meshing pairs by improving the hydrodynamic lubrication characteristics of meshing pairs through profile optimization and realizing the suspension meshing of meshing pairs on the basis of dynamic torque and resistance torque balance matching. Obviously, the design concept of the column envelope profile and the curved envelope profile is consistent with this conclusion.

With the presentation of various SSC meshing pair profiles, the problem of meshing pair abrasion had been solved, and the performance of SSC had been greatly improved. Among these profiles, the curved envelope profile shows great performance advantages. However, there are still several problems in its application. There is no complete design method of curved envelope profile, which limits its application under different scenarios. Meanwhile, the star wheel tooth angle and the screw groove bottom angle in the curved envelope meshing pair are not only difficult to process, but also easy to interfere under extreme conditions, resulting in the failure of the profile at the top of the meshing pair and the deterioration of the thermal and mechanical properties. Besides, the contact line of the star wheel tooth top and the contact line of the star wheel tooth side are discontinuous, resulting in large leakage clearance in the junction area between the star wheel tooth angle and the screw groove bottom angle. Based on these, in this paper, the design method of the coupling profile of the curved envelope meshing pair of the SSC was proposed, and the structure of the curved envelope meshing pair was optimized to ensure that good meshing characteristics were maintained under different conditions, which provided a theoretical basis for the design and improvement of the curved envelope meshing pair of the SSC.

2. Design of curved envelop meshing pair profile

2.1. Design principle of the curved envelop meshing pair profile

In order to obtain a curved envelope profile with excellent performance, the following design criteria should be followed. First, ensure excellent hydrodynamic lubrication characteristics to prevent friction between the meshing pair. Second, guarantee great sealing performance to ease the leakage problem between the star wheel tooth and the side of the screw groove. Third, ensure good processing technology to make the processing flow simple and the processing accuracy high.

The profile design must obey the conjugate meshing principle. Figure 2 shows the relative motion relationship of the star wheel-screw meshing pair, where O_1 and O_3 are the fixed coordinate systems of the star wheel and the screw, respectively, representing the starting position of the star wheel and the screw; O_2 and O_4 are fixed on the star wheel and screw respectively and rotate with the rotor. The angular velocities of the screw and the star wheel are ω_{sr} and ω_{sw}, respectively, and the corresponding screw rotation angle and the star wheel rotation angle are α_{sr} and α_{sw}, respectively.

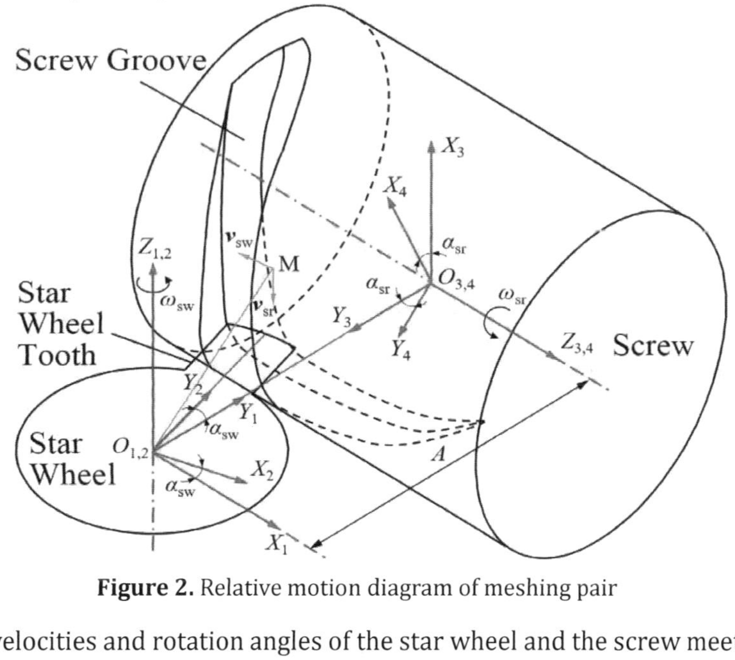

Figure 2. Relative motion diagram of meshing pair

The angular velocities and rotation angles of the star wheel and the screw meet Formula (1):

$$\frac{\omega_{sw}}{\omega_{sr}} = \frac{\alpha_{sw}}{\alpha_{sr}} = \frac{1}{\varepsilon} \tag{1}$$

where ε is the tooth ratio, that is, the ratio of the number of star wheel teeth to the number of screw grooves.

The rotation angular velocity of the star wheel can be calculated according to the tooth ratio ε and the compressor rotation speed *n*, as listed in Equation (2):

$$\omega_{\mathrm{sw}} = \frac{2\pi n}{60\varepsilon} \tag{2}$$

In the star wheel-screw meshing pair, the side of the screw groove is the envelope surface of the star wheel tooth surface, and the two form a continuous tangent conjugate surface under the law of relative motion [11]. The normal vector ***n*** and the relative velocity ***v*** at any meshing point on the conjugate surface are perpendicular to each other, which satisfies the conjugate relationship described by Formula (3):

$$\boldsymbol{v} \cdot \boldsymbol{n} = 0 \tag{3}$$

In the SSC, the flank of the star wheel tooth is usually the mother side. When its contact line equation is determined, the conjugate contact surface on the screw groove can be obtained according to the conjugate meshing relationship of Formula (3). The coordinates of any meshing point *M* on the flank of the star wheel tooth can be described by Formula (4):

$$\begin{cases} x_2 = x(t,\tau) \\ y_2 = y(t,\tau) \\ z_2 = z(t,\tau) \end{cases} \tag{4}$$

where *t* and τ are two parameter variables of the surface equation.

In order to obtain the relative velocity ***v*** between the star wheel and the screw, the coordinate transformation method should be used to convert the coordinates of the *M* point to coordinate system O_1, as shown in formula (5):

$$\begin{cases} x_1 = x_2 \cos\alpha_{\mathrm{sw}} - y_2 \sin\alpha_{\mathrm{sw}} \\ y_1 = x_2 \sin\alpha_{\mathrm{sw}} + y_2 \cos\alpha_{\mathrm{sw}} \\ z_1 = z_2 \end{cases} \tag{5}$$

Then by solving the product of the angular velocity and the rotation radius of point *M*, the velocity of the point in coordinate system O_1 can be obtained, as shown in Formula (6):

$$\boldsymbol{v}_1^{\mathrm{II}} = -\omega_{\mathrm{sw}} y_1 \boldsymbol{i} + \omega_{\mathrm{sw}} x_1 \boldsymbol{j} \tag{6}$$

where the superscript II represents the absolute velocity of point *M* on the star wheel; ***i*** and ***j*** are the unit vectors in the X and Y directions, respectively.

After coordinate transformation, the velocity of point *M* in coordinate system O_3 can be calculated, as shown in Formula (7):

$$\boldsymbol{v}_3^{\mathrm{II}} = -\omega_{\mathrm{sw}} \left(x_2 \cos\alpha_{sw} - y_2 \sin\alpha_{sw} \right) \boldsymbol{j} - \omega_{\mathrm{sw}} \left(y_2 \cos\alpha_{sw} + x_2 \sin\alpha_{sw} \right) \boldsymbol{k} \tag{7}$$

where ***k*** is the unit vector in the Z direction.

The velocity at point *M* on the side of the screw groove can be described by Formula (8):

$$\boldsymbol{v}_3^{\mathrm{I}} = -\omega_{\mathrm{sr}} y_3 \boldsymbol{i} + \omega_{\mathrm{sr}} x_3 \boldsymbol{j} \tag{8}$$

The coordinates of the meshing point in coordinate system O_3 can be expressed by Formula (9):

$$\begin{cases} x_3 = z_1 \\ y_3 = A - y_1 \\ z_3 = x_1 \end{cases} \tag{9}$$

where A is the center distance between the star wheel and the screw.

Therefore, the relative velocity of the star wheel and the screw at point M can be expressed as Formula (10):

$$\boldsymbol{v}_3 = \boldsymbol{v}_3^{\mathrm{II}} - \boldsymbol{v}_3^{\mathrm{I}} \tag{10}$$

Combine formulas (3),(9) and (10) and transform the coordinates, the coordinates of the meshing point M in coordinate system O_4 can be obtained by equations (11):

$$\begin{cases} x_4 = x_3 \cos\alpha_{sr} + y_3 \sin\alpha_{sr} \\ y_4 = -x_3 \sin\alpha_{sr} + y_3 \cos\alpha_{sr} \\ z_4 = z_3 \end{cases} \tag{11}$$

Besides the constraints of the above profile design criteria and conjugate meshing principle, it is necessary to ensure that the position of the contact line changes continuously in a large meshing area as much as possible.

2.2. Equations of Curved Envelope Profile

Figure 3 shows the profile and section of the curved envelope meshing pair in detail. In the curved envelope meshing pair, the meshing surface is a continuous and smooth surface in space, which can ensure that the position of the contact line changes continuously in the meshing area with the rotation of the rotor, disperse the meshing point to the greatest extent, and improve the meshing performance of the star wheel-screw meshing pair. The vertical section of the star wheel tooth is made at any tooth height parameter, and the profile on the section can be divided into three parts: L_{curve} is the curve part of the meshing line (the envelope curve), L_{up} and L_{down} are the tangents of the upper and lower bounds of the meshing line, respectively.

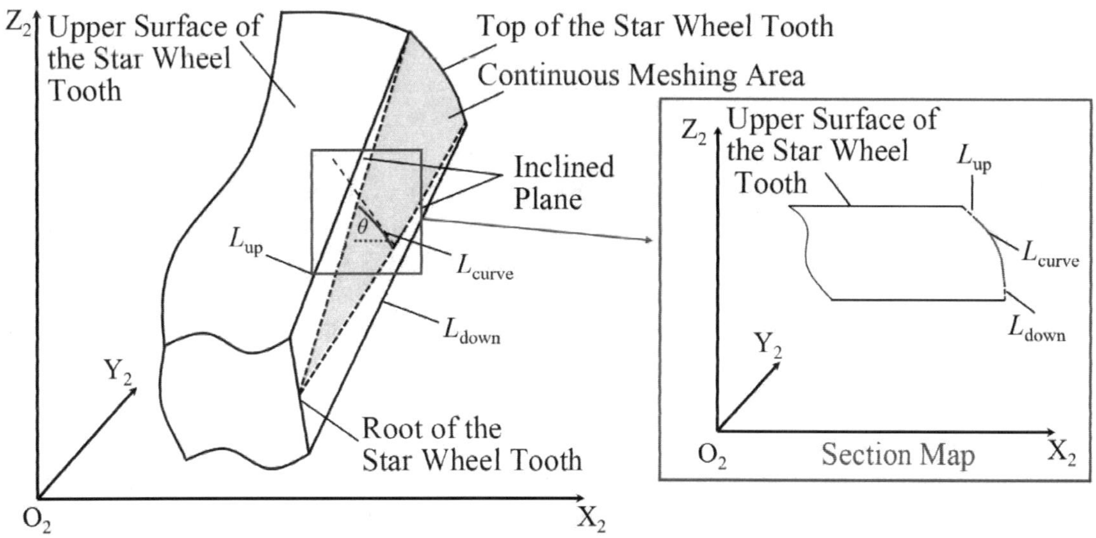

Figure 3. The feature map of the back side of the tooth and section map

The coordinates of the meshing point of any section of the star wheel tooth in coordinate system O_2 satisfy Formula (12):

$$\begin{cases} X_2 = x \\ Y_2 = y \\ Z_2 = z(x,y) \end{cases} \tag{12}$$

where $z(x, y)$ is the parametric equation of the envelope curve.

The normal vector of the contact point is shown in Equation (13):

$$\boldsymbol{n}_2 = -\frac{\partial z}{\partial x}\boldsymbol{i} - \frac{\partial z}{\partial y}\boldsymbol{j} + \boldsymbol{k} \tag{13}$$

After coordinate transformations of the relative velocity equation, the relative velocity of the contact point in coordinate system O_2 can be described as Equations (14):

$$\begin{cases} v_{X_2} = \omega_{sw}\left(-Y_2 + \varepsilon Z_2 \sin\alpha_{sw}\right) \\ v_{Y_2} = \omega_{sw}\left(X_2 + \varepsilon Z_2 \cos\alpha_{sw}\right) \\ v_{Z_2} = \omega_{sw}\varepsilon\left(A - Y_2\cos\alpha_{sw} - X_2 \sin\alpha_{sw}\right) \end{cases} \tag{14}$$

Substituting Formula (13) and Formula (14) into Formula (3), Formula (15) can be obtained.

$$\varepsilon\left(A - Y_2\cos\alpha_{sw} - X_2\sin\alpha_{sw}\right) + \frac{\partial z}{\partial x}(Y_2 - \varepsilon Z_2 \sin\alpha_{sw}) - \frac{\partial z}{\partial y}(X_2 + \varepsilon Z_2\cos\alpha_{sw}) = 0 \tag{15}$$

Combining Equation (12) and Equation (15), the meshing contact condition between the envelope curve on the star wheel tooth and the side of the screw groove is shown in Equation (16):

$$x = f\left(y, \alpha_{sw}\right) \tag{16}$$

The inclination angle θ of the side of the screw groove can be obtained according to Formula (17):

$$\theta = \arctan\left(\frac{\varepsilon\left(A - Y_2\cos\alpha_{sw} - X_2\sin\alpha_{sw}\right)}{Y_2 - \varepsilon Z_2 \sin\alpha_{sw}}\right) \tag{17}$$

Continuous curves such as hyperbolic, parabolic and elliptic can be used to form the basic curve of the envelope surface. Among them, the eccentricity of the ellipse is between 0 and 1, which makes it more convenient to control the curvature change of the envelope curve, and is beneficial to the gap adjustment of the leakage channel. Therefore, the ellipse is used as the basic curve of the envelope surface to solve the profile equations.

Figure 4 shows the envelope ellipse in coordinate system O_2. Let the angle between the X_2 axis and the major axis of the ellipse be φ, the centrifugal angle be ϕ, and the length of the major and minor axes of the ellipse be m and n, respectively. The center of the enveloping ellipse at different tooth heights changes continuously along axis L, forming a continuous enveloping surface. The angles between axis L and $X_2O_2Y_2$ and $X_2O_2Z_2$ are η and β, respectively. The mathematical relationship between the envelope ellipse characteristic parameters and the geometric parameters of the meshing pair can be obtained according to Equation (18):

$$\tan\phi = -\frac{1}{\varepsilon B_3 A_1 + B_6 A_3 + B_9 A_5}\begin{bmatrix}(\varepsilon B_3 A_2 + B_6 A_4 + B_9 A_6) + (\varepsilon B_1 A_1 + B_4 A_3 + A_5 B_7)\sin\phi\ \tan\phi \\ +(\varepsilon B_2 A_2 + B_5 A_4 + B_8 A_6)\cos\phi \\ +(\varepsilon B_2 A_1 + \varepsilon B_1 A_2 + B_4 A_4 + B_5 A_3 + B_7 A_6 + B_8 A_5)\sin\phi\end{bmatrix} \tag{18}$$

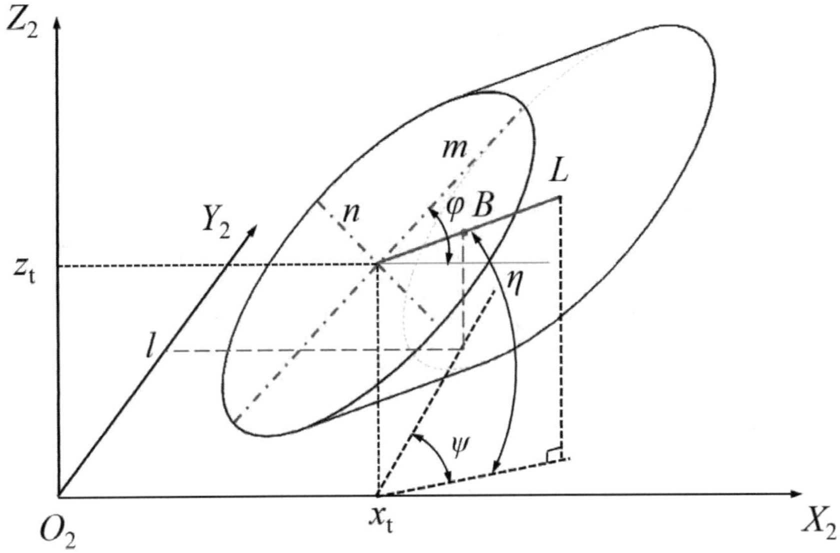

Figure 4. Envelope ellipse in the coordinate system O_2

The coefficients in the above equations satisfy Formula (19) and Formula (20) respectively.

$$\begin{cases} A_1 = m\cos\varphi \\ A_2 = n\sin\varphi \\ A_3 = (\sin\alpha_{sw}\tan\varphi\cos\psi + \cos\alpha_{sw}\tan\varphi\sin\psi + \cos\alpha_{sw}\tan\eta)m\dfrac{\cos\varphi}{\cos\psi} \\ A_4 = (\cos\alpha_{sw}\tan\varphi\tan\eta - \sin\alpha_{sw}\cos\psi - \cos\alpha_{sw}\sin\psi)n\dfrac{\cos\varphi}{\cos\beta} \\ A_5 = (\sin\alpha_{sw}\tan\eta - \cos\alpha_{sw}\tan\varphi\cos\psi + \sin\alpha_{sw}\sin\psi)m\dfrac{\cos\varphi}{\cos\psi} \\ A_6 = (\cos\alpha_{sw}\cos\psi - \sin\alpha_{sw}\sin\psi + \sin\alpha_{sw}\tan\varphi\tan\eta)n\dfrac{\cos\varphi}{\cos\psi} \end{cases} \tag{19}$$

$$\begin{cases} B_1 = n\sin\alpha_{sw}\sin\varphi \\ B_2 = -m\sin\alpha_{sw}\cos\varphi \\ B_3 = l\left(\sin\alpha_{sw}\tan\psi - \cos\alpha_{sw}\right) + A - x_t\sin\alpha_{sw} \\ B_4 = \left(\cos\alpha_{sw}\tan\varphi - \varepsilon\right)n\cos\varphi \\ B_5 = -\left(\cos\alpha_{sw} + \varepsilon\tan\varphi\right)m\cos\varphi \\ B_6 = l\left(\cos\alpha_{sw}\tan\psi + \sin\alpha_{sw} - \varepsilon\dfrac{\tan\eta}{\cos\psi}\right) - x_t\cos\alpha_{sw} - \varepsilon z_t \\ B_7 = n\sin\alpha_{sw}\sin\varphi \\ B_8 = -m\sin\alpha_{sw}\cos\varphi \\ B_9 = -l\cos\alpha_{sw} + l\sin\alpha_{sw}\tan\psi - x_t\sin\alpha_{sw} \end{cases} \tag{20}$$

The ϕ values corresponding to different star wheel rotation angles and star wheel tooth heights can be solved by numerical solution method, and the coordinates of each meshing point on the envelope surface are obtained according to ϕ. Then the coordinates of the meshing points are converted to coordinate system O_4 by coordinate transformation, and the screw groove profile equations are obtained, as shown is Equastions (21):

$$\begin{cases} x_{sr} = \left(\cos\alpha_{sr}\sin\varphi - \sin\alpha_{sr}\sin\alpha_{sw}\cos\varphi\right)m\cos\phi \\ +\left(\sin\alpha_{sw}\sin\alpha_{sr}\sin\varphi + \cos\alpha_{sr}\cos\varphi\right)n\sin\phi \\ +\left(l\sin\alpha_{sw}\tan\psi - l\cos\alpha_{sw} - x_t\sin\alpha_{sw} + A\right)\sin\alpha_{sr} \\ +\left(z_t + l\dfrac{\tan\eta}{\cos\psi}\right)\cos\alpha_{sr} \\ y_{sr} = \left(-\cos\alpha_{sr}\sin\alpha_{sw}\cos\varphi - \sin\alpha_{sr}\sin\varphi\right)m\cos\phi \\ +\left(\cos\alpha_{sr}\sin\alpha_{sw}\sin\varphi - \sin\alpha_{sr}\cos\varphi\right)n\sin\phi \\ +\left(l\sin\alpha_{sw}\tan\psi - l\cos\alpha_{sw} - x_t\sin\alpha_{sw} + A\right)\cos\alpha_{sr} \\ -\left(z_t + l\dfrac{\tan\eta}{\cos\psi}\right)\sin\alpha_{sr} \\ z_{sr} = m\cos\alpha_{sw}\cos\varphi\cos\phi - n\cos\alpha_{sw}\sin\varphi\sin\phi + \left(x_t - l\tan\beta\right)\cos\alpha_{sw} - l\sin\alpha_{sw} \end{cases} \tag{21}$$

It can be seen that when designing the curved envelop profile, it is only needed to determine the mathematical equations of the envelope to solve the coordinates of the contact points and meshing points corresponding to any star wheel rotation angle and star wheel tooth height. Then, smooth the surface according to the obtained three-dimensional point set of the rotor, draw the spatial surface, and the curved envelop surface of the meshing pair is finally obtained.

2.3. Standard tooth width of curved envelop meshing pair

After establishing the profile equations of the curved envelope meshing pair, it is also necessary to determine the tooth width of the star wheel, so as to determine other geometric parameters of the meshing pair of the SSC according to the requirements of the exhaust volume. In the straight line envelope meshing pair, the width of the star wheel tooth is a fixed value, and the size of the fixed tooth width is usually solved according to Formula (22):

$$b = 2a \sin\frac{\gamma}{2} - \Delta b \cos\frac{\gamma}{2} \tag{22}$$

where a is the vertical distance between the outer edge of the screw and the center of the star wheel/m; γ is the star wheel indexing angle/rad, $\gamma=2\pi/z_{sw}$; Δb is the minimum axial thickness of the groove rib at the tooth root of the star wheel/m. The existing calculation and test results show that the thickness of the rib is not less than 2~3 mm to meet the engineering requirements.

However, in the curved envelope meshing pair, the geometric size in the width direction of the star wheel tooth is determined by the parameters of the curved envelope profile, which is not a fixed value. In fact, the meshing point of the curved envelope profile has a limited range of motion on the star wheel tooth section line, and the variation in the tooth width direction is usually within 1~3mm. Hence, the geometric parameters of the compressor can probably be calculated by an approximate tooth width instead of the tooth width of the curved envelope meshing pair. This approximate tooth width is defined as the standard tooth width of the curved envelope meshing pair. The selection method is as follows:

After the basic curve of the curved envelope profile is initially selected, measure the top tooth width b_1 and the root tooth width b_2 of the star wheel tooth on the high-pressure surface of the star wheel, and their average value is taken as the standard tooth width. The standard tooth width b' can be expressed by Formula (23) :

$$b' = \frac{b_1 + b_2}{2} \tag{23}$$

By comparing the screw groove volume calculated by the standard tooth width and obtained form three-dimensional model, it can be found that when the displacement is in the range of $1m^3 \cdot min^{-1}$~ $100m^3 \cdot min^{-1}$, the error of the total volume of the two and the error of the screw groove volume at each angle are less than 1 %, indicating that the calculation of other geometric parameters of the curved envelope meshing pair according to the standard tooth width will not cause obvious errors in engineering.

3. Star wheel tooth angle-screw groove bottom angle coupling profile

After completing the design of the flank profile of the curved envelope meshing pair, it is necessary to further optimize the geometric structure of the star wheel tooth angle and the screw groove bottom angle. The bottom angle of the screw groove and the tooth angle of the star wheel in the straight line envelope meshing pair are close to the right angle and can be processed by turning tool. However, milling cutter is usually used to process the curved envelop meshing pair. If the right angle design is maintained in the curved envelop meshing pair, problems such as large leakage channel and cutter interference will appear. In order to improve the problems, a design method of star wheel tooth angle-screw groove bottom angle coupling profile is proposed. The groove bottom angle is subjected to arc transition treatment to form an interface arc surface between the side and bottom of the groove; the coupling arc surface of the star wheel tooth angle satisfies the conjugate relationship with the transition arc surface of the screw groove bottom angle.

The transition arc profile equations of the groove bottom angle are mainly related to the rotation angle of the star wheel α_{sw}, the eccentricity angle of the tool α_{ct}, the angle between the tool symmetry line and the reference line α_{ch} and other parameters, as shown in Figure 5. In the above mentioned parameters, the angle between the tool symmetry line and the reference line changes with the position of the contact point between the tool and the flank of the screw groove. Thus, the position of the contact point, which is the point of tangency between the tool arc and the flank of the screw groove, should be determined first.

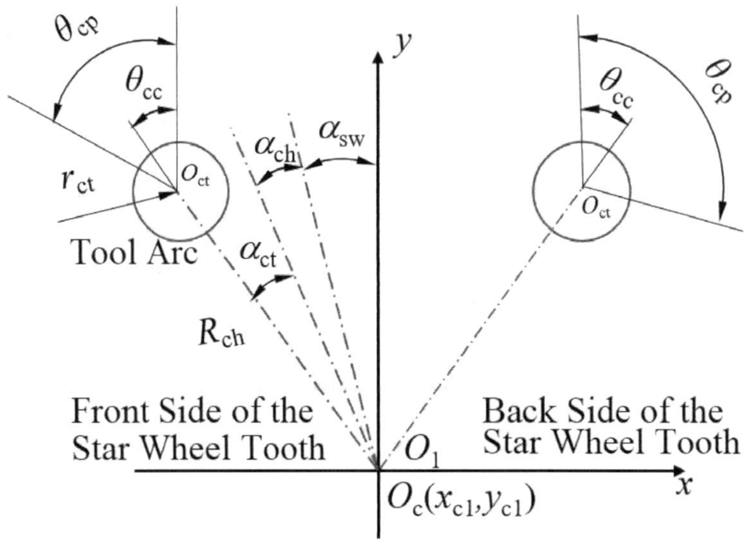

Figure 5. Transition arc diagram (in coordinate system O_1)

As shown in Figure 6, (x_{p1},y_{p1} are the coordinates of the contact point on the plane z_{p1}=0. To find the point, first an α_{ch} value is estimated and the corresponding position of the tool center is determined. Then, the normal line of the screw groove flank is established and moved along the flank of the screw groove. When the normal line passes the tool center, judge whether the distance from the flank of the screw groove to the tool center is equal to the tool radius r_{ct}. If it is, the estimated α_{ch} value is correct. If it is not, change the α_{ch} value and repeat the above steps until the tool circle is tangent to the flank of the screw groove. Mathematically, Formula (21 is converted into coordinate system O_1 and is combined with Formula (18 to form an iterative function, by solving which the coordinates of the contact point and the α_{ch} angle under different rotation angle of the screw can be obtained. Then the transition arc profile equations in coordinate system O_1 can be established according to the geometrical relation. After coordinate transformation, the transition arc profile equations in coordinate system O_4 are shown in Formula (24):

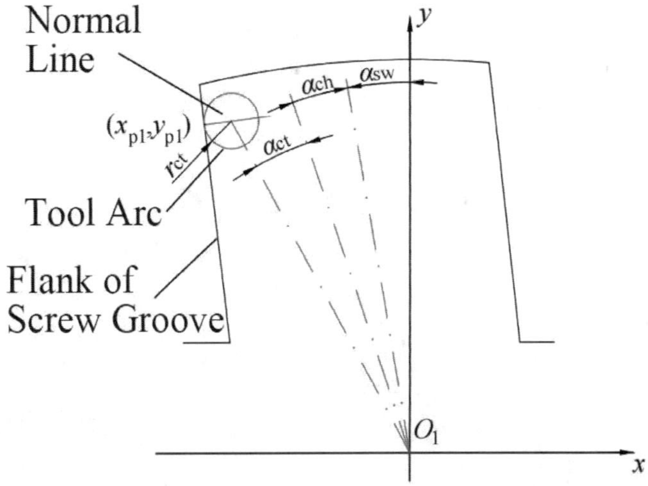

Figure 6. Diagram of the process of finding the contact point (in coordinate system O_1)

$$\begin{cases} x_{c4} = \left(A - R_{ch}\cos\left(\alpha_{ct} + \alpha_{ch} + \alpha_{sw}\right) - r_{ct}\cos\theta_{cc}\right)\sin\left(\varepsilon\alpha_{sw}\right) \\ y_{c4} = \left(A - R_{ch}\cos\left(\alpha_{ct} + \alpha_{ch} + \alpha_{sw}\right) - r_{ct}\cos\theta_{cc}\right)\cos\left(\varepsilon\alpha_{sw}\right) \\ z_{c4} = -R_{ch}\sin\left(\alpha_{ct} + \alpha_{ch} + \alpha_{sw}\right) \pm r_{ct}\sin\theta_{cc} \end{cases} \tag{24}$$

where α_{ct} is the eccentricity angle of the tool/rad; α_{ch} is the angle between the tool symmetry line and the reference line/rad; r_{ct} is the tool radius/m; θ_{cc} is the characteristic angle of any point on the transition arc/rad, taking the clockwise direction as the positive direction; '-' represents the front side of the tooth; '+' represents the back side of the tooth.

After obtaining the transition arc surface of the screw groove bottom angle, the envelope surface of the interface arc in the star wheel tooth angle can be obtained according to the meshing principle. This surface is actually a variable diameter arc surface, so it is defined as the coupling variable diameter arc surface of the star wheel tooth angle. According to the principle of conjugate meshing, Formula (25) can be obtained:

$$\theta_{cc} = \arctan\left(\frac{\varepsilon K_1 K_3 \sin(\varepsilon\alpha_{sw} - \varphi_{sr}) - K_2 K_3 + K_4 + K_1 K_5}{r_c K_3 \cos\theta_{cc}}\right) \tag{25}$$

where φ_{sr} is the screw angle when the screw envelops the star wheel/rad.

The coefficients in Formula (25) satisfy Formula (26):

$$\begin{cases} K_1 = A - R_{ch}\cos\left(\alpha_{ch} + \alpha_{ct} + \alpha_{sw}\right) - r_{ct}\sin\theta_{cc} \\ K_2 = -R_{ch}\sin\left(\alpha_{ch} + \alpha_{ct} + \alpha_{sw}\right) \\ K_3 = r_{ct}R_{ch}\sin\left(\alpha_{ch} + \alpha_{ct} + \alpha_{sw} - \theta_{cc}\right)\sin\left(\alpha_{sr} - \varphi_{sr}\right) + \varepsilon r_{ct}K_1\cos\left(\alpha_{sr} - \varphi_{sr}\right) \\ K_4 = \varepsilon r_{ct}K_1\left(A - K_1\cos\left(\alpha_{sr} - \varphi_{sr}\right)\right)\sin\theta_{cc} \\ K_5 = \left[r_{ct}R_{ch}\sin\left(\alpha_{ch} + \alpha_{ct} + \alpha_{sw} - \theta_{cc}\right)\cos\left(\alpha_{sr} - \varphi_{sr}\right) - \varepsilon r_{ct}K_1\sin\left(\alpha_{sr} - \varphi_{sr}\right)\right]\cos\left(\alpha_{sr} - \varphi_{sr}\right) \end{cases} \tag{26}$$

After obtaining the solution φ_{sr} (α_{sw}) of formula (26), the contact line between the star wheel tooth angle and the bottom angle of the screw groove under different α_{sw} and θ_{cc} can be iteratively calculated. Finally, the smooth surface is formed from the contact lines to form the coupling variable diameter arc surface of the star wheel tooth angle. Figure 7 shows the screw and star wheel designed and processed by the above-mentioned star wheel tooth angle-groove bottom angle coupling profile design method.

Figure 7. The physical diagram of screw and star wheel designed by coupling profile

4. Conclusion

In this paper, aiming at solving the problems of imperfect design theory of curved envelope meshing pair profile and large leakage gap between star wheel tooth angle and screw groove bottom angle, easy interference under extreme conditions and difficult processing of groove bottom right angle, the design principle of curved envelope profile was analyzed, the profile equation with ellipse as the basic envelope curve was established, the method of simplifying the design of curved envelope meshing pair by using standard tooth width was proposed, the coupling profile of star wheel tooth angle-groove bottom angle was invented, and equations of transition arc profile of groove bottom angle and the coupling variable diameter circular surface of star wheel tooth angle were established. The coupling profile design method of curved envelope meshing pair proposed in this paper would provide theoretical support for the design and improvement of curved envelope meshing pair of SSC, which would be helpful for the promotion and application of SSC in more fields such as high temperature heat pumps and steam compression.

Acknowledgement

This work was supported by National Natural Science Foundation of China [grant number 52076162], The Youth Innovation Team of Shaanxi Universities and Scientists and Engineers Team Construction of Qin Chuangyuan Platform (2024QCY-KXJ-004.

References

[1] Zimmern B and Patel G C 1972 Design and Operating Characteristics of the Zimmern Single Screw Compressor *International Compressor Engineering Conference* p 16

[2] Zimmern B 1975 Improvements in rotary positive-displacement machines for compression or expansion of a fluid UK Patent 1388537

[3] Feng Q *et al* 2017 New concept single screw compressors and their manufacture technology *IOP Conference Series: Materials Science and Engineering* **232** 012001.

[4] Li J *et al* 2012 A New Design of the Tooth Profile for Single Screw Compressors *International Compressor Engineering Conference* p 2070

[5] Post W W 1983 De hydrodynamische filmsmering in een globoïde wormcompresso *PhD Thesis* Eindhoven University of Technology, Eindhoven

[6] Post W and Zwaans M 1986 Computer Simulation of the Hydrodynamic Lubrication in A Single Screw Compressor *International Compressor Engineering Conference p* 536

[7] Jensen D 1998 A New Single Screw Compressor Design that Enables a New Manufacturing Process *International Compressor Engineering Conference* p 1306

[8] Shen Y, Luan H and Wang K 2010 Research on the Meshing Zones of Single Screw Compressor Formed by Cylindrical Milling *Manufacturing Technology & Machine Tool* pp 142-144.

[9] Huang R *et al* 2015 An optimization of the star-wheel profile in a single screw compressor *Proceedings of the Institution of Mechanical Engineers, Part A: Journal of Power and Energy* **229** pp 139-150.

[10] Huang R *et al* 2017 Characteristics between the meshing pairs with different envelope profile in single screw compressors *IOP Conference Series: Materials Science and Engineering* **232** 012071.

[11] Wu X 2009 *Gear Meshing Principle* (Xi'an: Xi'an Jiaotong University Press)

Impact of rotor geometry and fluid injection on screw compressor performance

A Kumar[1,2]**, A Kovacevic**[1] **and N Stosic**[1]

[1]Centre for Compressor Technology, Department of Engineering, City, University of London, Northampton Square, London, EC1V OHB, England, United Kingdom
[2]Kirloskar Pneumatic Company Limited, Hadapsar Industrial Estate, Pune, 411 013, Maharashtra, India

E-mail: abhishek.kumar.2@city.ac.uk

Abstract. Screw compressors are widely used in industrial and commercial applications due to their high efficiency, reliability, and durability. However, there is still room for improvement in terms of energy efficiency and performance. To address this, the current study investigates the impact of screw rotor geometrical parameters: wrap angle, relative length, and built-in volume ratio and fluid injection methods (oil, and water injection) on the performance of screw compressors. The research offers a detailed analysis of these parameters, with a focus on the empirical validation of water injection's influence through experimental testing on an industrial screw compressor. The findings reveal that water injection reduces specific power consumption by 4.78 % compared to oil injection, highlighting its potential for improving compressor efficiency. This work contributes valuable insights into optimizing screw compressor design for enhanced performance and energy efficiency.

Nomenclature

P	power (kW)	Q	volume flow rate (m^3/min)
SPC	specific power consumption ($kW/m^3/min$)	P_{dis}	discharge pressure (bar)
L/D	relative length	ϕ	wrap angle (deg)
VI	built-in volume ratio	W_{tip}	tip speed (m/s)

1. Introduction

Screw compressors have been a cornerstone of industrial and commercial applications since their inception in the mid-20th century. The innovative designs of Heinrich Krigar and Alf Lysholm laid the foundation for these efficient, reliable, and durable machines. Over the years, advancements in materials, manufacturing techniques, and computational methods have continued to refine screw compressor performance. As the demand for more energy-efficient and high-performance systems grows, optimizing screw compressors has become crucial. This optimization requires a comprehensive understanding of the various parameters influencing compressor operation, including profile clearances, geometrical configurations, thermodynamic behaviors, and fluid injection strategies. Each parameter plays a vital role in determining the overall performance and efficiency of the compressor [1][2][3].

For instance, managing profile clearances is critical to minimize gas leakage while maintaining mechanical robustness, as demonstrated by Dhayanandh et al. [4]. Optimizing oil injection parameters is also essential, as Dundagekar et al. showed that oil improves lubrication and thermal management [5]. Kumar et al. have demonstrated the effectiveness of optimizing design and fluid parameters using Bayesian and evolutionary algorithms to enhance the performance of screw compressors. Their studies, as reported in [6] and [7], highlight the potential of these advanced optimization techniques in improving the efficiency and reliability of screw compressors by systematically adjusting design and fluid parameters to achieve optimal performance.

Recent studies have focused on the impact of geometrical parameters on compressor efficiency. Tian et al. developed a thermodynamic model highlighting the role of rotor profiles and clearances in achieving optimal volumetric efficiency and pressure dynamics [8]. The influence of fluid injection, particularly water injection, has also been a key area of research. Li and Wang's theoretical work on water-injected compressors demonstrated how water enhances cooling and sealing, leading to improved compressor performance and longevity [8].

Experimental investigations, such as those by Dhayanandh, have validated these theoretical findings by empirically demonstrating the performance improvements brought by optimized oil and water injection techniques [4]. The interplay between thermodynamic and geometrical parameters has also been scrutinized, revealing complex dependencies that dictate the overall operational efficiency of screw compressors [9]. In addition to the screw parameters, the choice of rotor materials is crucial for optimizing performance. The material properties, such as strength, durability, and thermal conductivity, significantly impact the compressor's efficiency and reliability. For instance, Kumar et al. have investigated the influence of rotor materials on screw compressor performance, highlighting the importance of selecting materials that can withstand the high temperatures and pressures within the compression chamber [10].

The literature review underscores the necessity of optimizing screw compressor parameters. However, before optimization, it is crucial to understand the impact of these parameters—including rotor profiles, geometrical dimensions, and fluid injection methods—on compressor performance. This study focuses on analyzing the effects of key geometrical parameters, laying the groundwork for future optimization efforts. Additionally, the research explores the impact of water injection, as opposed to traditional oil injection, within the compression chamber, highlighting the thermodynamic improvements and potential challenges in developing a water-injected screw compressor.

The paper is structured as follows: Section 2 presents the modelling of the KAS-200 industrial screw compressor block, incorporating both experimental and analytical approaches using a chamber model-based methodology with oil and water as the injected fluids. Section 3 explores the impact of key geometrical parameters on compressor performance, as well as the thermodynamic effects of varying fluid injections. Concluding insights are provided in Section 4.

2. Experimental and analytical modelling of KAS-200 screw compressor

This study focuses on the KAS-200 industrial screw compressor block, manufactured by Kirloskar Pneumatic Company Limited (KPCL), Pune, India. The KAS-200 is an oil-flooded air screw compressor with a power rating of 22 kW, operating at a pressure ratio of 8.5, and a nominal tip speed of approximately 35 m/s, delivering a volume flow rate of around 4m^3/min. Key geometrical parameters of the 4/5 lobe combination screw rotors include:

- Rotor Length-to-Diameter Ratio (L/D) = 1.55

- Male Rotor Wrap Angle (ϕ) = 306.647°
- Built-in Volume Ratio (VI) = 4.6
- Working Fluid: Air
- Maximum Chamber Volume = 175,259 mm^3

Due to proprietary restrictions, detailed size and profile specifications are not disclosed. The KAS-200 compressor block was produced and assembled at KPCL for experimental validation using two fluid injection types—oil and water. An isometric view of the KAS-200 compressor block is shown in Figure 1.

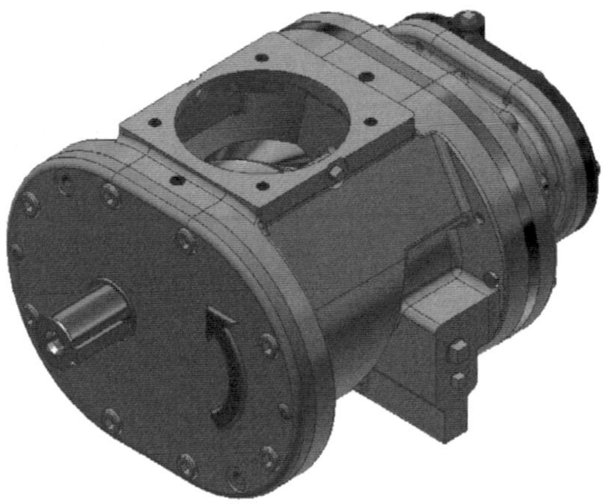

Figure 1: Isometric view of the KAS-200 compressor block

A customized testing package was developed for the KAS-200 compressor, featuring different separator tanks depending on the fluid i njection t ype. T he o il i njection s ystem u ses a n air-oil separator based on centrifugal and gravity principles, while the water injection system employs an air-water separator equipped with a demister pad to remove entrained liquid droplets effectively, as shown in Figure 2 . All components in direct contact with water, including the air-water separator tank, pipes, and small parts, were fabricated from corrosion-resistant stainless steel (SS304L) to ensure durability.

The compressor block, including the rotors and housing, remained consistent for both oil and water injection tests. The rotors are made from normalized medium carbon steel, and the housing is constructed from grey cast iron. Anti-friction bearings were used to ensure smooth operation.

The complete setup for the water-injected screw compressor is shown in Figure 2.

Following the experimental tests with both oil and water injections, the data obtained was utilized to develop an analytical model of the KAS-200 using SCORG (Screw Compressor Rotor Grid Generation) software from PDM Analysis. SCORG is a leading tool for grid generation and performance analysis of positive displacement screw machines, conducting thermodynamic calculations through a multi-chamber model [11] [6]. This analytical model was used to evaluate the impact of key geometrical parameters and to validate the experimental results of fluid injection on the screw compressor's performance, keeping the other input conditions and the rotor profile constant.

3. Results and Discussion

Following the completion of the experimental setup, tests were conducted under steady-state conditions at specified operating points, ensuring that the compressor reached stable parameters

Figure 2: Demister pad and complete package of the water-injected screw compressor

before recording data. For each steady-state condition, data was recorded for a minimum of 10 minutes. The results at each test point were averaged from 10 data samples and normalized to the required values of pressure ratios and speeds, in accordance with the ISO 1217 standard for displacement compressor testing.

The following sub-sections present the experimental results for the KAS-200 compressor using water and oil as the injected fluids.

3.1. *Experimental results*

The performance of the KAS-200 compressor was evaluated under varying operating conditions, focusing on key performance parameters such as discharge pressure, tip speed, power consumption, specific power consumption (SPC), and volumetric flow rate. Table 1 summarizes the testing results for both water and oil as the fluid injected into the compression chamber.

Table 1: Testing Results for KAS-200 Compressor Block

P_{dis} (bar a)	W_{tip} (m/s)	P (kW)	SPC (kW/m^3/min)	Q (m^3/min)
Fluid Injection: Water				
8.00	32	23.56	7.032	3.35
Fluid Injection: Oil				
8.00	32	24.74	7.385	3.35

The experimental results indicate that the water-injected screw compressor demonstrated an improvement in efficiency compared to the oil-flooded variant. Specifically, the water-injected configuration achieved a 4.78% reduction in specific power consumption, signifying a more energy-efficient operation under consistent operating conditions.

However, the introduction of water in a compressor traditionally designed for oil flooding resulted in notable challenges. After approximately 25 hours of operation, signs of corrosion, bacterial pollution and scaling were observed on the compressor components, as shown in Figure 3. These issues underscore the complexities involved in developing a reliable water-flooded screw compressor and highlight the necessity for further research and development to address these challenges effectively.

IOP Conf. Series: Materials Science and Engineering **1322** (2024) 012008 doi:10.1088/1757-899X/1322/1/012008

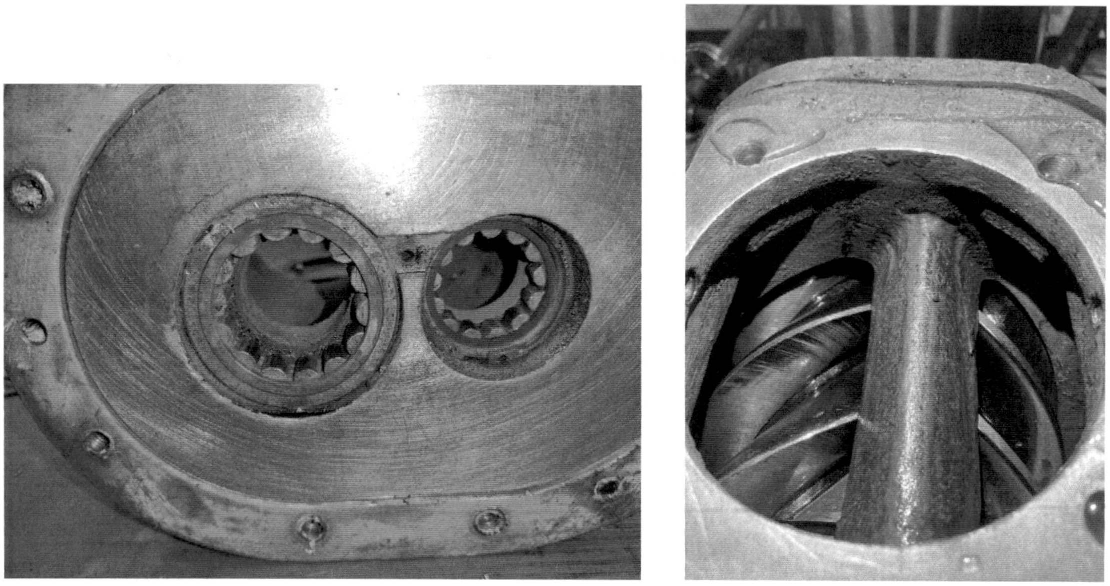

(a) Suction side view (b) Top view

Figure 3: Condition of the compressor block after 25 hours of operation

3.2. Analytical results

3.2.1. Effect of fluid injection To validate the experimental testing results with oil and water as fluid injection, analytical simulations were conducted using the SCORG software. The analytical model employed a chamber-based approach to evaluate the influence of fluid injection on the compressor's specific power consumption and discharge temperature.

The results, presented in Figures 4a and 4b, illustrate the impact of using water and oil in the compression chamber. The water-injected air screw compressor demonstrated lower specific power consumption compared to the oil-injected variant. Additionally, the discharge temperatures for the water-injected compressor were significantly lower, typically ranging between 45°C and 60°C, whereas the oil-injected compressor exhibited higher discharge temperatures.

The superior cooling performance of the water-injected system can be attributed to the higher specific heat capacity of water compared to oil. This results in more effective heat absorption, reducing the amount of fluid required for injection into the compression chamber. Furthermore, water's lower viscosity compared to standard compressor oils (such as ISO VG 68 Grade) leads to reduced viscous drag, which in turn lowers frictional power losses [12]. These factors collectively contribute to improved compressor performance and lower discharge temperatures.

3.2.2. Effect of geometrical parameters This research explores the effects of key geometrical parameters—such as the relative length of the rotor (L/D), wrap angle (ϕ), and built-in volume ratio (VI) on the performance of the KAS-200 compressor. These parameters were varied while keeping the rotor profile and clearances constant to achieve the desired volume flow rate at a fixed tip speed. The individual impacts of these parameters on compressor performance are discussed below:

Relative length of rotor: The length-to-diameter (L/D) ratio of the male rotor significantly influences compressor performance. Increasing the rotor length enhances the total displacement

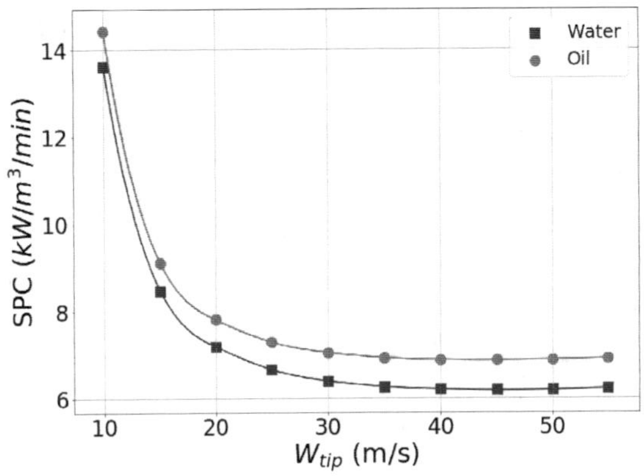

(a) SPC vs. W_{tip}

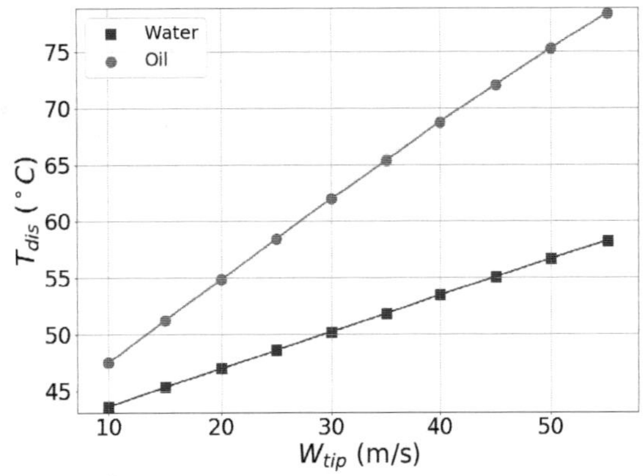

(b) T_{dis} vs. W_{tip}

Figure 4: Impact of fluid injection type on specific power consumption and discharge temperature.

volume, which generally improves performance. However, larger rotors are heavier, requiring more power to maintain the same RPM as smaller rotors, as shown in Figure 5. There are also practical constraints, such as the maximum power capacity of the compressor frame. A higher L/D ratio can lead to greater rotor deflection, potentially compromising the reliability of the compressor. Therefore, the L/D ratio must be chosen carefully, considering deflection analysis. A useful guideline is to limit the maximum rotor deflection to within one-third of the machine's radial clearance [13], as illustrated in Figure 6.

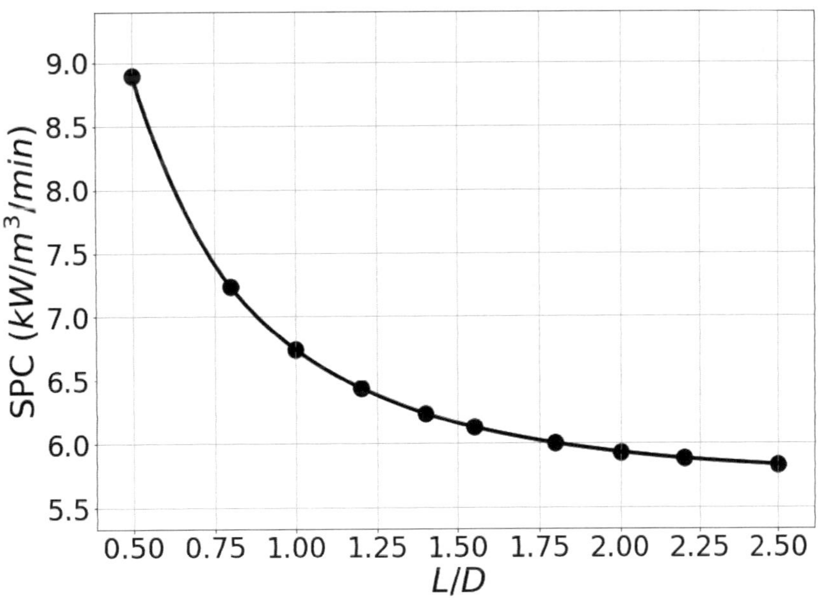

Figure 5: Effect of L/D ratio on KAS-200 compressor performance.

Wrap angle: The existing wrap angle of the male rotor in the KAS-200 compressor is 306.647°. Figure 7 shows the variation in specific power consumption as a function of the wrap angle. The

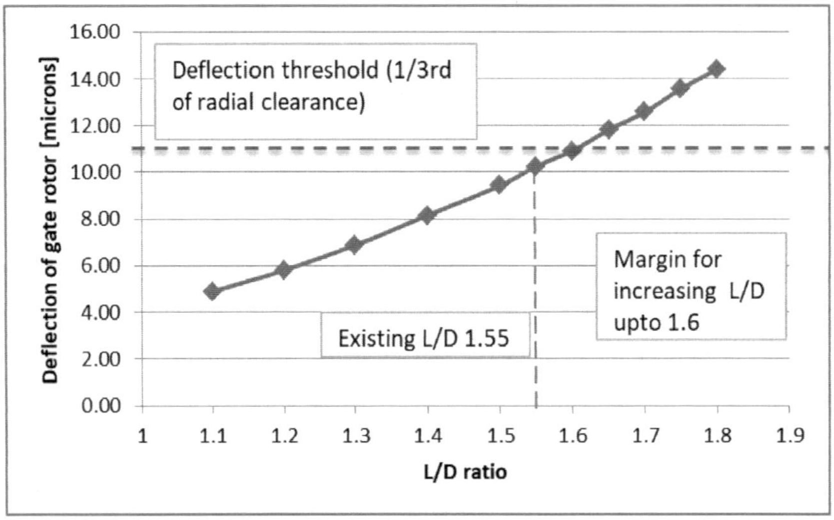

Figure 6: Effect of L/D ratio on deflections in the gate rotor at an 8.5 pressure ratio.

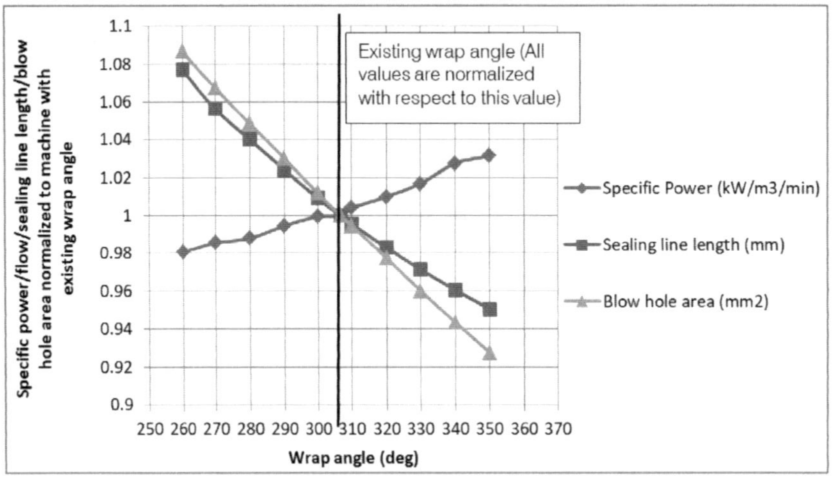

Figure 7: Variation in specific power and leakage areas with respect to wrap angle.

results indicate that specific power decreases as the wrap angle is reduced. Within the tested range of 350° to 200°, the specific power varies by 0.348 kW/m^3/min. However, reducing the wrap angle also increases leakage areas, which can negatively affect performance.

Other rotor characteristics, such as actual chamber volume and displacement volume, increase as the wrap angle decreases, counteracting the effects of increased leakage through blow holes and sealing line length (see Figure 8). While these changes can enhance performance under certain conditions, higher leakage rates could reduce efficiency, especially under high-pressure ratios or extreme working conditions. Deflection analysis of the rotors showed that changes in the wrap angle had an insignificant impact on deflection. The effective performance gain from reducing the wrap angle to 260° (beyond which further gains are minimal) is approximately 1.9%.

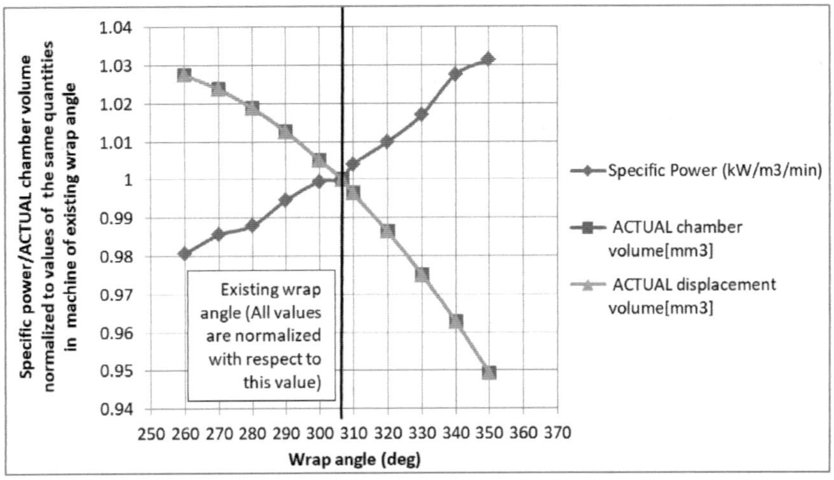

Figure 8: Variation in actual chamber volume and actual displacement volume with respect to wrap angle.

Built-in volume ratio: Figure 9 demonstrates that the specific power consumption increases when the built-in volume ratio deviates from the optimal value of 4.6. When the VI is decreased below 4.6, the internal compression is insufficient to match the external pressure ratio, leading to under-compression. This results in a need for additional external compression, increasing the overall power consumption and thus the SPC. On the other hand, increasing the VI beyond 4.6 causes over-compression, where the discharge pressure inside the compressor exceeds the external discharge pressure. This excess pressure leads to increased leakage losses and additional frictional power consumption, which also raises the SPC. Thus, a VI of 4.6 strikes an optimal balance by closely aligning the internal and external pressures, minimizing both under-compression and over-compression losses, and optimizing the compressor's efficiency. The ideal VI is specific to the application and depends on the targeted pressure ratios, as shown in Figure 9 [13].

This detailed assessment of the geometrical parameters in the KAS-200 screw compressor underscores the need for a balanced approach to optimizing performance while maintaining reliability across various operating conditions. Based on this research, the optimal geometrical parameters for the KAS-200 compressor were determined to be:

- Rotor Length-to-Diameter Ratio (L/D): 1.55
- Male Rotor Wrap Angle (ϕ): 306.647°
- Built-in Volume Ratio (VI): 4.6

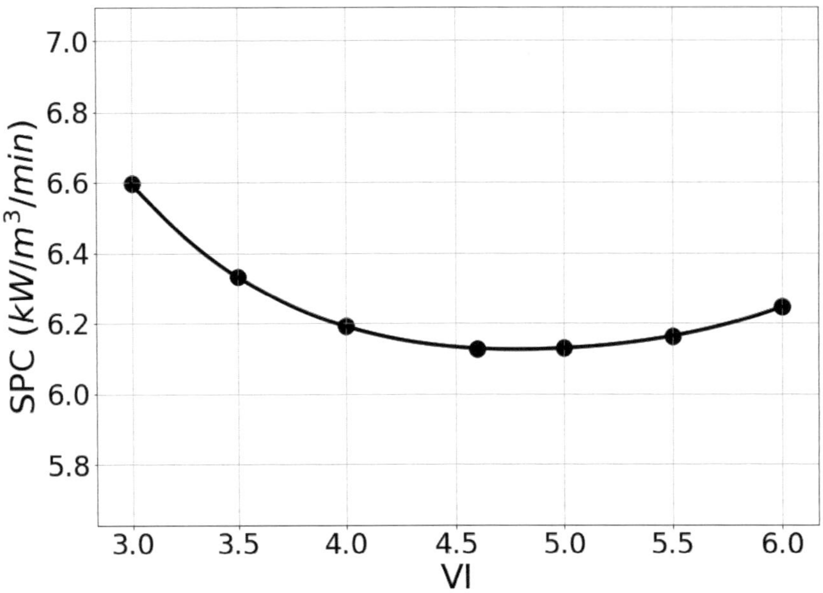

Figure 9: Effect of built-in volume ratio on KAS-200 compressor performance.

4. Conclusion

This study underscores the significant influence of rotor geometrical parameters and fluid injection on the performance of screw compressors. The analysis revealed that key geometrical parameters such as the built-in volume ratio, relative length, and wrap angle directly impact the compressor's specific power consumption. Optimizing these parameters is crucial for achieving maximum efficiency and performance. Additionally, the choice of fluid injection, whether oil or water, plays a pivotal role in determining compressor efficiency. The experimental results demonstrated that water injection led to a 4.78% reduction in specific power consumption compared to oil injection, highlighting its potential for improving compressor performance. This performance gain is attributed to the superior cooling properties of water and its lower viscosity, which reduce frictional losses within the compressor. While these findings provide valuable insights into optimizing screw compressor performance, future research should address potential challenges associated with water injection, such as corrosion, scaling, and bacterial contamination. Developing strategies to mitigate these issues will be essential for the long-term reliability and efficiency of water-injected screw compressors.

Declaration of competing interest

The authors declare no known competing financial interests or personal relationships that might have influenced the work reported in this paper.

Data availability

Data will be made available on request.

Acknowledgments

We gratefully thank Kirloskar Pneumatic Company Limited, Pune, India for sponsoring this research. We also want to acknowledge the Centre for Compressor Technology, City, University of London for their continued guidance and support in this research.

References

[1] Stosic N, Smith I and Kovacevic A 2005 *Screw compressors: mathematical modelling and performance calculation* Springer Science & Business Media

[2] Stosic N, Smith I K and Kovacevic A 2003 *Optimisation of screw compressors* Applied Thermal Engineering 23(10):1177–1195

[3] Stosic N and Hanjalic K 1997 *Development and optimization of screw machines with a simulation model – part i: Profile generation*

[4] Dhayanandh K K et al 2020 *Influence of oil injection parameters on the performance of diesel powered screw air compressor for water well application* Measurement 152:107323

[5] Dundagekar S P et al 2022 *Experimental investigation of the effect of oil injection flow rate on the performance of oil-injected twin-screw compressor*

[6] Kumar A et al 2022 *On performance optimisation for oil- injected screw compressors using different evolutionary algorithms* IOP Conference Series: Materials Science and Engineering volume 1267 page 012021 IOP Publishing

[7] Kumar A et al 2024 *Performance prediction and bayesian optimization of screw compressors using gaussian process regression* Engineering Applications of Artificial Intelligence 133:108270

[8] Li Y et al 2022 *Influence of water injection parameters on the performance of a water-lubricated single-screw air compressor* Journal of Mechanical Science and Technology 36(1):445–456

[9] Mustafin T N et al 2016 *Analysis of influence of screw compressor construction parameters and working condition on rotor temperature fields* Procedia Engineering 152:423–433

[10] Kumar A, Patil K, Kulkarni A and Patil S 2023 *Investigating alternative rotor materials to increase displacement and efficiency of screw compressor while considering cost and manufacturability* International Conference on Compressors and their Systems pages 115–125 Springer

[11] Kovacevic A, Rane S and PDM Analysis 2024 *Scorg - screw compressor rotor grid generation. Software Package for Design and Analysis of Positive Displacement Machines* **url**: https: //pdmanalysis.co.uk/scorg/ (accessed 16/08/2024).

[12] Madhav K V 2014 *The Commercial Viability of Water Injected Air Screw Compressor Systems* PhD thesis, City University London

[13] Mujic E 2009 *A numerical and experimental investigation of pulsation induced noise in screw compressors* PhD thesis, City University London

IOP Conf. Series: Materials Science and Engineering **1322** (2024) 012009 doi:10.1088/1757-899X/1322/1/012009

OilMixProp 1.0: Package for Thermophysical Properties of Oils, Common Fluids, and Their Mixtures

X Yang* **and M Richter**

Chemnitz University of Technology, Applied Thermodynamics, 09107 Chemnitz, Germany

*E-mail: xiaoxian.yang@mb.tu-chmenitz.de

Abstract. This work presents the first version of OilMixProp (short for oil mixture properties), which could be the first software package capable of calculating all the core thermophysical properties of fluids involving user-defined oils. Properties needed in thermodynamic cycle analysis are all included: density, phase equilibria, heat capacity, entropy, enthalpy, speed of sound, viscosity, and thermal conductivity. Approximately 632 pure fluids are available, and users can define their oil by determining its fluid constants using embedded fitting tools. Analogous to *refpropm* (Matlab interface of REFPROP 10.0), a function *OilPropm* is developed as the only interface for calculating thermophysical properties. Various input combinations (e.g., temperature and pressure, pressure and enthalpy, temperature and vapor fraction, etc.) are available to cater to the needs of thermodynamic cycle analysis. Sophisticated outputs are designed to deliver either complete information (phases and properties in each phase) or selected information about a fluid at the given condition. The package is written in Matlab and will be converted to other languages (e.g., C++) in the future.

1. Introduction

Oils are widely used in everyday life, e.g. for cooking (sunflower oil, olive oil, etc.) and for health care (fish oil, etc.) and are essential components of various industrial applications, e.g. as lubricants in automobiles, screw machines and in refrigeration technology. The oil products available on the market are usually composed of a base oil and a variety of additives and impurities. This makes modeling all essential thermophysical properties (including density, phase behavior, heat capacity, entropy, enthalpy, viscosity, and thermal conductivity) of an oil product very difficult, since conventional models for thermophysical properties, such as empirical multiparameter equations of state (EoS), can only be developed for pure fluids or mixtures with known components and composition. To tackle this challenge, a novel modeling approach was developed in our previous work [1] by treating an oil product as a quasi-pure fluid, setting up a simple set of equations for all the essential properties, and proposing a parameter-fitting procedure using a minimal set of experimental data (fewer than 20 and at least 12 data points). This approach can be easily extended for calculations of fluid mixtures, such as oils + refrigerants. For most properties, calculations using this approach generally agree with the experimental data within the same level of experimental uncertainty.

The basis of this modeling approach [1] is complicated as it involves the Patel-Teja-Valderrama (PTV) EoS [2,3] (more complicated than the famous Soave-Redlich-Kwong (SRK) [4] and Peng-Robinson (PR) [5] EoS) and the latest residual entropy scaling (RES) approaches developed by Yang et al. [6–9]. In addition, a robust flash algorithm for vapor-liquid equilibrium (VLE) is required for the calculation of gas solubility, which is an important property of mixtures containing oils. To facilitate the implementation and application of our modeling approach and to make the thermophysical properties of oils and their mixtures easily accessible to users, in this work, a software package, OilMixProp (short for oil mixture properties) was developed. The package is written in Matlab and will be converted to other languages (e.g., Python, C^{++} and Julia) in the future.

In this paper, we present the first version of OilMixProp. The main functions of OilMixProp 1.0 include: (1) parameter fitting package to obtain fluid constants of an oil from experimental data, (2) calculations of thermophysical properties of oils, common fluids, and their mixtures with various input combinations (e.g, temperature and pressure *T-p*, pressure and enthalpy *p-h*, pressure and vapor fraction *p-Q*, etc), and (3) plot various phase diagrams in VLE conditions, like pressure vs. composition, *p-x*. The primary feature of OilMixProp 1.0 is that all the core thermophysical properties of oils and their mixtures with other fluids can be calculated, which is not available in many other existing thermophysical property software packages, even in those commonly used reference ones (such as REFPROP 10.0 [11], TREND 5.0 [12] and CoolProp 6.4.1 [13]). Furthermore, when oils are involved, the fluid mixture is typically in a two-phase region: an oil-rich liquid phase dissolved with other components and a gas phase with a trace amount of oil. To study such systems well, a robust VLE flash algorithm was developed as the core of OilMixProp. Last but not least, a sophisticated design of inputs and outputs is implemented. Various input combinations are available, which are analogous to the function *refpropm*, the Matlab interface of REFPROP 10.0 [11]. The output is either a class (a datatype in Matlab) of a fluid (a pure fluid or a mixture) with all information available or a specified property at each phase. The output information could include compositions and all the core thermophysical properties in each phase, as well as the phase fraction. Such an output design is different to *refpropm*, which might only output a single density value (when density is the requested output), while according to the inputs, the fluid is obviously in the two-phase region (see more details in Section 4.1).

Finally, it is important to mention that there are about 632 pure fluids, unlimited user-defined oils, and their mixtures available in OilMixProp 1.0. OilMixProp does not aim to become a reference software package (such as REFPROP 10.0 [11]) for high-accurate calculations of thermophysical properties of pure fluids and mixtures. OilMixProp aims to be a reliable solution when the fluids to be studied (e.g. fluids containing oils) are not available in those reference software packages, e.g., so far only approximately 150 pure fluids and their mixtures are available in REFPROP 10.0. Compared to REFPROP 10.0, OilMixProp 1.0 has a lower accuracy for pure fluids and in single phases. However, when it comes to the VLE calculation of mixtures, OilMixProp 1.0 could be as accurate as REFPROP 10.0 and even provide more reliable results in some situations (see Section 4.1). Part of this effort is funded within subproject 3 of the KETEC (Research Platform Refrigeration and Energy Technology) project [10].

2. Fundamentals

2.1 Models and fluid constants

Thermophysical properties calculated with OilMixProp mainly include (but not only include): density, phase behavior, heat capacity, entropy, enthalpy, speed of sound, viscosity and thermal conductivity. These properties are typically required in the analysis of thermodynamic cycles. An overview of the properties and the used calculation models are given in Figure 1. In total seven Cubic EoS [PTV [2,3], SRK [4], PR [5], Yang-Frotscher-Richter (YFR) [14], Peng-Robinson-Stryjek-

Vera (PRSV) [15], Wilson-Redlich-Kwong (WRK) [16] and Redlich-Kwong (RK)[17]] are available for calculations of density, phase behavior and the residual properties. Here YFR EoS [14] is a new cubic EoS recently developed within our group (Applied Thermodynamics, Chemnitz University of Technology) aided with symbolic regression [18]; it has better accuracy in liquid phase property calculation. Due to the lack of parameters, PRSV is not recommended, and due to the lack of accuracy, WRK and RK should not be used. With an additional equation for the ideal gas isobaric heat capacity as a linear function of temperature [linear-$c_p^o(T)$], heat capacities, entropy, enthalpy and speed of sound can be obtained. Adopting the latest RES approach developed by Yang et al. [6–9], viscosity and thermal conductivity can be calculated. The fundamentals of these properties' calculation from this model set [cubic EoS + linear $c_p^o(T)$ + RES] has already been given in the supporting information of our previous paper [1]. For mixtures, the van der Waals mixing rule is used in Cubic EoS and the mixing rules developed by Yang et al.[8,9] are used for both viscosity and thermal conductivity.

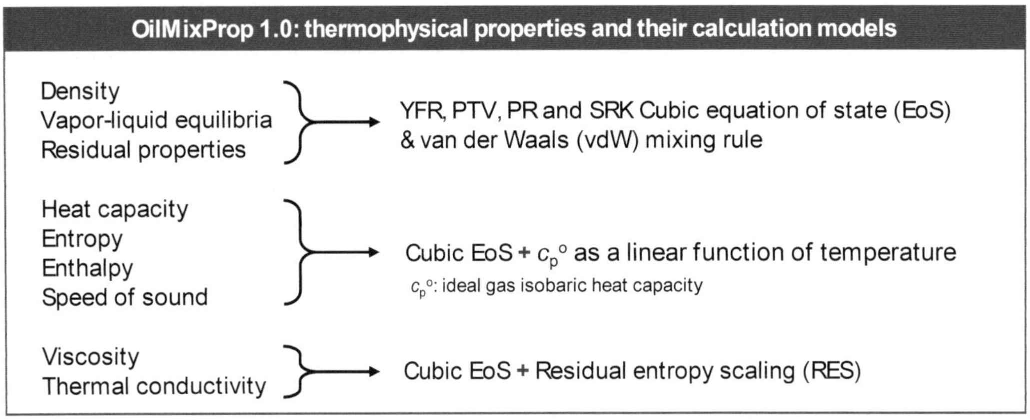

Figure 1. Available key thermophysical properties in OilMixProp 1.0 and the calculation models.

With the model set of cubic EoS + linear-$c_p^o(T)$ + RES, fluid characteristic constants of a pure fluid or a quasi-pure oil that are needed include: molar mass M, critical temperature T_c, critical density ρ_c, critical pressure p_c, acentric factor ω, two parameters k_0, k_1 in the linear-$c_p^o(T)$, Lennard–Jones (L-J) pair potential parameters ε/k_B and σ, RES fitted parameters for viscosity $n_{\mu k}$ (k = 1,2,3,4) and for thermal conductivity $n_{\lambda k}$ (k = 1,2,3,4), and three parameters for the critical enhancement of thermal conductivity (φ_0, Γ, q_D). For the 151 pure fluids (see *Classes/Fluid_Constants.txt* in the package) available in REFPROP 10.0 [11], the fluid characteristic constants are obtained by fitting to calculations of REFPROP (k_0 and k_1), from Yang et al's RES approach[6–9] ($n_{\mu k}$ and $n_{\lambda k}$), or from REFPROP directly (all other constants). For remaining pure fluids, the constants M, T_c, ρ_c, p_c, and ω are obtained from various sources mainly from DWSIM [19], $n_{\mu k}$ are from Yang et al's RES approach[6–9], and the remaining properties are constants being the average of those 151 fluids in REFPROP. Therefore, one should expect a lower accuracy when calculating properties of these fluids. For an oil product, the determination or estimation methods for these fluid characteristic constants have been given in our previous work [1]. In brief, Raoult's law of boiling point elevation (RBPE) is used for determining the average M, the modified Rackett equation (mRE) for critical point information (T_c, p_c and ρ_c), PTV or YFR EoS for ω, linear-$c_p^o(T)$ for k_0 and k_1, RES approaches of Yang et al. [6–9] for $n_{\mu k}$ and $n_{\lambda k}$ (k = 1,2,3,4). Parameters ε/k_B, σ, φ_0, Γ and q_D have negligible impact on the result and are set as constants. An overview of these constants and their relations with the models are given in Figure 2.

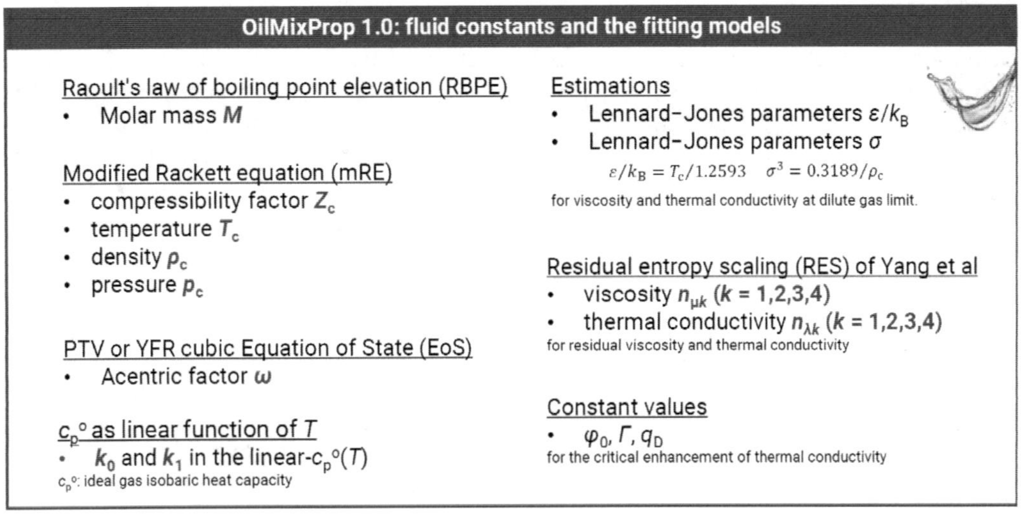

Figure 2. Fluid constants of a quasi-pure oil in OilMixProp 1.0, and their determination methods.

2.2 Vapor-liquid equilibria

Although the fundamentals of VLE calculation are available in literature [20], to explain the self-developed VLE flash algorithm as clearly as possible in Section 3.1, the generalized fundamentals of phase equilibria are described here briefly. Assuming a mixture with n components stays in equilibrium in m phases, the relations among temperature T, pressure p, overall composition $\boldsymbol{Z} = [z_1, z_2, \ldots z_n]$ where z_i is mole fraction of component i, compositions $\boldsymbol{Z}_j = [z_{j1}, z_{j2}, \ldots z_{jn}]$ in the jth phase, phase fraction $\boldsymbol{Q} = [q_1, q_2, \ldots q_m]$, fugacity coefficients $\boldsymbol{\varphi}$ (an $n \times m$ matrix, with φ_{ji} being that of the ith component in the jth phase) are:

$$\sum_{i=1}^{n} z_i = 1 \quad (1)$$

$$\sum_{i=1}^{n} z_{ji} = 1, (j = 1,2, \ldots m) \quad (2)$$

$$\sum_{j=1}^{m} q_j = 1 \quad (3)$$

$$\sum_{j=1}^{m} z_{ji} \cdot q_j = z_i, (i = 1,2, \ldots n) \quad (4)$$

$$\varphi_{ji} = func(T, p, \boldsymbol{Z}_j), (i = 1,2, \ldots n, \text{and } j = 1,2, \ldots m) \quad (5)$$

$$\varphi_{ji} \cdot z_{ji} = \varphi_{ki} \cdot z_{ki} \; (i = 1,2, \ldots n; \; k \text{ and } j = 1,2, \ldots m, \text{but } k \neq j) \quad (6)$$

Here, Eq. (1) is not an independent equation as it can be derived from Eqs. (2) to (4). Eqs. (5) require additional fluid constants which are typically known. Therefore, there are in total $2+n+m+m\times n\times 2$ variables with $1+n+m+m\times n+(m-1)\times n$ equations; and with $n+1$ variables given, equilibrium conditions could be solved. The most typical scenario is that, the known variables are temperature T, pressure p and overall composition $\boldsymbol{Z}$, which happen to be $n+1$ variables (only n-1 independent variables in $\boldsymbol{Z}$). This can be solved by the VLE flash algorithm to be discussed in Section 3.1. Another typical scenario is to calculate the phase transition temperature (or pressure), with overall composition $\boldsymbol{Z}$, pressure (or temperature) known and one of the phase fractions q_j equals 0. This corresponds to the bubble-point and dew-point calculations in VLE.

IOP Conf. Series: Materials Science and Engineering **1322** (2024) 012009 doi:10.1088/1757-899X/1322/1/012009

3. Algorithms

The calculations of most of the thermophysical properties are pretty straightforward according to models described in section 2.1. Here, in this section, we only describe important calculations that require iterations: the self-developed flash algorithm for VLE calculation, the phase transition point (bubble-point and dew-point) determination, and the calculation with one input existing in the two-phase region.

3.1 Flash algorithm

For VLE conditions, there exist only 2 phases (m = 2), therefore, the generalized phase equilibria described in section 2.2 can be specified and simplified as follows. Some variables can be specified as: $\boldsymbol{X} = [x_1, x_2,...x_n]$ and $\boldsymbol{Y} = [y_1, y_2,...y_n]$ are molar fractions of components in the liquid and vapor phases, respectively; $\boldsymbol{\varphi}_\mathrm{L} = [\varphi_{\mathrm{L}1}, \varphi_{\mathrm{L}2},... \varphi_{\mathrm{L}n}]$ and $\boldsymbol{\varphi}_\mathrm{V} = [\varphi_{\mathrm{V}1}, \varphi_{\mathrm{V}2},... \varphi_{\mathrm{V}n}]$ are fugacity coefficients of components in the liquid and vapor phases, respectively; and Q is the vapor fraction. New variables $\boldsymbol{K} = [K_1, K_2,...K_n]$ are defined:

$$K_i = y_i/x_i, (i = 1,2, ... n) \tag{7}$$

as typically seen in literature [20]. The Rachford-Rice equation [21]:

$$\sum_{i=1}^{n} x_i - y_i = 0 \tag{8}$$

is adopted to ensure that Eqs. (2) are all satisfied. Then Eq. (7) is applied to Eqs. (4) and to Eqs. (6) respectively to yield

$$x_i = \frac{z_i}{1 + Q(K_i - 1)}, (i = 1,2, ... n) \tag{9}$$

$$y_i = K_i x_i, (i = 1,2, ... n) \tag{10}$$

$$K_i = \varphi_{\mathrm{L}i}/\varphi_{\mathrm{V}i}, (i = 1,2, ... n) \tag{11}$$

With these new variables and equations, and with T, p and $\boldsymbol{Z}$ known, the VLE condition ($\boldsymbol{X}$, $\boldsymbol{Y}$ and Q) can be solved using the self-developed flash algorithm, which is illustrated in Figure 3 and described as follows. (I) At given T, p and $\boldsymbol{Z}$, with an initial guess of $\boldsymbol{K}$ according to Wilson's approximation

$$K_i = \frac{p_{\mathrm{c},i}}{p} \exp[5.373(1 + \omega_i)(1 - \frac{T_{\mathrm{c},i}}{T})], (i = 1,2, ... n) \tag{12}$$

Q can be solved according to Eqs. (8) to (10) using the Newton-Raphson (N-R) method and meanwhile $\boldsymbol{X}$ and $\boldsymbol{Y}$ are calculated. (II) With $\boldsymbol{X}$ and $\boldsymbol{Y}$ known and according to Eq. (5), $\boldsymbol{\varphi}_\mathrm{L}$ and $\boldsymbol{\varphi}_\mathrm{V}$ can be obtained, and new $\boldsymbol{K}$ is calculated using Eq. (12). (III) iterating the Q and $\boldsymbol{K}$ calculations until the converge condition is met.

To improve the robustness of the algorithm, a trick was implemented in the flash algorithm: five different initial guesses (0.001, 0.1, 0.5, 0.9 and 0.999) of Q needed for the N-R method are used. This trick, when convergence fails with all initial guesses of Q, also provides useful and reliable information to judge if the fluid is in a homogeneous liquid or vapor phase: If more iterated Q (the values Q_e in Figure 3) satisfies $Q \geq 1$ than $Q \leq 0$, the fluid is in the vapor phase, otherwise in the liquid phase. There are also tricks in improving calculation efficiency: e.g., once the looping Q deviates further away from the VLE range ($0 \leq Q \leq 1$), the Q (Q_0 in Figure 3) jumps to the next initial guess.

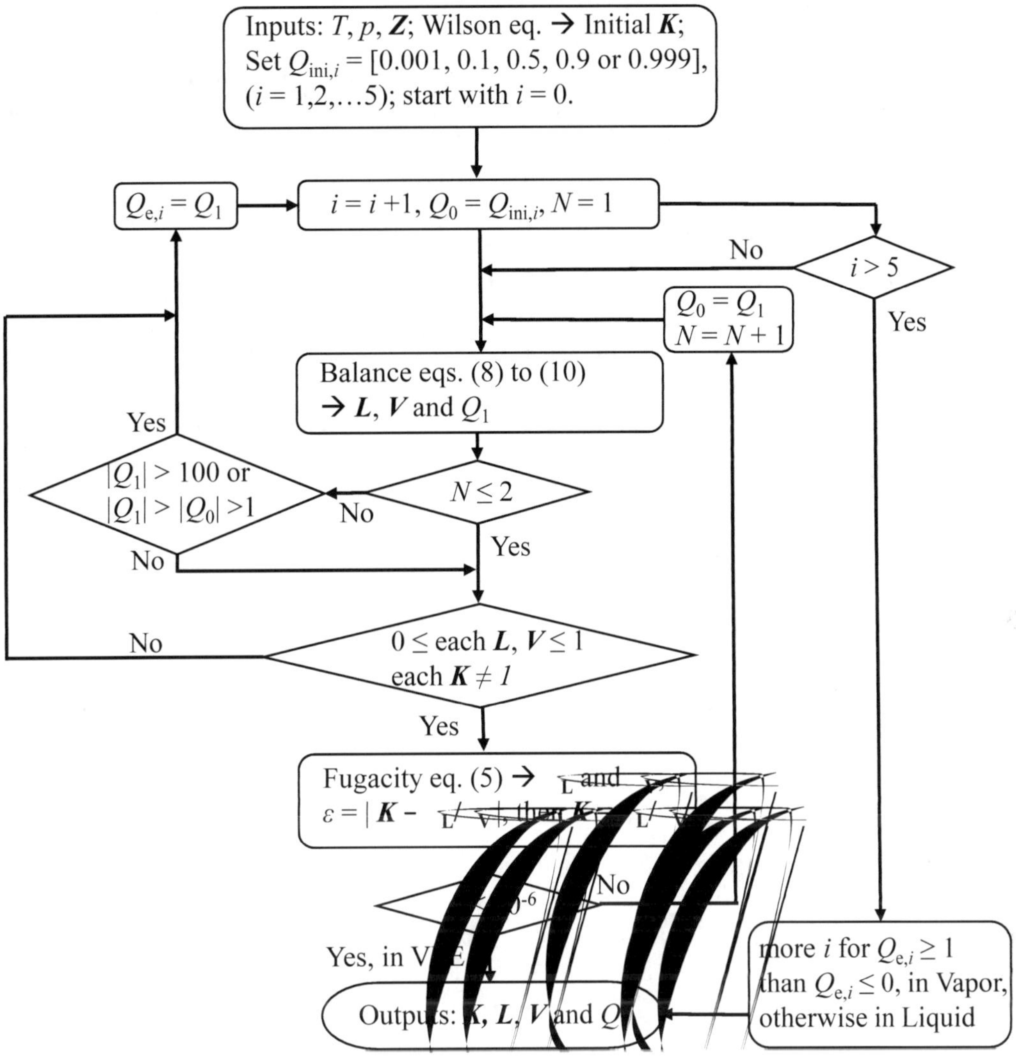

Figure 3. The vapor- liquid-equilibrium (VLE) flash algorithm implemented in OilMixProp 1.0.

As a result, this algorithm is reliable and robust for almost all the pure fluids and mixtures tested. One unresolved issue lies in the phase identification when the mixture is in the supercritical region: so far the output is the liquid phase in the denser region and the vapor phase vice versa. This issue can be solved by adding the function of calculating mixture critical point, which will be part of the future OilMixProp 2.0.

3.2 Algorithms for other key calculations

Except for the flash algorithm, there are also other calculations that require iterations, i.e., phase transition point (bubble-point and dew-point) determination, and VLE calculation with one input only exists in the two-phase region.

In phase transition point calculation, Q equals 1 or 0, which makes $\boldsymbol{K}$ the only values that need to be solved iteratively (compared to the flash algorithm, see Figure 3). It seems easier but since one of the T and p is unknown, an initial guess of T or p is always needed and should be reliable otherwise the iteration will fail or converge to a wrong value. Except for suggesting users try to provide an initial guess as accurate as possible, an algorithm is implemented in OilMixProp 1.0 to guarantee convergence to the right value successfully. For example, to calculate the dew-point temperature at a known pressure, the algorithm will first determine the phase condition (V for vapor, L for liquid, and VLE for vapor-liquid equilibrium) using the flash algorithm at the guessed temperature given by the user. If, for example, the fluid is in VLE condition, 5 K will be added to the guessed temperature and the phase condition is determined again; this step will go on until reaching the vapor phase. Then an initial guess of temperature being the average of the last two temperatures is used for the dew point temperature calculation according to Eqs. (5) and (8) to (11). If no initial guess is given by the users, valid initial guesses, such as the molar average of the transition point of each pure component, will be used.

There are various input combinations available in OilMixProp 1.0 (see all in section 4.2). Sometimes it is necessary to calculate fluid mixture properties with one of the input values only existing in a two-phase region. For example, in heat pump or refrigerant cycle analysis, at the outlet of the valve, the known variables are pressure and enthalpy and the fluid is in a two-phase region. This is a scenario of calculating temperature from pressure and enthalpy. The algorithm will solve this function

$$h_{\text{input}} = Q * h_{\text{V}} + (1 - Q) * h_{\text{L}} \tag{13}$$

using the N-R method with an initial guess of temperature either provided by the user or being those implemented (e.g., a molar average of pure component solutions) in the algorithm. In Eq. (13), h_{input} is the molar enthalpy being the input value, h_{V} and h_{L} are molar enthalpies in the vapor and liquid phase, respectively. The values of h_{V}, h_{L} and Q are obtained using the flash algorithm discussed in section 3.1. If the fluid is in a single phase, Q = 1 or 0.

4. Functions and examples

The core of OilMixProp 1.0 is to calculate the thermophysical properties of fluids, and this will be described in section 4.1. To determine fluid constants of a custom-defined oil so that this oil and its mixture can also be calculated, fitting tools are developed, see section 4.2. Last but not least, visualization tools are implemented so that important phase diagrams can be plotted, see section 4.3.

4.1 Property calculation function OilPropm

Analogical to the function *refpropm*, which is the Matlab interface of REFPROP 10.0 [11], a function *OilPropm* in OilMixProp 1.0 was developed as the 'only' interface that users need to call. Examples of using this function are given in *FluidCalc.m* in the package. An overview of the function *OilPropm* is given in Figure 4.

As shown in Figure 4, in total night input values are needed to call *OilPropm*:

(1) The output indicator (input 1). Except for specifying a property to calculate, *OilPropm* could also deliver a class (a datatype in Matlab, using output indicator 'all') that contains all information, compositions and all the core thermophysical properties in each phase as well as the phase fraction, of the fluid at the defined condition.

(2) First input property (inputs 2 and 3): indicator and its value. So far only 'P' for pressure in kPa and 'T' for temperature in K are available. For any calculations, at least one of these two variables has to be known.

(3) Second input property (inputs 4 and 5): indicator and its value. Six indicators are available: 'P', 'T', 'D' for density in kg·m^{-3}, 'Q' for vapor fraction in mass basis, 'H' for enthalpy in J·kg^{-1} and 'S'

for entropy in $J \cdot kg^{-1} \cdot K^{-1}$. Together with the first indicator, these combinations should cover all the needs in typical thermodynamic cycle analysis.

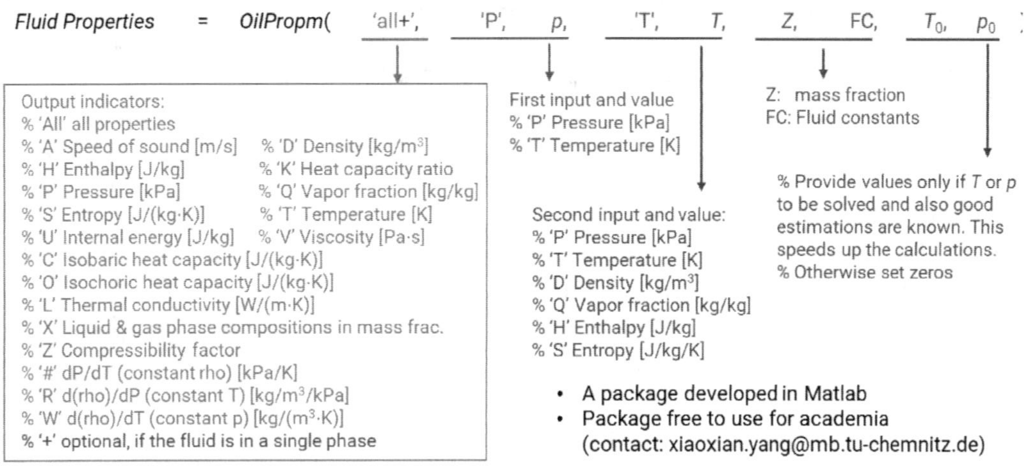

Figure 4. Overview of the function *OilPropm* in OilMixProp 1.0.

(4) Fluid and model information (inputs 6 and 7): Z and FC. Here Z is the overall composition (all phases included) in mass fraction and is an array in column rather than in line, e.g., Z = [0.3, 0.7]' rather than [0.3, 0.7] in Matlab. FC is a class (a datatype in Matlab) containing all fluid constants and model information. A typical example of generating FC of the R1233zde + Emkarate RL32 mixture using PTV EoS is given as follows

```
Fluids = {'R1233zde','Emkarate RL32'};
AllEOS = {'PR','SRK','PTV','YFR'};  CubicEOS = AllEOS{3};
FC = GetGlobals(CubicEOS, Fluids);
```

This design is different from *refpropm* whose inputs are names of fluids. Every time *refpropm* is called, fluid constants have to be read once; this wastes time if the same fluids are calculated several times or in loops.

(5) Initial guesses of temperature T_0 or pressure p_0 if they are unknown (inputs 8 and 9). They could be set to zero; however, we strongly recommend users provide guesses as accurate as possible, because this will significantly speed up the calculation, and avoid failed convergence or convergence at a wrong value. Here is the reason. If one of the T or p is unknown, the OilMixProp will first determine this unknown T or p by adopting the user's guess as the initial value for the equation solving using the N-R method. If the equation-solving fails or yields a wrong value, various initial guesses (e.g., molar average of pure component solutions) for the N-R method will be applied until the right solution is found. In the thermodynamic cycle analysis, such good guesses are always available, e.g., based on the properties of the previous state.

For the output, *refpropm* (Matlab interface of REFPROP 10.0) indicates it specifically, e.g., 'D' for density and the output is usually a single value. This might mislead users who have no sophisticated knowledge about thermophysics and the mixtures under study. For example call *refpropm* at an VLE condition:

refpropm('D','T',303.15,'P',1e3,'water','nitrogen',[0.5 0.5])

but the output is a single value of 21.924 kg·m^{-3} without indicating this value is in the liquid phase, vapor phase or overall phases. With OilMixProp 1.0, such a problem could be avoided. For the above mixture calculation (water and nitrogen mixture with an equal mass fraction), the output (with indicator 'all') of *OilPropm* using the YFR EoS is illustrated in Figure 5. From this comprehensive output, users will see that the liquid phase is almost pure water with a trace amount of nitrogen dissolved, and the gas phase is mainly nitrogen with little amount of water. As for the result and accuracy, the vapor phase density (11.113 kg·m^{-3}, see Figure 5) is very close to that (11.129 kg·m^{-3}) of pure nitrogen at the same T and p condition calculated with *refpropm*, and the liquid phase density (979.698 kg·m^{-3}, see Figure 5) agree within 2 % of pure water at the same T and p condition calculated with *refpropm*. The density output (21.924 kg·m^{-3}) of *refpropm* for this mixture is likely the value of the mixture considering all phases.

```
0.50 Water + 0.50 Nitrogen in mass fraction with YFR EoS
 T:              303.150000   K
 P:             1000.000000   kPa
 MM:                 21.929   g/mol
 Phase:                 LV

Phase behavior in mass fraction.  Vapor Fraction: 0.501418
                   Water       Nitrogen
Liquid:          0.999999      0.000001
Vapor:           0.002829      0.997171

        ---- Properties in each phase ----
                    Liquid          Vapor
 Z:               0.007296       0.998578
 MM:                18.015         27.970    g/mol
 rho:              979.698         11.113    kg/m3
 Entropy:           -3.165          1.516    kJ/K/kg
 Enthalpy:       -2393.055         28.871    kJ/kg
 cp:                 4.166          1.061    kJ/K/kg
 cv:                 2.961          0.749    kJ/K/kg
 SoS:             4513.070        356.844    m/s
 vis:             0.785165       0.017813    mPa s
 TC:              0.687892       0.027020    W/m/K
 Kappa            1.407187       1.416654
 d(rho)/dP:       0.000094       0.011125    kg/m3/kPa
 d(rho)/dT:      -0.771048      -0.037613    kg/m3/K
 dp/dT_rho:    6366.504562       3.380864    kPa/K

  -- Properties with all phases combined --
 rho_all:                  21.915          kg/m3
 Entropy_all:              -0.818          kJ/K/kg
 Enthalpy_all:          -1178.658          kJ/kg
 cp_all:                    2.609          kJ/K/kg
 cv_all:                    1.852          kJ/K/kg
 Kappa_all               1.409107
```

Figure 5. An example result of *OilPropm* using output indicator 'all'.

Such output design of *OilPropm* also significantly reduces the computational time if multiple properties at a defined condition are needed (typically scenarios in thermodynamic cycle analysis) as *OilPropm* only needs to be called once. The order of properties to be calculated is: (1) phase and volumetric information (density, volume, or compression factor) in each phase using the VLE flash algorithm, (2) caloric properties (heat capacities, entropy, enthalpy and speed of sound) in each phase, and (3) transport properties (viscosity and thermal conductivity) in each phase. If one of the T and p is unknown, T or p should be determined in the very beginning. Here, the time-consuming part is the first step (VLE flash calculation), while the last two steps are very quick as no iterations are needed. If users know for sure the fluid is in a single phase, adding the optional output indicator '+' will skip the VLE flash calculation and thus significantly reduce the calculation time. Also, if only a single property is of interest, the user could specify this output property, e.g., 'S' for entropy, to speed up calculation; this is the same output design as *refpropm*.

4.2 Fitting tools for oil constants

There are thousands of oil products available in the market. OilMixProp 1.0 offers fitting tools to determine the fluid constants of any oil, so that the thermophysical properties of this oil and its mixtures with other fluids can be calculated in OilMixProp 1.0. The required experiment and experimental data are described in our previous work [1] and summarized in Figure 6.

OilMixProp 1.0: fluid constants and the needed experiments

Less than 20 (at least 12) experimental points are needed → all constants of an oil

4 (0) → two masses and two boiling point temperatures → M

0 (0) → Z_c fixed at a reasonable value (e.g., 0.2563) → Z_c

4 (2) → four, could be down to two, points of (p = 1 atm, T, ρ) → T_c, p_c, and ω

4 (2) → four, could be down to two, points of (p = 1 atm, T, c_p) → k_0 and k_1

4 (4) → four points of (p = 1 atm, T, μ) → $n_{\mu k}$ (k = 1,2,3,4)

4 (4) → four points of (p = 1 atm, T, λ) → $n_{\lambda k}$ (k = 1,2,3,4)

Note: Temperature suggested to be (278.15 to 368.15) K or smaller
Pressure can be another value rather than 1 atm and does not have to be constant.

Figure 6. Fluid constants in OilMixProp 1.0 and the needed experiments to fit them of oils. Means of the parameters can be seen in Figure 2.

Aided with Figure 6, the needed experimental points for fitting the fluid constants of an oil in OilMixProp 1.0 can be described as follows.

(1) Optionally, two mass and two boiling point temperature measurements are needed to determine the average M. If data are not available, M is a defaulted value of 200 g·mol^{-1}, or users can provide a better estimation.

(2) Z_c is fixed at 0.2563, or users can provide a better estimation.

(3) Mandatorily, at least two sets of temperature, pressure and density (T, p, ρ) data are needed to fit T_c, p_c, and ω.

(4) Optionally, at least two heat capacity (T, p, c_p) data sets are needed to fit k_0 and k_1; if data are not available, k_0 and k_1 are defaulted values being average of those of the 151 fluids in REFPROP 10.0.

(5) Optionally, at least four viscosity (T, p, μ) data sets are needed to fit $n_{\mu k}$ (k = 1,2,3,4); if data are not available, $n_{\mu k}$ are defaulted values being global parameters of group 6 according to Yang et al. [8,22].

(6) Optionally, at least four thermal conductivity (T, p, λ) are needed to fit $n_{\lambda k}$ (k = 1,2,3,4); if data are not available, $n_{\lambda k}$ are defaulted values being global parameters of group 6 according to Li et al. [23].
(7) Optionally, for a binary system, one data point of mixture bubble-point pressure is needed to determine the binary interaction parameter (BIP) in the vdW mixing rule; otherwise, BIP = 0.

As a result, with a minimal amount of two (T, p, ρ) data sets, an oil could be fitted and implemented into OilMixProp 1.0. This is of course not recommended as the accuracy of other properties will be very low. We recommend collecting data (either from experiment or literature) as many as possible for each needed property (see Figure 6) and carrying out the fit. In OilMixProp 1.0, users can choose a few data for fitting and the remaining for confirming the fitting. An example result is given in Figure 7, where only data in full symbols are used for fitting.

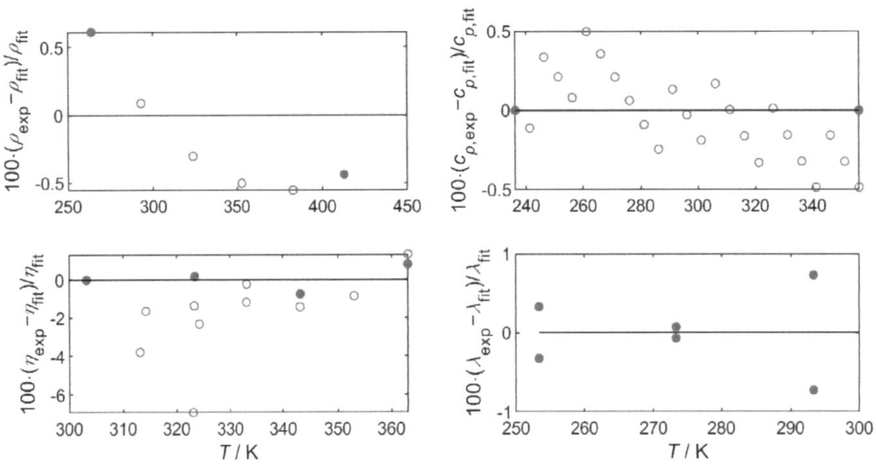

Figure 7. Example of the fluid constants fitting of PAG68: relative deviation of the experimental data [1] from the model predictions. Only the data in filled symbols are used for fitting.

OilMixProp 1.0 has included many examples to fit oil constants and the BIP from experimental data. The code for fitting oil constants is *Fitter_PureOil.m* and that for BIP is *Fitter_CKij.m*. The experimental data of these examples are given in the folder 'ExpData'. It is important to follow the examples to prepare the experimental data files, which are of course very easy to understand and prepare.

4.3 Phase diagrams

When studying a fluid mixture with unfamiliar components, such as oils, it is important to visualize its phase behavior by plotting phase diagrams of pressure vs. composition (p-x), temperature vs. composition (T-x), or pressure vs. enthalpy (p-h), etc. In OilMixProp 1.0, the p-x diagram at constant temperature is calculated by running a flash calculation at each pressure point, and the T-x diagram is calculated similarly. For potentially azeotrope mixture, calculating dew-point and bubble-point pressures (or temperatures) at each composition are also implemented to complete the whole phase diagram. Examples of a p-x diagram of a zeotrope mixture of propane + dodecane at 419.15 K (with experimental data from [24]) and a T-x diagram of an azeotrope mixture of carbon dioxide + ethane mixture at 3.0 MPa are given in Figure 8 and Figure 9, respectively. Many tricks, e.g., well-defined pressure (or temperature) points, are adopted to optimize the calculation speed while providing the completed phase diagram. More details can be seen from the code *PhaseDiag_px.m*, and users are encouraged to modify and improve the plotting code by themselves.

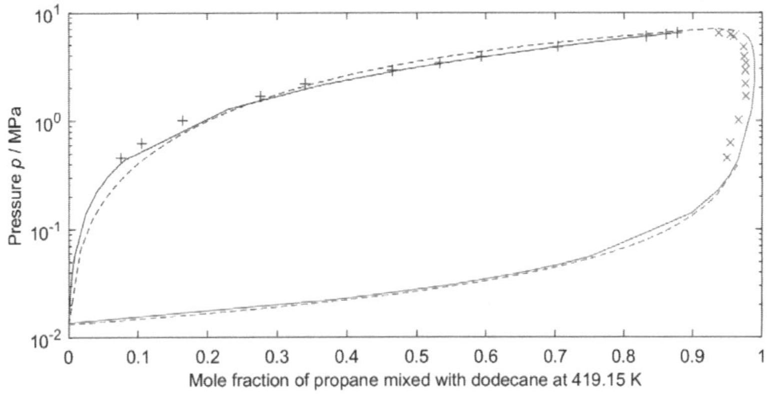

Figure 8. Pressure vs. composition diagram for the propane + dodecane mixture at 419.15 K. Symbols: experimental data [24]; solid curves: OilPropMix 1.0; dashed curves: REFPROP 10.0.

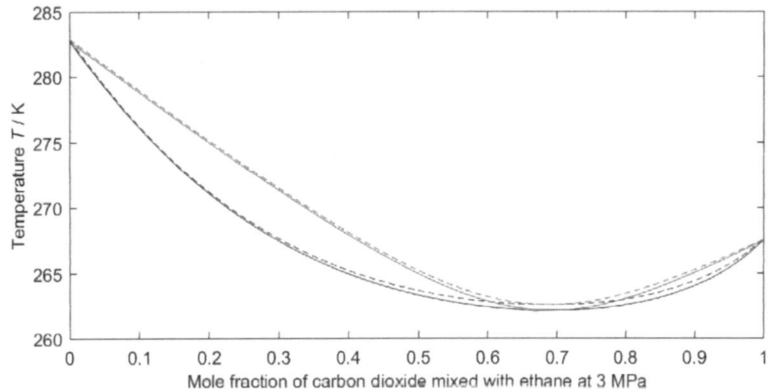

Figure 9. Temperature vs. composition diagram for the carbon dioxide + ethane mixture at 3 MPa. Solid curves: OilPropMix 1.0; dashed curves: REFPROP 10.0.

As an example, the *p-h* phase diagrams of pure propane and the propane + HATCOL4467 mixture at a propane mass fraction of 0.995 are plotted in Figure 10. The HATCOL® 4467 is mainly composed of ISO VG 68 polyol ester and is a compressor's oil miscible to propane. With only 0.005 mass fraction of HATCOL4467 added to propane, the dew point curve significantly shifts to low pressure; this is expected as one can easy imagine that oil condenses so easily. While the bubble point curve shifts towards higher pressures. This might help improve the coefficient of performance (COP) of a heat pump (HP) system as more heat needs to be transferred in the heat exchangers.

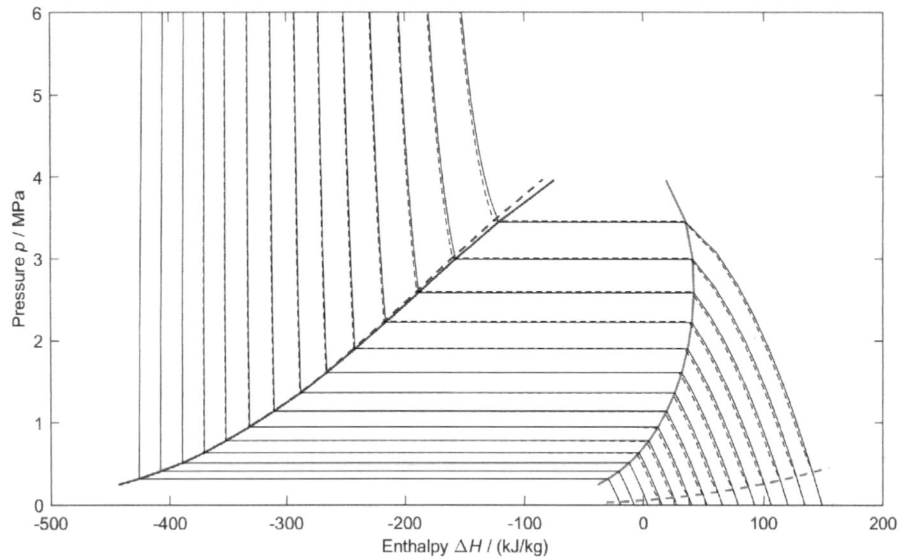

Figure 10. Pressure vs. enthalpy diagrams of pure propane (solid curves) and the propane + HATCOL4467 mixture at a propane mass fraction of 0.995 (dashed curves).

5. Conclusion and Outlook

In summary, we present here a software package OilMixProp 1.0, which calculates all the core thermophysical properties of oils, common fluids, and their mixtures. Properties that are needed in thermodynamic cycle analysis are all included: density, phase behavior, heat capacity, entropy, enthalpy, speed of sound, viscosity, and thermal conductivity. This package could be the first tool for comprehensive thermophysical property calculations of fluids involving user-defined oils. Using a minimal set of experimental data (fewer than 20 and at least 12 data points), fluid constants of an oil and its mixture with other fluids can be fitted with OilMixProp 1.0, and thus predictions can be achieved at larger temperature and pressure ranges. The models for property calculations in OilMixProp 1.0 include the cubic equation of state (EoS) for density, phase behavior, and residual properties, an additional equation for the ideal gas isobaric heat capacity c_p^o as a linear function of temperature T for caloric properties, and the residual entropy scaling developed by Yang et al. [6–8] for transport properties. Here, the latest developed Yang, Frotscher, and Richter (YFR) EoS with improved liquid phase accuracy is first implemented in a software package.

Based on a self-developed robust flash algorithm, reliable calculation of mixtures at vapor-liquid equilibrium (VLE) can be achieved in OilMixProp 1.0. The function *OilPropm* is the interface for property calculations. It was developed analogously to *refpropm*, the Matlab interface of REFPROP 10.0, but with improved input and output designs. The input improvement includes: (1) using fluid constants instead of fluid names as input, which avoids repeating reading fluid constants, and (2) allowing users to provide initial guesses of temperature and pressure (if they are unknown) as optional inputs, which could potentially significantly reduce computational time. The output of *OilPropm* could be a class (a data type in Matlab) containing all information (phase fraction, as well as compositions and properties in each phase) of the fluid at the defined condition. This avoids the ambiguous output as delivered in some cases by *refpropm,* and in most cases reduces the calculation time.

The package is written in Matlab, and is free for academic institutions; please contact the authors to acquire it. Future developments of OilMixProp include: (1) adding functions to calculate critical points of mixtures; (2) updating constants of some fluids needed for transport property calculations based on the RES approach; (3) enabling more phase diagram plots; (4) applying solid constants to more fluids so that solids are considered and the VLE phase diagram could be updated to SVLE phase diagram; (5) development of a graphical user interface (GUI) to facilitate the use of the package for none professional Matlab programmer; (6) converting the package to other languages, e.g., python, so that more it could be accessed by more users. Some of these developments have already been carried out, and version 2.0 is estimated to be released in the middle of 2025.

Acknowledgments

The realization of the project and the scientific work was supported by the German Federal Ministry of Education and Research on the basis of a decision by the German Bundestag (funding code 03SF0623A). The authors gratefully acknowledge this support and carry the full responsibility for the content of this paper.

References

[1] Yang X, Hanzelmann C, Feja S, Trusler J M and Richter M 2023 Thermophysical Property Modeling of Lubricant Oils and Their Mixtures with Refrigerants Using a Minimal Set of Experimental Data *Ind. Eng. Chem. Res.* **62** 18736–49

[2] Patel N C and Teja A S 1982 A new cubic equation of state for fluids and fluid mixtures *Chem. Eng. Sci.* **37** 463–73

[3] Valderrama J O 1990 A Generalized Patel-Teja Equation of State for Polar and Nonpolar Fluids and Their Mixtures *J. chem. eng. Jpn.* **23** 87–91

[4] Soave G 1972 Equilibrium constants from a modified Redlich-Kwong equation of state *Chem. Eng. Sci.* **27** 1197–203

[5] Peng D-Y and Robinson D B 1976 A new two-constant equation of state *Ind. Eng. Chem. Fundam.* **15** 59–64

[6] Yang X, Xiao X, May E F and Bell I H 2021 Entropy Scaling of Viscosity—III: Application to Refrigerants and Their Mixtures *J. Chem. Eng. Data* **66** 1385–98

[7] Yang X, Kim D, May E F and Bell I H 2021 Entropy Scaling of Thermal Conductivity: Application to Refrigerants and Their Mixtures *Ind. Eng. Chem. Res.* **60** 13052–70

[8] Yang X, Xiao X, Thol M, Richter M and Bell I H 2022 Linking viscosity to equations of state using residual entropy scaling theory *Int. J. Thermophys.* **43** 183

[9] Yang X, Xiao X, Thol M, Richter M and Bell I 2023 A Residual Entropy Scaling Approach for Viscosity of Refrigerants, other Fluids and their Mixtures *NIST*

[10] Urbaneck T, Matthes M, Richter M, Hempel O, Safarik M, Franzke U, Noack R, Honke M, Müller M, Henninger S, Schossig P, Henning H-M, de Oliveira H T, Morgenstern A, Nienborg B, Gschwander S, Engelmann P and Benndorf G 2022 Research Platform Refrigeration and Energy Technology (KETEC)

[11] Lemmon E W, Bell I H, Huber M L and McLinden M O 2018 *NIST Standard Reference Database 23: Reference Fluid Thermodynamic and Transport Properties-REFPROP, Version 10.0, National Institute of Standards and Technology*

[12] Span R, Beckmüller R, Hielscher S, Jäger A, Mickoleit E, Neumann T and Pohl S 2020 *TREND. Thermodynamic Reference and Engineering Data 5.0.* (Bochum, Germany: Lehrstuhl für Thermodynamik, Ruhr-Universität Bochum)

[13] Bell I H, Wronski J, Quoilin S and Lemort V 2014 Pure and Pseudo-pure Fluid Thermophysical Property Evaluation and the Open-Source Thermophysical Property Library CoolProp *Ind. Eng. Chem. Res.* **53** 2498–508

[14] Yang X, Ophelia Frotscher and Richter M 2024 A new cubic equation of state developed with symbolic regression to improve liquid density calculation accuracy *Ind. Eng. Chem. Res.* (to be submitted)

[15] Stryjek R and Vera J H 1986 PRSV: An improved Peng—Robinson equation of state for pure compounds and mixtures *Can. J. Chem. Eng.* **64** 323–33

[16] Wilson G M 1964 Vapor-liquid equilibria, correlation by means of a modified Redlich-Kwong equation of state *Advances in Cryogenic Engineering* (Springer) pp 168–76
[17] Redlich O and Kwong J N S 1949 On the thermodynamics of solutions. V. An equation of state. Fugacities of gaseous solutions *Chem. Rev.* **44** 233–44
[18] Viktor Martinek 2023 Thermodynamics-informed symbolic regression (TiSR). A tool for the thermodynamic equation of state development. *Zenodo*
[19] Daniel Medeiros DWSIM - Open-Source Chemical Process Simulator
[20] Ritschel T K S and Jørgensen J B 2017 *Computation of Phase Equilibrium and Phase Envelopes* (DTU Compute)
[21] Rachford H H Jr and Rice J D 1952 Procedure for Use of Electronic Digital Computers in Calculating Flash Vaporization Hydrocarbon Equilibrium *J. Petrol. Tech.* **4** 19–3
[22] Martinek V, Yang X, Bell I H, Herzog R and Richter M 2024 Entropy scaling of viscosity IV - application to 124 industrially important fluids *J. Chem. Eng. Data* (to be submitted)
[23] Li Z, Yang X and Duan Y 2024 Linking Thermal Conductivity to Equations of State Using Residual Entropy Scaling Theory *Int. J. Thermophys.* (to be submitted)
[24] Gardeler H, Fischer K and Gmehling J 2002 Experimental Determination of Vapor–Liquid Equilibrium Data for Asymmetric Systems *Ind. Eng. Chem. Res.* **41** 1051–6

IOP Conf. Series: Materials Science and Engineering **1322** (2024) 012010 doi:10.1088/1757-899X/1322/1/012010

One-dimensional investigations of the periodic liquid-injection in twin-screw compressors

M Heselmann, T Monden and A Brümmer

TU Dortmund University, Chair of Fluidics, Dortmund, Germany

E-mail: matthias.heselmann@tu-dortmund.de

Abstract. The injection of liquid fluid (mostly oil) in twin-screw machines brings several advantages like sealing the gaps, at least partially, and absorbs some of the compression heat. To predict the machine behavior, it is essential to consider the injection process, because it determines the amount of liquid within the working chambers. This in turn determine the absorption of compression heat, the degree of sealing the gaps and the hydraulic losses. Therefore, in this study the transient liquid injection is investigated using a one-dimensional method of characteristics within the injection nozzle. The theoretical results are compared with the actual liquid mass flow rate of an oil-injected twin-screw compressor.

Nomenclature

a	propagation velocity	(m/s)
a	parameter	
A, B	base points	
$\bar{\bar{A}}$	matrix of coefficients	
c	contraction	
c	velocity	(m/s)
d	diameter	(m)
f_x	specific friction	(m^2/s^2)
hyd	hydraulic smooth	
inj	injection	
i, j	counting index	
$\bar{\bar{I}}$	identity	
K_s	compression module	(Pa)
$\dot{m}$	mass flow rate	(kg/s)
n	nozzle	
n	compressor speed	(Hz)
p	pipe	
p	pressure	(Pa)
Re	Reynolds number	
t	time	(s)
$\vec{u}$	variable vector	
x	location	(m)
x	measurement device	
α	contraction angle	(°)
γ	roughness ratio	
Δ	difference	
ζ	pressure loss coefficient	
η	dynamic viscosity	(Pas)
λ	pipe friction coefficient, eigenvalues	
ρ	density	(kg/m^3)
Φ	general parameter	
ϕ	rotation angle	(°)

1. Introduction

Screw machines are among the most frequently used compressor concepts in industry. This applies to the areas of compressed air generation, compression of process gases and use as refrigerant compressors [1]. Above all, screw compressors with liquid injection (usually oil) have established themselves. This is largely due to the sealing and cooling effect of the injected liquid. Especially in times of global warming, optimisation with regard to the reduction of CO2 emissions is of particular interest. The industrial compression of gases accounts for one of the highest CO2 emissions globally [2]. In addition, a 2017 study shows that oil-injected screw compressors and vane compressors account for up to 42% of global energy consumption in the field of industrial gas compression [3].

Research into injected screw machines generally takes the form of two disciplines: research into fluid distribution and research into fluid action (hydraulic losses and heat transfer). The two mechanisms are usually considered separately from each other. For heat transport, for example, it is assumed that the injected fluid exists in the form of droplets in the working chamber during the entire compression process [4]. On the other hand, the viscosity, for example, is kept constant for the distribution of the fluid in the working chamber. In addition to the two disciplines mentioned above, two main calculation methods are used to investigate injected screw machines. On the one hand, these are three-dimensional Computational Fluid Dynamics (CFD) simulations [5]. However, they require a large amount of time to generate a suitable computational mesh and corresponding computing capacity to solve the system of partial differential equations. This method is therefore more suitable for recalculating individual operating points. On the other hand, there is a zero-dimensional chamber model simulation. Zero dimensional because only mass and energy conservation have to be solved. Here, the geometry of the screw machine must be abstracted in such a way that all relevant variables (volume, gap areas, etc.) are represented by a time-dependent variable (usually the angle of rotation). The solution of the system of equations is then very fast compared to the CFD calculations, with comparable accuracy [6]. This is partly due to the fact that submodels can be specifically selected to represent individual mechanisms (gap flow, heat transport, etc.). The significantly lower computing capacity means that this method is also used for extensive variation calculations.

In the numerical simulation of injected screw machines with CFD or chamber models, the first challenge is to determine the amount of fluid injected. This is due to the fact that injection usually takes place in a working chamber. This means that the injection nozzle is periodically closed, at least partially, by the passing rotor lobe. Furthermore, injection takes place into an encapsulated working chamber, which means that the pressure difference from the oil reservoir into the working chamber is not constant. It is therefore a transient flow. In this work, precisely this unsteady flow through the injection nozzle is analysed. The aim is to determine the amount of injected oil so that this can be incorporated into a suitable sub-model for oil injection useable in CFD or chamber model simulations.

For this purpose, a geometrically similar model of the injection nozzle of a commercially available screw compressor is additively manufactured (3D-printed) on a scale of 1:2. Water then flows through the nozzle at steady state to determine the pressure loss coefficient of the nozzle. In order to capture the transient process of the rotor lobe passing by, the rotor geometry is also positioned geometrically similar (includes lobe contour and housing gap) at different rotation angles behind the nozzle. The pressure loss coefficients of the measured rotation angles are then mapped into a continuous function using a curve fit. This means that a pressure loss coefficient is available for each angle of rotation (point in time). In order to transfer the pressure loss coefficients of the nozzle through which water flows to those in the commercial compressor, the physical similarity must also be ensured. For this purpose, the pressure loss coefficient of the actual nozzle (without rotor lobe) is divided into individual pressure loss coefficients (cross-sectional changes, pipe sections), which can be found in the literature. In this way, the physical similarity can be guaranteed by a constant Reynolds number by adapting the pressure difference to the medium. For the numerical method used (method of characteristics) to calculate the transient injection, it does not matter at which point the calculated pressure loss coefficient is taken into account. Therefore, the pressure loss coefficient is summarised again and taken into account at the outlet of the nozzle. This allows the injection mass flow rates determined by simulations to be compared directly with the measured injection mass flow rates.

2. Numerics

The method used for the numerical calculation of the flow through the injection nozzle is described in this chapter. The flow is considered one-dimensional (along the flow direction), compressible, unsteady, friction-bound and adiabatic without supply or removal of work. The conservation equations for mass and momentum are then for constant cross sections:

$$\frac{\partial \rho}{\partial t} + \frac{\partial(\rho \cdot c)}{\partial x} = 0 \tag{1}$$

$$\frac{dc}{dt} + \frac{1}{\rho} \cdot \frac{\partial p}{\partial x} + f_x = 0 \quad \text{with} \quad f_x = \frac{\lambda}{d} \cdot \frac{c \cdot |c|}{2} \quad \text{and} \quad \lambda = \begin{cases} 64/Re & \text{if} \quad Re \leq 2320 \\ 0.316/\sqrt[4]{Re} & \text{if} \quad Re \leq 10^5 \end{cases} \tag{2}$$

The continuity equation (Eq. 1) describes the relationship between the local mass change and the convective mass transport. In the conservation of momentum (Eq. 2), the total momentum change (local and convective) is balanced with the acting pressure forces and the specific frictional forces (f_x) resulting from the shear stresses [7]. The influence of possible weight forces resulting from the acceleration due to gravity is neglected here. To solve the system of partial differential equations, these are converted into a linear, hyperbolic system of differential equations. With the help of the mathematical characteristics method, this can in turn be converted into a total differential equation system and solved explicitly using the finite difference method [8].

2.1. *Method of characteristics*

The application of the method of characteristics transforms the partial system of equations consisting of equations 1 and 2 into a system of total differential equations. To do this, the partial system of differential equations of the homogeneous form

$$\vec{u}_t + \bar{\bar{A}} \cdot \vec{u}_x = \vec{0} \tag{3}$$

have real eigenvalues λ,

$$\det\left(\bar{\bar{A}} - \lambda \cdot \bar{\bar{I}}\right) = 0 \tag{4}$$

which are also referred to as characteristics [9]. The indices on the variable vector $\vec{u}$ represent the temporal (index t) and spatial (index x) derivatives of the variables.

$$\text{vector of variables:} \quad \vec{u} = \begin{pmatrix} p \\ c \end{pmatrix} \quad \text{matrix of coefficients:} \quad \bar{\bar{A}} = \begin{pmatrix} c & \rho a^2 \\ \frac{1}{\rho} & c \end{pmatrix} \tag{5}$$

wherein

$$a = \sqrt{\frac{dp}{d\rho}} = \sqrt{\frac{K_s}{\rho}} \tag{6}$$

is the propagation velocity solving Laplaces' equations [7]. There are two eigenvalues for the present system of equations, for each of which an eigenvector can be determined. Subsequently, the homogeneity of the system of differential equations in the form of the specific frictional force ($f_x > 0$ points in x-direction) is to be taken into account when determining the normal form via a corresponding matrix transformation [9]. The determined normal form of the system of differential equations can be converted into total differentials along the respective characteristics:

$$\frac{dp}{dt} + \rho a \frac{dc}{dt} - \rho a f_x = 0 \quad \text{for} \quad \lambda_A = \frac{dx}{dt} = c + a \tag{7}$$

$$\frac{dp}{dt} - \rho a \frac{dc}{dt} + \rho a f_x = 0 \quad \text{for} \quad \lambda_B = \frac{dx}{dt} = c - a \tag{8}$$

These total differentials are only valid along their characteristic directions of propagation dx/dt. The gradient of the characteristics in the local time plane is not constant, but depends on the local, time-dependent flow and propagation velocities c and a. The characteristics correspond to the propagation of an outgoing (A) and returning (B) sound wave. These propagate relative to the moving medium with the instantaneous, local propagation velocity a.

2.2. *Linearisation and discretisation*

For the characteristic differentials, however, the fluid states at the base points Φ_b of the characteristics (A and B) must be taken into account in order to apply the method.

$$\frac{d\Phi}{dt} = \frac{\Phi_i^{j+1} - \Phi_b^j}{\Delta t} \tag{9}$$

However, since only the fluid states at the nodes are known in the respective time steps Δt, these must be interpolated via individual nodes. The position of the base point of a characteristic between two nodal points can be determined via the gradient of the respective characteristic. Figure 1 visualise this method in a coordinate system of space x and time t. Once all the quantities for determining the two characteristic state variables of the new time step j+1 are known, they can be determined explicitly:

$$p_i^{j+1} = \frac{1}{2}((p_A + p_B) + (\rho a)_i^j (c_A - c_B)) \quad (10)$$

$$c_i^{j+1} = \frac{1}{2}\left(\frac{1}{(\rho a)_i^j}(p_A - p_B) + (c_A + c_B)\right) + f_{x,i}^j \Delta t \quad (11)$$

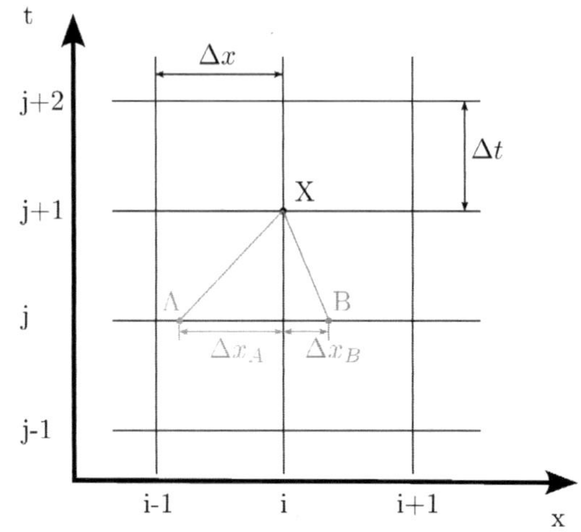

Figure 1. Discretised space-time grid.

The propagation velocity a, which is also required, is determined via Eq. 6. The method of characteristics to be used here is consistent [10]. In order for the method to converge with increasing resolution, a corresponding stability criterion must be met. This states that the minimum spatial step size of an equidistant space-time grid depends on the maximum gradient of the characteristics and the temporal step size:

$$\Delta x \geq (|c| + a) \cdot \Delta t \quad (12)$$

If this criterion is not met, stability problems may occur and the convergence of the numerical scheme cannot be guaranteed [10].

2.3. *Boundary conditions*

The method of characteristics used here can be applied to any pipe sections with constant cross-section. At the edges or system boundaries, however, only one characteristic is available for determining the primitive variables (pressure p and velocity c). In order to obtain a specific system of equations, Dirichlet boundary conditions are used for the investigations to be carried out here [11]. A boundary condition must be specified at a boundary by explicitly specifying a time functional for the pressure $p(t)$ or the velocity $c(t)$.

2.4. *Connection of elements*

In order to simulate a system with several different cross-sections and intermediate fixtures, additional transition conditions must be implemented. Only couplings of two elements (e.g. 1 and 2) are considered here. Whereas in the numerical scheme for a continuous pipe section, two unknown characteristic variables must be determined at each node, in the case of a transition condition between two elements, two unknowns must be calculated at each edge of the elements - i.e. a total of four unknowns. Accordingly, in addition to the two characteristics, two further equations must be solved. For this purpose, the general conservation equations (Eqs. 1 and 2) are discretised according to the numerical conventions. Those are the conservation of mass

$$\rho_1^j \cdot A_1 \cdot c_1^{j+1} = \rho_2^j \cdot A_2 \cdot c_2^{j+1} \quad (13)$$

and the conservation of momentum

$$p_1^{j+1} + \frac{\rho_1^j}{2} \cdot (c_1^j)^2 = p_2^{j+1} + \frac{\rho_2^j}{2} \cdot (c_2^j)^2 + \zeta_1 \cdot \frac{\rho_1^j}{2} \cdot (c_1^j)^2 \quad . \quad (14)$$

The formulations of the conservation equations have been chosen in such a way that an explicit linear system of equations can be set up. For this purpose, the local density is taken into account from the known time step j in each case. The incoming and outgoing mass flow rates are balanced directly with each other and local density changes within a control volume are not taken into account. The conservation of momentum used corresponds to a discretised expression of the conservation of momentum along a streamline [12]. In order to obtain a linear system of equations, the quadratic terms for the dynamic pressure are mapped via the local flow velocity from the known time step j. The expressions for the pressure loss is related to the edge of element 1 (upstream) as an example.

3. Experiments with water

The injection nozzle, which is inserted in a glass housing during the operation of a commercial compressor, is used to investigate the injection. The compressor has a nominal power of 55 kW, lobe combination of 5+6 and a main rotor wrap angle of 300°. Technical data of the compressor have been published by Heselmann et al. [13]. The geometry of the nozzle is shown in Fig. 2. It consists of four characteristic areas. The connection to the pipe, two pipe sections and a continuous reduction in cross-section. The flow direction is from left to right. To keep the experimental setup as simple as possible, the experiments

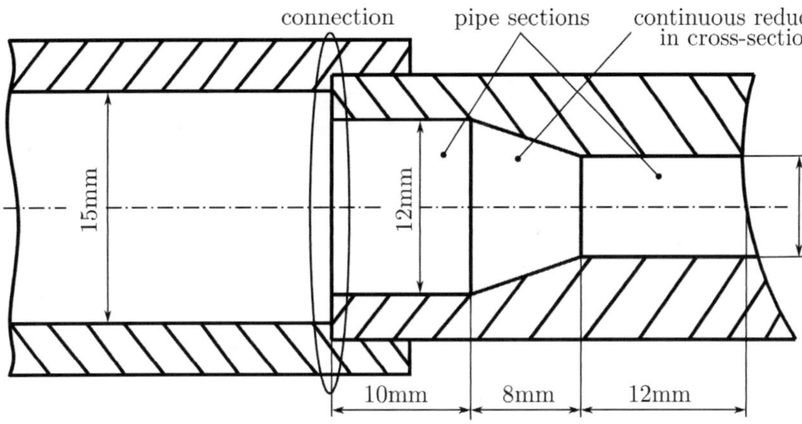

Figure 2. Geometry of the nozzle used in the commercial compressor with glass housing.

are performed with water to validate the numerical method. The water is taken from the public water network through a standard tap and flows through the nozzle. Since the mass flow rate is limited, the nozzle was halved for the experiments with water. This makes it possible to achieve meaningful pressure differences across the nozzle. A pressure gauge is placed directly in front of the nozzle to measure the pressure difference with respect to atmospheric outlet conditions. The water mass flow rate is determined using a stopwatch, a container in which the water is collected, and a scale. The pressure loss coefficient can then be determined integrally from the measured pressure difference and the measured mass flow rate. Table 1 lists the used instruments to gain the values described before. The container is listed here including measurement uncertainty as mass losses have occurred due to splashing when collecting the water masses. Since the mass flow rate with the measured variables used here is a composite measured variable, the sensitivities of the composite measured variable to the measured variables are formed according to Ratcliffe [14]:

$$\Delta \dot{m} = \sqrt{\sum_i \left(\frac{\partial \dot{m}}{\partial x_i} \Delta x_i \right)^2} \quad (15)$$

Where $\Delta \dot{m}$ is the measurement uncertainty of the mass flow rate, x_i is the i-th measuring device and Δx_i is the measurement uncertainty of the measuring device.

Table 1. Listing of measuring instruments and the associated measurement uncertainties.

measurement device	range	uncertainty
stopwatch	0 - 60 s	± 0.3 s (o.r.)
scale	0 - 10 kg	± 0.5 % (o.r.)
pressure gauge	0 - 0.1 MPa	± 1.5 % (o.f.s)
container	0 - 0.01 m^3	± 1.6 % (o.r.)

3.1. Transient pressure loss coefficient

In order to map a periodic injection, it is necessary to take the passing rotor lobe into account. This blocks a different proportion of the nozzle outlet opening depending on the rotational angle (time). Just like the nozzle, the gate rotor is scaled geometrically similar (halved) and placed in front of the nozzle at different angles of rotation. The angle of rotation is defined to the best knowledge so that the rotor lobe in the zero position ($\phi = 0$) is centred in front of the nozzle opening. According to this definition, the rotation angles $\phi = -5°$, $\phi = -2°$, $\phi = 2°$ and $\phi = 5°$ are also manufactured and measured. The actual measurement is carried out under stationary conditions. Figure 3 shows the measured pressure loss coefficient for the outflow from the nozzle into the container past the rotor lobe. The graph in Fig. 3, which approximates the measured points for the transient pressure loss coefficient $\zeta(\phi)$, referred to the velocity at the nozzle opening, can be described by the following function depending on the gate rotor rotation angle ϕ:

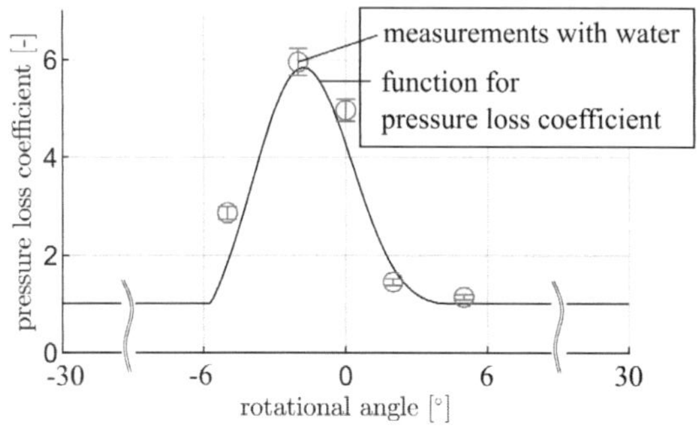

Figure 3. Pressure loss coefficient for a rotor lobe passing the nozzle opening over the rotational angle of the gate rotor.

$$\zeta(\phi) = \begin{cases} a \cdot \exp\left(-\left(\frac{\phi+b}{c}\right)^2\right) + \sin(d \cdot \phi) + e & \text{if} \quad -5.682 < \phi < 4.374 \\ 1 & \text{if} \quad \text{otherwise} \end{cases} \tag{16}$$

$$\text{a} = 5.5194 \quad \text{b} = 1.8153 \quad \text{c} = 2.9913 \quad \text{d} = 0.1 \quad \text{e} = 0.5$$

As Fig. 3 show, the maximum pressure loss coefficient is just before the defined zero position. However, as this is easy to identify in the later evaluation and is more decisive for a zero position, the dimension-less representation of the angle of rotation (time) is related to the extreme points of the pressure loss coefficient. This is shown in Fig. 4. With six rotor lobes, one period corresponds to a rotation angle of $\phi = 60°$. With the results of the measurements carried out, a stationary pressure loss coefficient is now available for each angle of rotation. In the following, this flow coefficient is applied to the unsteady flow through the injection nozzle outlined above.

4. Consideration of similarity

In order to transfer the pressure loss coefficients of the nozzle through which the water flows to the commercial compressor, similarities must be observed. This applies to both geometric and physical similarity. For this purpose, the pressure loss coefficient used in Eq. 14 is decomposed into its individual components. These are the previously described time-dependent (rotation angle dependent) outflow loss from the nozzle and the pressure loss within the actual nozzle. According to Fig. 2, the pressure loss coefficient for the nozzle consists of 4 characteristic areas. For each area a pressure loss coefficient is given below, which can be found in the literature. In this way, the physical similarity can be ensured by a constant Reynolds number by adapting the pressure difference to the medium.

4.1. Pressure loss coefficients

To calculate the pressure loss coefficients, the injection nozzle is assembled as a combination of piping elements. The nozzle can then be regarded as a piping system. The pressure loss coefficients ζ given below apply to quasi-one-dimensional, friction-loaded, stationary and incompressible flows through completely filled piping systems. They are proportional to the dynamic pressure $\rho c^2/2$, whereby the velocity c refers to the outlet cross-section of the element under consideration [15]. The first characteristic area is the connection of a tube ($d_1 = 15$ mm) to the nozzle ($d_2 = 12$ mm). This is treated as an abrupt contraction of the cross-section, index c. The pressure loss coefficient can be calculated according to [16] using the following equation:

equation:

$$\zeta_c = 0.578 + 0.395\left(\frac{d_2}{d_1}\right) - 4.538\left(\frac{d_2}{d_1}\right)^2 + 14.243\left(\frac{d_2}{d_1}\right)^3 - 19.222\left(\frac{d_2}{d_1}\right)^4 + 8.54\left(\frac{d_2}{d_1}\right)^5 \tag{17}$$

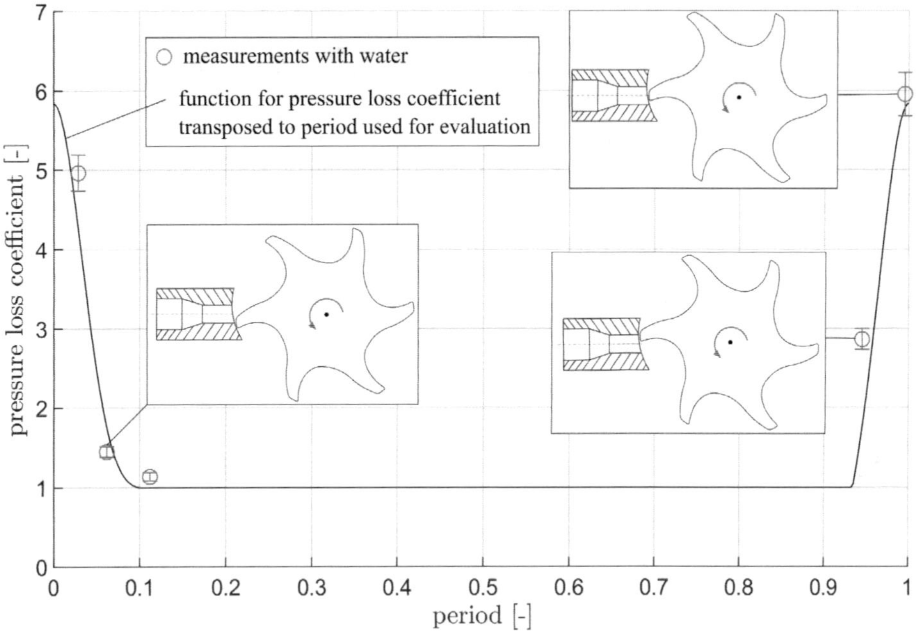

Figure 4. Pressure loss coefficient over dimensionless time (period) for a rotor lobe passing the nozzle opening. (One period equals 60° of gate rotor rotation.)

According to the direction of flow, index 1 indicates the left-hand side and index 2 the right-hand side of the change in cross-section. The pressure loss coefficient for the sections with a constant cross-section is calculated according to the pipe friction law:

$$\zeta_p = \lambda_{hyd} \cdot \frac{l}{d} \quad \text{with} \quad \lambda_{hyd} = \begin{cases} 64/Re & \text{if} \quad Re \leq 2320 \\ 0.316/\sqrt[4]{Re} & \text{if} \quad Re \leq 10^5 \end{cases} \tag{18}$$

Here l is the length of the corresponding pipe section and d the corresponding diameter. λ_{hyd} is the pipe friction coefficient, which is a function of the Reynolds number. The index hyd indicates that the surface is hydraulically smooth. In addition to the Reynolds number, the calculation of the pipe friction coefficient depends on the flow structure, which is differentiated on the basis of the Reynolds number (laminar $Re \leq 2320$ or turbulent $Re \in (2320;\ 10^5])$ [7]. The semi-empirical formula for λ_{hyd} can be used to determine the pressure loss coefficient ζ_p for hydraulically smooth pipe sections. The Reynolds number

$$Re = \frac{\rho \cdot c \cdot d}{\eta} = \frac{4 \cdot \dot{m}}{\pi \cdot d \cdot \eta} \tag{19}$$

relates the inertial forces to the frictional forces and is therefore dependent on the density ρ, the flow velocity c, a characteristic length d and the dynamic viscosity η. With the conservation of mass $\dot{m} = \rho c A$ and the diameter as the characteristic length for pipes, the Reynolds number can also be specified as a function of the mass flow rate $\dot{m}$, the diameter and the dynamic viscosity. In a continuous reduction in cross-section, the fluid retarded by the wall friction is constantly given new drive by the existing pressure gradient, so that the forward movement is also maintained in the friction layer close to the wall. As the cross-section narrows, the velocity profile therefore becomes more and more complete. As a result of the wall friction, the actual flow velocity in the smaller cross-section is slightly lower than the theoretically calculable velocity with frictionless flow [15]. According to Zoebl [17], the pressure loss coefficient is therefore calculated from a combination of the pipe friction coefficient and a correction factor a, which takes into account both the diameter ratio and the angle of contraction α, given in table 2. Within the table, the value for the current pairing of diameter ratio and contraction angle is determined by linear interpolation.

$$\zeta_n = a \cdot \lambda_{hyd} \tag{20}$$

In order to match the theoretical pressure loss coefficients to the measured pressure loss coefficients, a parameter is needed to do just that. Two of the three models for the pressure loss coefficient include the Blasius pipe friction coefficient. Since this applies to hydraulically smooth pipes, it is used here. In the case of the 3D-printed nozzle, it can be assumed that both the diameter and the surface quality will deviate from the specified values. These deviations between theory and practice are represented in the parameter γ as the ratio of the actual pipe friction coefficient to the hydraulically smooth pipe friction coefficient:

$$\gamma = \frac{\lambda}{\lambda_{hyd}} \approx 7.18 \tag{21}$$

Table 2. Factors a for continuously narrowing cross-sections [17].

	values a for contraction angles α			
d_1/d_2	4°	6°	8°	20°
1.2	9.3	6.2	4.6	1.8
1.4	6.7	4.5	3.3	1.3
1.6	5.1	3.4	2.5	1.0
1.8	4.1	2.7	2.0	0.8
2.0	3.4	2.3	1.7	0.7

5. Periodic liquid-injection

The 1D simulation domain is represented as shown in Fig. 2, but halved in size to match the experiments with water. The continuous reduction in cross-section is represented by several segments of constant cross-section to account for inertial effects within the calculation method used. Whether the pressure loss coefficient is applied in the domain where it physically occurs, or whether it is summed and applied at the end of the entire simulation domain, the mathematical result is the same. For this reason, and because Blasius' theory is valid only for stationary flows, the pressure loss coefficient is applied at the end of the simulation domain. In order to be able to assess the plausibility of the results, similarity relationships are used to transfer the calculations of the model to the nozzle actually installed (scale 1:2). These are, on the one hand, the geometric similarity, which requires that the model and reality are similar by a length scale. And on the other hand the physical similarity. The Reynolds number is used here so that the Reynold's dependent pressure loss coefficients for the model and reality match. In the following, the results are shown for the values of the original size nozzle. The measurement results shown refer to a compression ratio of the compressor of $\Pi \approx 7$. The pressure gauge directly upstream of the injection nozzle delivers an averaged value of $p_{inj} \approx 6$ bar. This value is further used for the simulations as high pressure. As the machine is not indicated, the internal pressure within the chambers of the screw compressor and therefore the pressure difference across the injection nozzle is not known precise.

5.1 Transient simulation results

At the beginning of the calculation, the fluid in the calculation domain is at rest (Fig. 5, time = 0 ms). As time progresses, the flow velocity at the outlet of the nozzle changes according to the pressure difference and the previously described and experimentally determined pressure loss coefficient of the nozzle. Although the simulations are time-dependent, a recurring pattern emerges after a few periods. The period here refers to the repetitive pattern of the rotor lobe passing the nozzle opening (Fig. 5). Based on this period, the variations performed can be interpreted alongside the integrally calculated mass flows. To get transferable results, the calculated outflow velocity $c(t)$ is normalised with the stationary, frictionless flow velocity $c_{Bernoulli}$ (from the pipe to the nozzle opening):

$$\frac{c(t)}{c_{Bernoulli}} = \frac{c(t)}{\sqrt{\frac{2\Delta p}{\rho\left(1-\frac{d^4_{outlet}}{d^4_{inlet}}\right)}}} \tag{22}$$

This means that the interpretation is independent of the dimensions of the problem under consideration. Furthermore, a quasi-stationary velocity can be calculated for any point in time with a known pressure loss coefficient (from the pipe to the container):

$$c_{stationary} = \sqrt{\frac{2\Delta p}{\rho\left(\varsigma(t) - \frac{d^4_{outlet}}{d^4_{inlet}}\right)}} \tag{23}$$

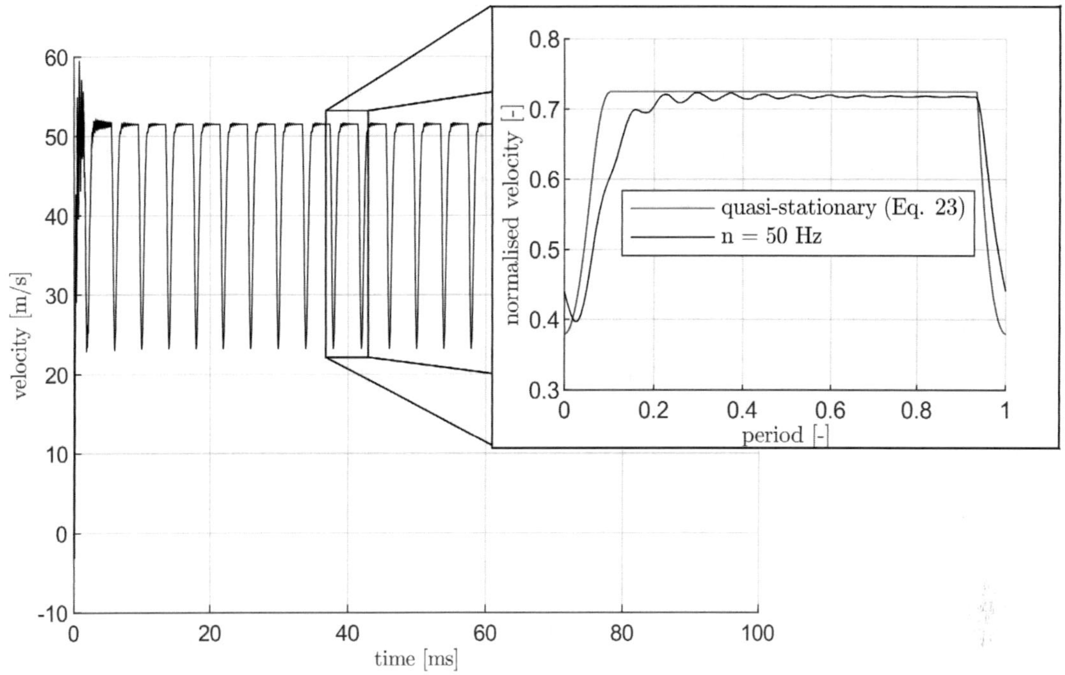

Figure 5. Simulated velocity over time as well as the normalised velocity (Eq. 22) over the period for analysis purposes. ($\eta = 30$ mPas, $\Delta p = 5$ bar, n = 50 Hz)

This corresponds to the flow velocity without inertia. The periods shown start at the rotor position at which the largest of the nozzle cross-sections is covered. This results in the maximum pressure loss coefficient but inertia of the flow shifts the minimum flow velocity. As the rotor does not completely block the opening and there is a housing gap, the liquid flows past the rotor. When the rotor continues to turn, the long rotor flank releases the opening coverage again, which increases the flow velocity. On the other side of the period, the flow is slowed down much faster by the short rotor flank.

5.2. Variation of compressor speed

In this chapter, the compressor speed and thus the absolute period duration will be varied first. Figure 6 shows the mass flow rate through the nozzle over the speed of the main rotor for different pressure differences across the nozzle as well as the injected oil mass flow rates recorded in the experiment with the commercial compressor. For the simulations, the dynamic viscosity of the oil was set to $\eta = 30$ mPas. First of all, it is noticeable that the measurement results are very well represented by the simulations with a pressure difference of $\Delta p = 5$ bar. This is surprising at first, as the pressure difference corresponds to that of the pressure gauge within the injection line just upstream of the injection nozzle, but the internal pressure in the compressor must be greater than 1 bar. Furthermore, the dynamic viscosity in the experiment is by no means constant but depends on the amount of heat introduced by the compression. The more heat is added by the compression process, the cooler the oil must be in order to keep the temperature on the high-pressure side constant. This increases the dynamic viscosity, the influence of which is considered in the following chapter. Figure 7 shows the nozzle exit velocity related to the loss-free, quasi-stationary velocity (Bernoulli) over the period for the compressor speeds n = 15 Hz, n = 50 Hz and n = 100 Hz. The normalised velocity (Eq. 22) shows that the profiles hardly differ and thus the mass flow rate through the nozzle is almost independent of the compressor speed. The influence of the transient inertia can be seen in both the fluid acceleration and the deceleration and thus almost balances out. The blue curve shows the stationary solution for the specified pressure loss coefficient of the nozzle. Deviations from this curve show both the inertia effects as described above and acoustic effects that are particularly noticeable where the nozzle is not covered by the rotor.

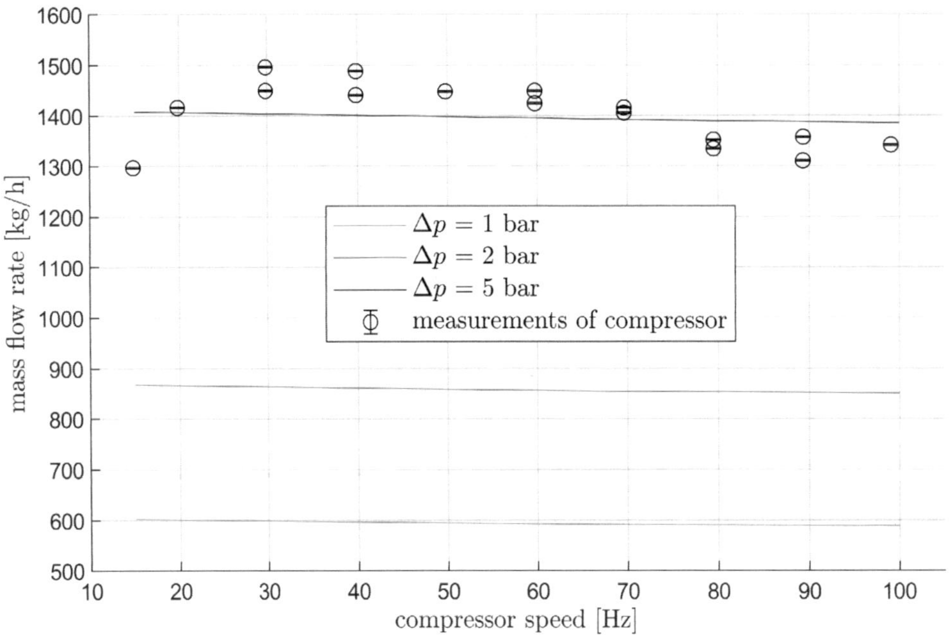

Figure 6. Injection mass flow rate as a function of the speed of the main rotor for different pressure differences. ($\eta = 30$ mPas)

5.3. Variation of pressure ratio

This chapter examines the influence of the pressure ratio. It is calculated here as the ratio of the outlet pressure to the inlet pressure and therefore within the range of 0 to 1 ($p_{outlet}/p_{inlet} \in (0\ 1)$). Using the pressure ratio makes it possible to transfer the results to similar systems. It is therefore used here although according to the Bernoulli (Eq. 23) the pressure difference determines the mass flow rate for liquid flows. Figure 8 shows the calculated mass flow rate over the pressure ratio of the nozzle for different oil viscosities. The viscosity is also varied at this point, as it is influenced by temperature, oil specifications (additives, ISO VG), etc. As the pressure difference decreases with increasing pressure ratio, the mass flow rate through the nozzle decreases accordingly. Apart from the geometric parameters not considered here, the pressure ratio has by far the greatest influence on the mass flow rate. The course of the mass flow rate for different viscosities is similar at first glance: greater viscosity leads to more friction and thus to decreasing mass flow rates. However, it should be noted that at some points depending on viscosity and pressure ratio, the flow structure changes from turbulent to laminar (from left to right). To visualise that phenomenon Fig. 8 shows the Colebrook diagram as used in section 4.1 to calculate the pressure loss coefficient for the nozzle. Beside the laminar and turbulent pipe friction loss coefficient, the actually pressure loss coefficients for the outlet cross section is shown according to the calculated mass flow rate. The discontinuities in the curves (called transition zone) occur precisely when the critical Reynolds number ($Re_{crit} = 2320$) is reached. In this region, according to the theory used, it is not possible to say with certainty which

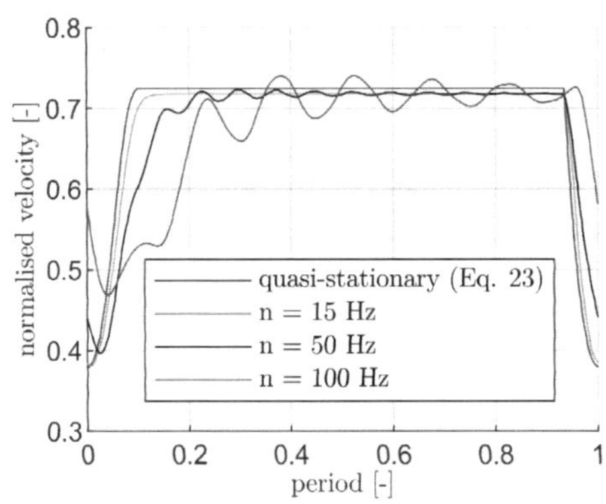

Figure 7. Normalised velocity over dimensionless time (period) for different compressor speeds. ($\eta = 30$ mPas, $\Delta p = 5$ bar)

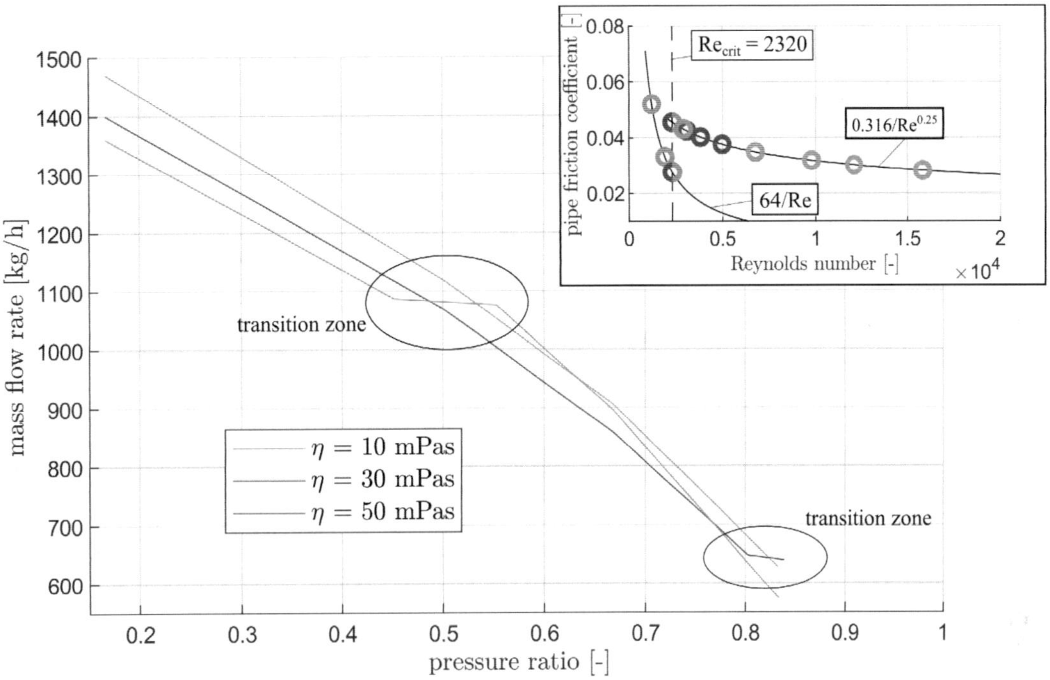

Figure 8. Mass flow rate as a function of the pressure ratio across the nozzle for different dynamic viscosities of the oil. Pipe friction coefficient as a function of the Reynolds number to illustrate that the simulated mass flow rates are in the transition zone between laminar and turbulent flow structures. (n = 50 Hz)

flow structure is present. Therefore, the mathematical model is not continuous. Here, too, the normalised velocity profile can be viewed over a period. Figure 9 shows this for different pressure differences and constant compressor speed and viscosity. It is noticeable that the normalisation with Bernoulli (Eq. 22) leads to the fact that the three curves shown hardly differ and reach about 70 % of the steady, frictionless flow velocity at the peak. When varying the compressor speed, it has already been shown that there

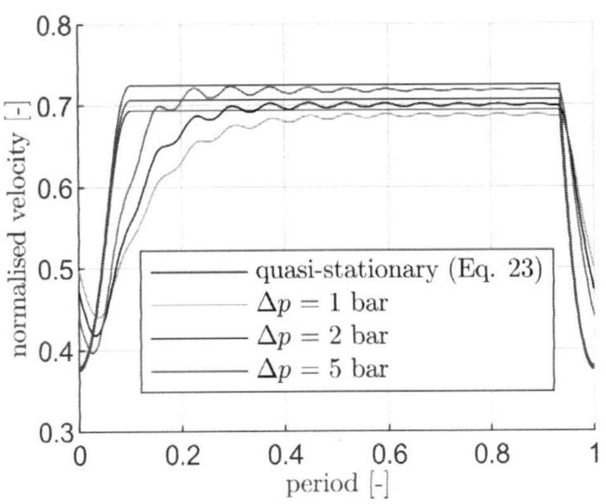

Figure 9. Normalised velocity over dimensionless time (period) for different p resssure differences. (η = 30 mPas, n = 50 Hz)

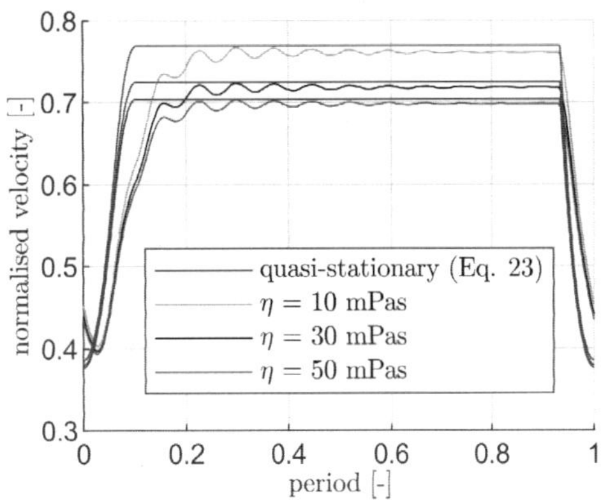

Figure 10. Normalised velocity over dimension-less time (period) for different dynamic viscosities.(Δp = 5 bar, n = 50 Hz)

is hardly any influence on the mass flow rate through the nozzle. The quasi-stationary velocities over a period are also shown here as a function of the pressure loss coefficient. The mass flow rate can be mapped very well with these. Figure 10 shows the normalised velocity for different viscosities and here too the mass flow rate can be determined almost exactly by the curve of the quasi-stationary calculated mass flow rate.

5.4. Influence of inertia

As already shown, the maximum flow velocity, is almost independent of the boundary conditions (compressor speed, pressure difference, viscosity), and in the range of 70 % of the frictionless calculated flow velocity. The quasi-stationary calculated flow velocity at the outlet of the nozzle is also shown. These curves show an associated flow velocity for each pressure loss coefficient $\zeta(t)$. Inertial effects are not considered and therefore the computational effort is completely saved. Figure 11 shows the simulated, transient calculated mass flow rates over the quasi-stationary calculated mass flow rates. This type of illustration is very useful to show deviations between two methods. If both methods calcu-late the same value, it is on the diagonal. A trend channel is also very easy to plot. Here the trend channel represents a deviation from the quasi-stationary calculated mass flow rate of ±3 %. An upward deviation means that the quasi-stationary calculated mass flow rate is less than the transient calculated mass flow rate and vice versa. The deviations of all cal-culated points are less than $|\Delta\ \dot{m}| < 3\%$. The mean value is $\Delta\ \dot{m} \approx -1.4\%$. In summary, it can be concluded that the inertia of the injected oil has hardly any effect o n t he m ass fl ow ra te in the system simulated here. The computational effort of a transient c alculation c an be s aved for nozzle mapping.

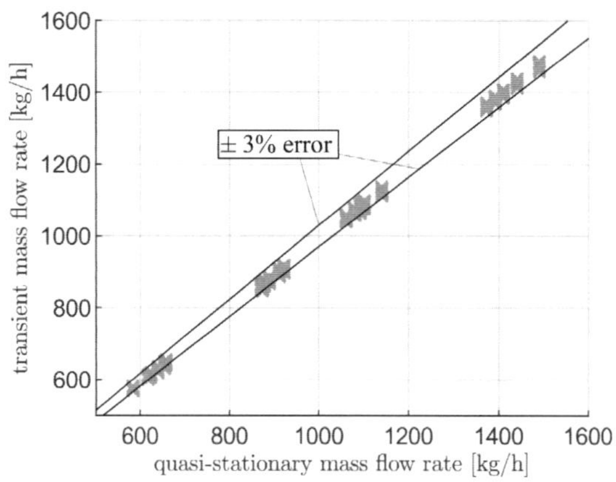

Figure 11. Transient vs quasi-stationary calculated mass flow rates.

6. Conclusion

In this work, the transient flow through the injection nozzle of a twin-screw compressor is analysed. For this purpose, a model is developed that can be used to calculate the pressure loss coefficient for the nozzle in combination with the gate rotor lobe. The nozzle is treated as a piping system and a function is determined for the rotor lobe travelling past the opening. The transient flow is calculated using the method of characteristics. The investigation shows, in good agreement with measurements, that the mass flow rate is almost independent of the compressor speed. The influence of viscosity is also small, although unpredictable effects can occur if the flow structure cannot be clearly determined (e.g. Re = 2320). The influence of the pressure ratio or pressure difference is the most pronounced. However, since the pressure difference and the pressure loss coefficient are known for each rotor position, a quasi-stationary calculated flow velocity through the nozzle can be used to determine the mass flow rate in good agreement with measurements and the method of characteristics. The mean deviation between the quasi-stationary calculated mass flow rate and the transient calculated mass flow rate is about $|\Delta\dot{m}| \approx 1.4\%$. This shows that the inertia of the oil has no effect on the injection mass flow rate in the studies conducted here. Further work could investigate the geometry of the nozzle and the influence of the piping upstream. Also of interest is the question of the shape and length of the injection jet to contribute to the distribution of the liquid fluid in the machine.

Acknowledgments

The authors gratefully acknowledge the German Research Foundation (DFG) for financially supporting this work within the priority program 2403 "Carnot Batteries: Inverse Design from Markets to Molecules" under project number 526153353.

References

[1] Utri M 2021 *Potenzial von nicht-konstanter Rotorsteigung für Schraubenkompressoren* Schriftenreihe des Fachgebietes Fluidtechnik (Logos Verlag Berlin) ISBN 9783832553340 URL `https://books.google.de/books?id=TVo9EAAAQBAJ`

[2] Radgen P and Blaustein E 2001 *Compressed Air Systems in the European Union. Energy, Emissions, Saving Potential and Policy Actions* (Stuttgart: LOG_X Verlag GmbH)

[3] Vittorini D and Cipollone R 2016 Energy saving potential in existing industrial compressors *Energy* **102** 502–515

[4] Persson J G 1987 Heat-exchange in liquid-injected screw-compressors *Schraubenmaschinen '87. Schraubenverdichter, Schraubenmotoren*

[5] S R Rane, A Kovacevic and N Stosic 2016 Cfd analysis of oil flooded twin screw compressors *International Compressor Engineering Conference* 2392

[6] Grieb M and Brümmer A 2022 Thermodynamic simulation and experimental validation of an oil-free twin-screw expander *IOP Conference Series: Materials Science and Engineering* **1267** 012015 DOI 10.1088/1757-899x/1267/1/012015

[7] Sigloch H 2012 *Technische Fluidmechanik* (Springer Berlin Heidelberg) ISBN 9783642228452 DOI 10.1007/978-3-642-22845-2

[8] Courant R, Isaacson E and Rees M 1952 On the solution of nonlinear hyperbolic differential equations by finite differences *Communications on Pure and Applied Mathematics* **5** 243–255 ISSN 1097-0312 DOI 10.1002/cpa.3160050303

[9] LeVeque R 2002 *Finite Volume Methods for Hyperbolic Problems* Cambridge Texts in Applied Mathematics (Cambridge University Press) ISBN 9780521009249 URL `https://books.google.de/books?id=QazcnD7GUoUC`

[10] Peussner M 2004 *Entwicklung und Implementierung eines numerischen Verfahrens zur Berechnung der instationären Gasströmung in Rohrleitungen mit Kolbenverdichtern* Master's thesis University Osnabrück

[11] Ferziger J, Perić M and Street R 2019 *Computational Methods for Fluid Dynamics* (Springer International Publishing) ISBN 9783319996936 URL `https://books.google.de/books?id=i9CpDwAAQBAJ`

[12] Tetenborg P 2018 *Entwicklung einer adaptiven und passiven Pulsationsdämpfereinheit* Schriftenreihe des Fachgebietes Fluidtechnik (Logos Verlag Berlin) ISBN 9783832547455 URL `https://books.google.de/books?id=bzyEDwAAQBAJ`

[13] Heselmann M, Dämgen U and Brümmer A 2024 Experimental investigation of the distribution of two-phase flow in oil-injected twin-screw compressors *13th International Conference on Compressors and Their Systems* ed Read M, Rane S, Ivkovic-Kihic I and Kovacevic A (Cham: Springer Nature Switzerland) pp 61–76 ISBN 978-3-031-42663-6

[14] Ratcliffe C and Ratcliffe B 2015 *Doubt-Free Uncertainty In Measurement: An Introduction for Engineers and Students* (Springer International Publishing) ISBN 9783319120638 DOI 10.1007/978-3-319-12063-8

[15] Truckenbrodt E 1989 *Fluidmechanik: Band 1: Grundlagen und elementare Strömungsvorgänge dichtebeständiger Fluide* (Springer Berlin Heidelberg) ISBN 9783662072752 URL `https://books.google.de/books?id=TQQHkAEACAAJ`

[16] Benedict R P, Carlucci N A and Swetz S D 1966 Flow losses in abrupt enlargements and contractions *Journal of Engineering for Power* **88** 73–81 ISSN 0022-0825 DOI 10.1115/1.3678482

[17] Zoebl H and Kruschik J 1978 *Strömung durch Rohre und Ventile: Tabellen und Berechnungsverfahren zur Dimensionierung von Rohrleitungssystemen* (Springer-Verlag/Wien)

IOP Conf. Series: Materials Science and Engineering **1322** (2024) 012011 doi:10.1088/1757-899X/1322/1/012011

A Bayesian-interference approach to quantify degradation parameters in a water-cooled variable speed screw compressor chiller

A J Hoess*, J Ma, E A Groll, J E Braun and D Ziviani

Ray W. Herrick Laboratories, Purdue University, West Lafayette, USA

*E-mail: ahoess@purdue.edu; dziviani@purdue.edu

Abstract. Variable-speed screw chillers can provide effective capacity modulations in commercial buildings to reduce energy consumption, match the load requirements, and contribute to lower equivalent carbon emissions. However, variable-speed operation can also yield to degradation phenomena inside the compressor. Assessing the real-time health status of chillers and detecting anomalies are important aspects to ensure long-term operation and improve reliability. To this end, an automated accelerated life test (ALT) procedure was developed and applied to a 513 kW (145.9 RT) water-cooled variable-speed screw chiller in a laboratory test facility to experimental assess performance degradation over time. After every 1,000 operating hours of ALT, steady-state performance tests at 30%, 50%, 75% and 100% load were conducted according to AHRI Standard 550/590. Additional operating conditions were included while conducting the tests to provide a more complete assessment of the degradation trends compared to the initially measured baseline. The experimental data set was then used to quantify the performance degradation based on the impact on the screw compressor operation. A Bayesian-inference identification approach has been developed to identify changes of sensitive parameters that statistically provide evidence for the likelihood of a performance degradation of the compressor. The information was used to implement the parameters in a dynamic model which was used to perform studies to predict the degradations of a chiller. The outcomes of this study help to improve the maintenance schedule of a compressor and even the whole chiller so that the system can be operated more economically and unplanned downtime can be minimized.

1. Introduction

The building sector is the largest energy consumer in the United States, accounting for 37% of the country's energy consumption and 74% of its total electricity use buildings use [1, 2]. Within this sector, HVAC&R equipment is the largest consumer group, responsible for 57% of the primary energy being used in commercial buildings [3]. The performance degradation of components in vapor compression cycles (VCC) can have a serious impact on a system's efficiency and energy consumption, thereby affecting the entire building sector. In 2023, about 60% (2,505,249 MWh) of the energy in the U. S. originated from fossil fuels. This makes it a crucial task gain knowledge about the performance degradation of VCC components to improve their design and maintenance schedules for an improved efficiency and reduced environmental impact [4]. When operating over extended periods, mechanical components of thermal systems can experience a reduced system performance and life time due to mechanical wear, overheating due to friction as well as insufficient component maintenance. Furthermore, the performance degradation can go along with reduced product safety and even environmental damage following partial or total loss of refrigerant. A better understanding of the performance degradation process and the ability to detect failures and system changes early on can be used to avoid these issues and safe the environment.

Therefore, fault detection and diagnostic (FDD) tools play an important role in the early identification of system changes and can help to combat negative impact on the functionality of components such as the compressor. Published work in this field includes either model based [5, 6, 7] or data-driven [8, 9, 10] approaches. That means, that either a calibrated transient model is used to train the FDD tool and its control logic or that a machine learning algorithm is trained directly on a recorded data set. Both methods have in common, that their accuracy relies significantly on the quality and size of the dataset that was used to train or calibrate the initial transient model, or the machine learning algorithm.

Accelerated life testing (ALT) is a promising test method to generate such a data set that gives insight in the performance changes of a VCC within a reduced amount of time. There are two major approaches in ALT: one is, to bring the system to a fast break down to evaluate which are the weakest parts and what could be improved, the second one is to investigate the behaviour of a system or component performance over an estimated life time. In the second test approach, a breakdown can happen but is not the designated target of the test. Since the given test requirements in this investigation are, that the system should not be manipulated and that the operating modes should be selected in a way that they could possibly appear in real life operation, the second test approach was selected. With the help of published work about conducted ALT on vapor compression systems [11, 12, 13], potential test modes were evaluated and the results are used to calibrate a transient model.

Since previous studies already covered the impact of system parameters such as refrigerant loss, oil distribution, and heat exchanger fouling [14, 15, 16, 17], the main focus of this study lies on the performance degradation of the variable-speed twin screw compressor.

2. Test Setup

The primary focus of the experimental investigation was on the evaluation of performance degradation of a screw compressor in long-term operation. Two key requirements for the tests were, that no modifications could be made on the vapor compression systems itself and its controls and that the operating conditions during the accelerated life testing should be picked in a way, in which they could also come up in a real-life situation.

To ensure the results were representative, an off-the-shelf water-cooled chiller using R-134a as refrigerant was selected for testing. The chiller is equipped with a semi-hermetic variable-speed twin screw compressor and has a rated performance of 513.11 kW (145.9 RT) in the cooling mode with an inlet water temperature level of 12.22 °C (54 °F) and 6.67 °C (44 °F) outlet temperature on the evaporator and a water inlet temperature of 29.44 °C (85 °F) on the condenser side. The compressor speed is controlled by an internal chiller control with a frequency inverter as variable speed drive (VSD).

An existing performance test stand for a 316.52 kW (90 RT) centrifugal chiller [14] was redesigned and commissioned to conduct performance tests according to the AHRI 550/590 Standard [18] and for automated 24/7 ALT operation. The redesign included measures to enhance the test rigs performance limit for chillers, reduce the water consumption and increase the energy efficiency of the operation. The test stand (Figure 1) consists out of two main water loops, the condenser and the evaporator cycle, which are connected through two internal heat exchangers, that allow the use of the condenser heat as load for the evaporator cycle. A flow regulated bypass allows the adjustment of different load stages based on the system temperatures. The condenser cycle is as well connected to a chilled water system that enables the dissipation of compressor heat through two additional shell-and-tube heat exchangers. The heat dissipation rate through the chilled water cycle is the major factor to impact the system temperatures. An auxiliary cycle that is connected to the laboratories steam heating system, allows additional heat input if necessary.

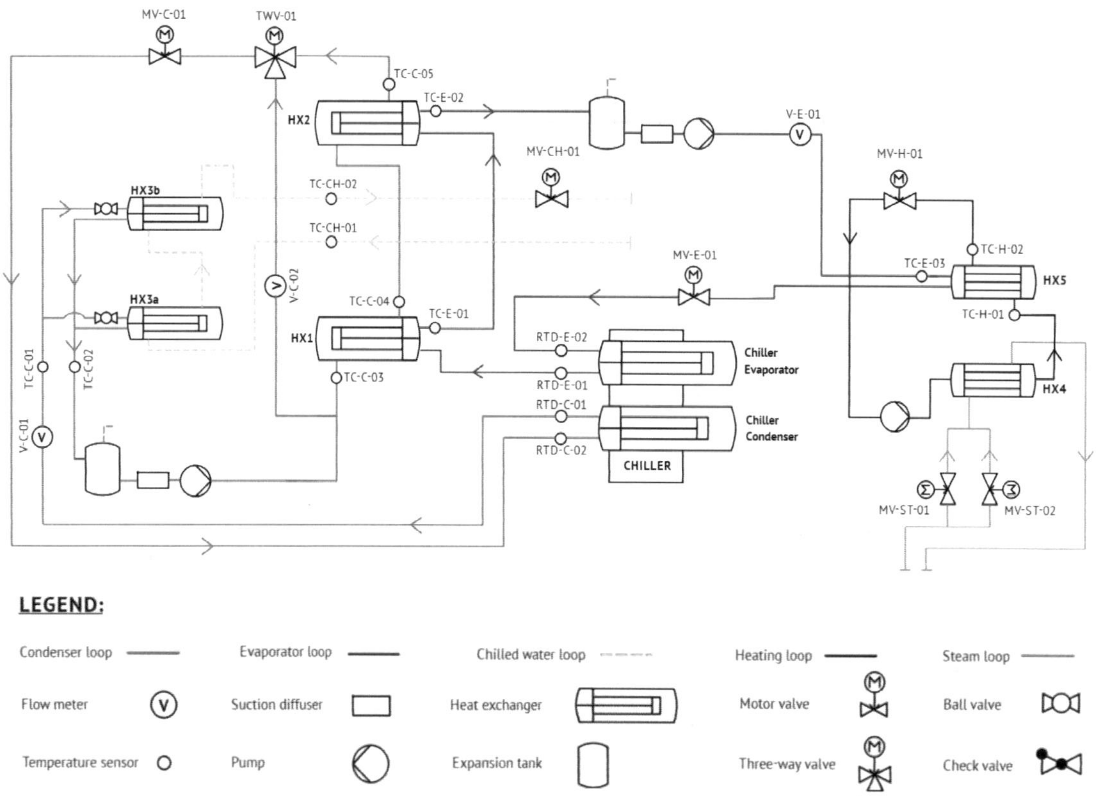

Figure 1. Test stand for chiller performance degradation testing [19]

Initial performance tests [20] indicated that steady state operation could be achieved for a performance of 439.61 kW (125 RT). This is lower than the rated 513.11 kW of the chiller but it was not seen as a problem since the same chiller system is also on the market for a rated 439.61 kW without any differences in the system setup. After additional smaller modifications to ensure standard conform performance tests, such as reduction of the pressure loss and new water pumps, the test rig was ready to perform performance tests and ALT operation. Further system details about the chiller and the test stand can be found in previous published literature [19, 20, 21].

3. Test Methodology

3.1 Performance Testing

Before the ALT was started, a baseline test was conducted to evaluate the performance of the chiller under the operating conditions given in the previously described test bed. After the baseline tests were finished, the performance tests were repeated at similar conditions (30%, 50%, 75%, and 100% load) every 1,000 hours of ALT operation until the current state of 4,000 operating hours (May 2024). The chiller performance was tested based on the test conditions in the AHRI 550/590 Standard [18]. Since the controls of the chiller trigger a safety shutdown for loads lower than 30%, the lowest test range was not selected at 25% but at 30%. For deviations to the required 25% rate, the efficiency can be extrapolated [18, 20]. Table 1 shows the test matrix that was set up for the chiller, based on the standard requirements and an evaporator water flow rate of 18.87 kg/s.

Table 1. Test matrix for AHRI 550/590 conform chiller test [18]

Target Load (%)	Target Capacity (kW)	Evaporator Inlet Water Temperature (°C)	Evaporator Outlet Water Temperature (°C)	Condenser Inlet Water Temperature (°C)
30	131.88	8.33	6.67	18.33
50	219.81	9.44	6.67	18.33
75	329.71	10.83	6.67	23.89
100	439.61	12.22	6.67	29.44

3.2 ALT Test Modes

Based on initial measurements and evaluations of system stress as well as based on practice-oriented considerations, two test modes were developed and applied to the system in an alternating pattern: (1) the high load/low head mode (HLLH) and (2) the low load/high head (LLHH) mode.

(1) The HLLH operates, as the name indicates with high cooling load, at a low pressure head. The idea behind this mode is, that the compressor is exposed to a high cooling demand and therefore has to operate at a high RPM while the oil pressure is kept low. This is possible, since the oil pressure depends on the differential pressure between evaporator and condenser level. This mode could appear in real life operation with a high cooling demand due to a manufacturing process, while outdoor temperatures are very low.

(2) The LLHH mode operates at low load demand with high pressure head. The thought behind this mode is, to increase the stress on the compressor with a high differential pressure and at the same time, keep the RPM low to increase the torque acting on the rotors. The mode could appear in real life application, during a period with high outdoor temperature, low cooling temperature and low cooling demand due to a downtime. The ALT operation proofed that test as less impactful; this is why the operation hours in this mode were reduced over time.

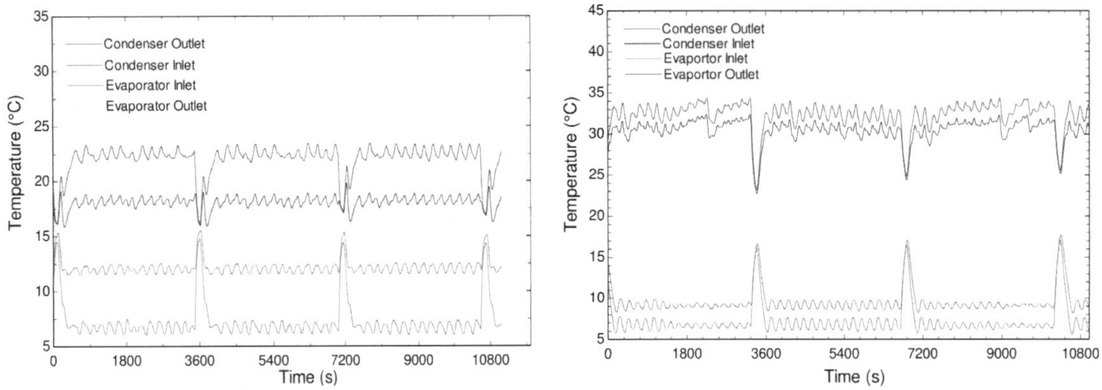

a) High load/low head operation b) Low load/high head operation

Figure 2. Operating Temperatures during ALT modes [21]

Both modes have in common that they were frequently interrupted by a start/stop interval in which the compressor was shut down to increase the mechanical stress on the motor and the rotors. After gaining some initial operating experience and to increase the stress on the system, the start/stop intervals were reduced from initially 60 to later 12 minutes. The ALT operation was carried out automatically in a 24/7 operation. A detailed description of the test development and implementation as well as information about the safety measures was published by Hoess et al. [19, 21]. Figure 2 shows the water temperatures in the chiller loops during continuous operation while Table 2 lists the target temperatures during for each ALT mode.

Table 2. Operating Temperatures during ALT [21]

Test Mode	Target Flow Rate (kg/s)	Target Evaporator Inlet Water Temp. (°C)	Target Evaporator Outlet Water Temp. (°C)	Target Condenser Inlet Water Temp. (°C)
HLLH	18.87	12.22	6.67	15.56
LLHH	18.87	7.61	6.67	32.22

4. Experimental Results and Discussion

During the performance tests, a total of 42 steady state points were recorded. To be considered as steady state according to the standard [18], a system must operate constant for at least 15 minutes without interruption. The temperatures can vary maximum ±0.28 °C (±0.50 °F) during this time. Figure 3 shows, an efficiency trendline that was established for the values that were evaluated in the baseline measurements. This characteristic efficiency line represents the efficiency of the system as it came from the manufacturer and serves as comparison for all following test results. All efficiencies were calculated the ratio of electricity consumed by the compressor in kW over the provided cooling capacity in RT as

$$\text{Efficiency} = \frac{\dot{W}_{\text{Comp}}}{\dot{Q}_{\text{E}}} \tag{4.1}$$

It should be mentioned, that the efficiency value that is calculated in Equation 4.1 is higher, the less efficient the chiller is. While this seems counterintuitive in the first place, it simply means, that more electrical power $\dot{W}_{Comp}$ is needed to achieve the same system capacity. The cooling capacity in RT for Equation 4.1 was calculated as function of water mass flow through the evaporator in kg/s, specific heat in kJ/(kW-K) and the temperature gradient between water input and output in K, converted from kW to RT with a factor following Equation 4.2.

$$\dot{Q}_E = \dot{m} \cdot c_p \cdot (T_{in} - T_{out}) \cdot 0.2843 \tag{4.2}$$

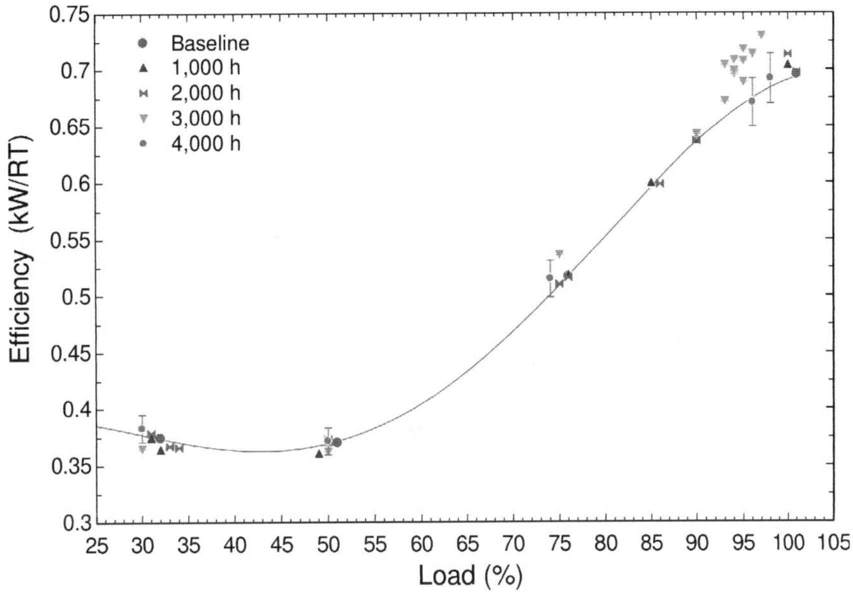

Figure 3. AHRI 550/590 Performance Measurements in 1,000-h Increments

The first repeating performance test was conducted after 1,000 h of ALT. At this stage, just slight deviations from the baseline could be detected on the upper and lower load range. However, these divergencies should be categorized irrelevant, based on the uncertainty of the measurement. The 2,000-h repetitive performance tests proofed these assumptions since here, the data points are even closer to the baseline. It should be mentioned here, that after the 2,000-h measurement, maintenance was conducted on the chiller in which the shell and tube heat exchangers were cleaned mechanically. Even though the heat exchangers showed no visible pollution or build-up, a second test set was collected after cleaning, to detect possible changes. After the cleaning, the water in the system was treated chemically to avoid microbiological growth and corrosion. This was not a problem per se since both water loops are closed loops but it made sure, that a potential heat exchanger fouling could be minimized on the water side. The 3,000-h test set was showed the first clear deviation of the efficiency for a range above 50% load. It is noteworthy that it was not possible to reach a 100% load during this set of tests and that all tests above 90% load showed deviations larger than the tolerance of ±0.28 °C (±0.50 °F) defined in the AHRI Standard 550/590 [18]. To still be able to identify a trend, the decision was made to increase the tolerance to ±0.56 °C (±1.00 °F). This measure allowed, to use 11 data points between 90% and 97%. Initially, this performance degradation was assumed to be related to the performance degradation of the twin screw compressor but a more in-depth investigation of the performance related operational data brought up an increased superheat and discharge superheat. This led to the conclusion, that a low refrigerant charge could be a cause of the performance loss. A similar case was investigated by Comstock [14]. Before the 4,000-h repetitive performance test was conducted, 16.3 kg (36 lbs) of the refrigerant R-134a were charged to the system to allow a full load operation again. This accounts for a charge loss of about 12% of the complete system charge according to the manual. The total recharge amount was evaluated based on the superheat temperatures. After recharging the system, the 4,000-h performance test was conducted. This test proved that the major performance loss at 3,000 h was charge related. However, there were measurable efficiency losses on the system at the lowest and the higher load tests. Compared to the baseline test, the efficiency loss at the minimum load point of 25% was 3.12%. The 50% load point showed almost no decrease with a 0.26% difference to the baseline point and could be assumed as almost steady. The 75% value showed a 3.46% efficiency decrease and the full load efficiency indicated a loss with -3.57% efficiency. There can be concluded that after a total of 4,026 h of operation at rough operating conditions and 12,124 start-stop intervals a deterioration of about 3-4% in all load ranges besides 50% appeared.

Even though, the deviation is still within the calculated range of uncertainty (error bars), it is important to state that the quality of repeated measurements proofed very little deviation and the thermocouples were calibrated on a regular base. Table 3 shows the normalized efficiency values of the performance tests. Furthermore, the efficiency was used to calculate the integrated part load value (IPLV). The IPLV is one standardized parameter to express the overall efficiency of a system over the whole load spectrum. It is a way to normalize the efficiency values of a system by their impact on the annual power consumption. The IPLV is calculated with the efficiency in kW/RT for the steady state test points of 100% (A-value), 75% (B-value), 50% (C-value) and 25% (D-value) according to AHRI standard 550/590 [18] with Equation 4.3.

$$\mathrm{IPLV.IP} = \frac{1}{\frac{0.01}{A} + \frac{0.42}{B} + \frac{0.45}{C} + \frac{0.12}{D}} \tag{4.3}$$

Using Equation 4.3, the IPLV values were calculated for all performance test sets. The 25% value can be extrapolated since the chiller operation was just possible to a lowest capacity of 30% [18, 20].

Table 3. Trend of IPLV and normalized Efficiency over 4,000 h of Operation

	Efficiency and IPLV				
Load (%)	**25**	**50**	**75**	**100**	**IPLV**
Baseline Efficiency (kW/RT)	0.3639	0.3705	0.5106	0.6882	0.4199
1000 h Efficiency (kW/RT)	0.3652	0.3675	0.5125	0.7042	0.4190
2000 h Efficiency (kW/RT)	0.3541	0.3732	0.5103	0.7133	0.4199
3000 h Efficiency (kW/RT)	0.3569	0.3624	0.5363	0.7530	0.4238
4000 h Efficiency (kW/RT)	0.3756	0.3714	0.5289	0.7137	0.4275

The values for 50% and 75% were calculated with linear interpolation for the recorded measurements that were not exactly on the load range. Table 3 shows the trend of the IPLV values over the complete test time. Compared to the baseline test, the IPLV and therefore the efficiency over all load ranges decreased by 1.78%.

However, to further evaluate the roots of potential system efficiency degradations and impact of the test modes on the hardware, especially on the compressor, additional compressor related data was visualized and investigated. Several system and compressor related parameters were investigated, such as the discharge and economizer superheat, system oil pressure and oil differential pressure, compressor speed, and motor temperature. The trends that can be seen in the diagrams for discharge temperature and oil pressure (Figure 4) all followed the distribution of the efficiency (Figure 3) so that no clear indicator for the performance loss could be indicated in an initial step. While the discharge temperatures show a larger change, the spread of these values is larger in general. The oil pressure barely changes and mostly follows the established trendline.

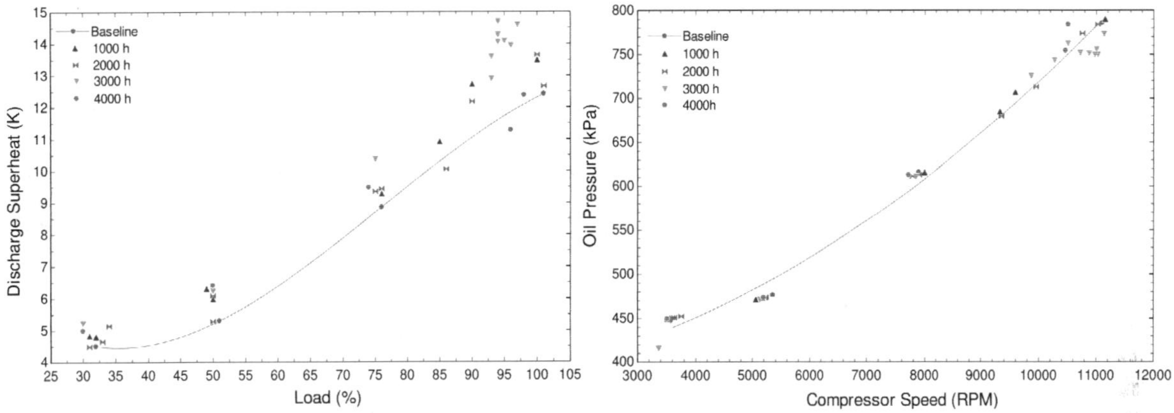

a) Load vs. Discharge Superheat b) Compressor Speed vs. Oil pressure

Figure 4. Analysis of system parameters

5. Model Calibration with reduced- dimension Bayesian Optimization Framework

The initial system analysis could not lead to a clear indication of parameters for performance loss. To evaluate these so-called sensitive parameters and to be able to realistically replicate system trends and behaviours, a transient chiller model was set up (Figure 5) using the software Dymola [22]. A Bayesian optimization (BO) framework will be applied in the next step to calibrate the transient model based on the recorded test data from the performance tests. The described approach follows the example described by Ma et al. [23] and is visualized in Figure 6.

In preliminary simulations, the existing physics-based model generally showed a good performance and accuracy in steady state. However, to align the model predicted outcomes with the generated physical test data and increase the accuracy in transient mode, multiple parameters must be fine-tuned. In the case of the investigated chiller, a variety of system and component data and geometry is unknown due to a lack of given technical specification but crucial for the investigation of the performance degradation. The measurable outputs of the test system were collected in a time-series sequence as $y_{0:T} = [y_0 \; y_1 \; \cdots \; y_T] \; y_t \in Y \subset \mathbb{R}^m$ where t= 0, 1, 2, ..., T (s) for the test stand data and t = 0, 5, 10, ..., T (s) for chiller system data. Each vector y_t represents a measurement at a specific time. Following that logic, an abstracted way to represent the VCC of the chiller is:

$$\hat{y}_{0:T} = F(\theta) \qquad \hat{y}_t \in Y \subset \mathbb{R}^m \tag{5.1}$$

That means that each vector for the recorded time sequence can be represented as a function F of the vector θ that denotes the parameter of interest for a certain time instance. For the calibration process it is necessary, to quantify the error of the transient prediction that is associated with each calculated output. This is done by calculating the normalized error residual (NER) where $y_{i,j}$ and $\hat{y}_{i,j}$ are the ith measured and predicted output variable at time j.

$$r_i = \frac{\sum_{j=0}^{T}(y_{i,j}-\hat{y}_{i,j})^2}{\sum_{j=0}^{T} y_{i,j}{}^2} \quad i=1, 2, 3, \ldots, m \tag{5.2}$$

The NER can be used to calculate the total prediction error as a set of calibration parameters using a weighted combination. This provides us the calibration objective function.

$$J(y_{0:T}, \hat{y}_{0:T}) = -\log\left(\sqrt{r(\Theta)^T W r(\Theta)}\right) \quad \text{where } r(\Theta) = [r_1, r_2, \ldots, r_m]^T \tag{5.3}$$

The vector r represents the NERs as calculated with Equation 5.2 and W is the diagonal weighting matrix.

The Gaussian Process (GP) regression is a Bayesian approach for curve fitting over a function space. Other than relying on a computationally expensive transient simulation, the GP regression provides a probabilistic frame work over a function space. With a set of input data points $\Theta_{1:n} = [\Theta_1 \;\; \Theta_2 \;\; \ldots \;\; \Theta_n]$ that are randomly generated samples from the parameter space it is possible to run transient simulations to feed the calibration objective function (Eqn. 5.3). Each data set of input parameters will generate a training data set $D = (\Theta_{1:n}, \tilde{J}_{1:n})$ with

$$\tilde{J}_{1:n} = [J(\Theta_1) \;\; J(\Theta_2) \;\; \ldots \;\; J(\Theta_n)]^T + \mathcal{E} \tag{5.4}$$

Being the noisy version of the true objective function and $\mathcal{E}$ being the additive independent identically Gaussian noise with zero mean and unknown variance.

Since every parametric evaluation would require a transient simulation and therefore a high computation capacity, the process will be optimized with the help of the Bayesian optimization process. Therefore, the results of the calibration objective function can be utilized for a GP model that uses a Bayesian paradigm for curve fitting over a function space. Based on the computed uncertainties, BO determines where to sample and aims for the solution in an iterative approach.

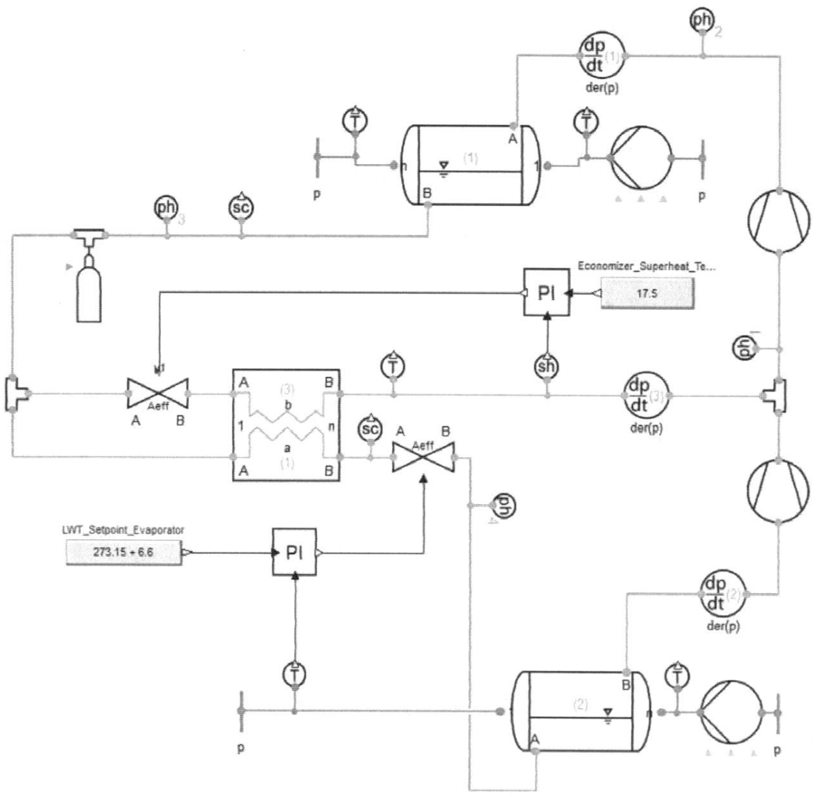

Figure 5. Transient Model Set up in Dymola [22]

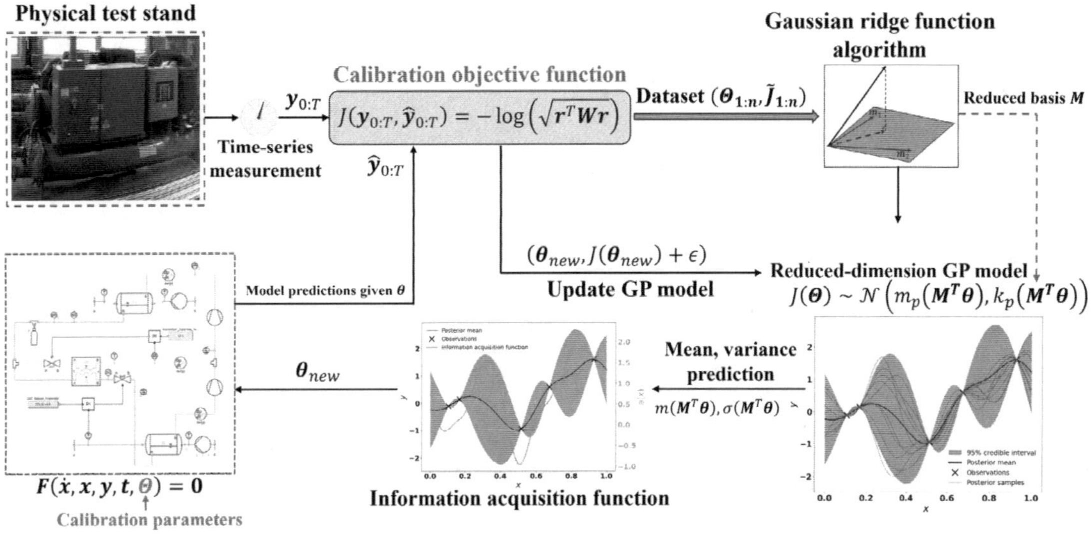

Figure 6. Overview of reduced-dimension BO calibration framework based on Ma et al. [23]

6. Conclusion and Future Work

Extensive testing was done to the chiller under controlled and equal operating conditions. In total, 4,026 h of ALT operation at demanding conditions and 12,124 start-stop intervals were applied to a commercial chiller with 513.11 kW cooling capacity that was integrated in an optimized test rig. These factors make it a quite unique long-term data set in the field of chiller operation. The accelerated life testing was interrupted every 1,000 h for a performance test of the chiller according to AHRI 550/590 [18]. The efficiency of the chiller shows a degradation of 3-4% at the 25%, 75% and 100% test points, while the efficiency at 50% remained almost constant. The IPLV reflected that trend and shows a reduction of 1.78% of the part load performance of the system. Even though slight performance degradation could be detected, it was not possible to clearly identify an individual or several sensitive parameters that show a change of the compressor performance in the initial step of the data set evaluation.

For this reason, the Bayesian optimization framework must be developed in the next data evaluation step. With the help of this process, variables can be identified that are sensitive to performance changes of the compressor and helps to calibrate the transient system model.

Acronyms

ALT	Accelerated Life Testing
BO	Bayesian Optimization
COP	Coefficient of Performance
EWT	Entering Water Temperature
FDD	Fault Detection and Diagnostics
GP	Gaussian Process
HVAC&R	Heating, Ventilation, Air Conditioning and Refrigeration
HLLH	High Load/Low Head
IPLV	Integrated Part Load Value
LLHH	Low Load/High Head
LWT	Leaving Water Temperature

NER	Normalized Error Residual
RTD	Resistance Temperature Detector
RPM	Revolutions per Minute
VSD	Variable Speed Drive
VCC	Vapor Compression Cycle

References

[1] DOE 2024 DOE announces $46 million to boost energy efficiency and slash emissions in residential and commercial buildings *Department of Energy* Available: https://www.energy.gov/articles/doe-announces-46-million-boost-energy-efficiency-and-slash-emissions-residential-and#:~:text=Residential%20and%20commercial%20buildings%20are,of%20its%20total%20carbon%20emissions (Accessed: Jul. 13, 2024)

[2] EIA 2024 How much energy is consumed in U.S. buildings? *U.S. Energy Information Administration* Available: https://www.eia.gov/tools/faqs/faq.php?id=86&t=1 (Accessed: Jul. 12, 2024)

[3] EIA 2024 2018 Commercial buildings energy consumption survey *U.S. Energy Information Administration* Available: https://www.eia.gov/energyexplained/use-of-energy/commercial-buildings.php (Accessed: Jul. 11, 2024)

[4] EIA 2024 Electricity data browser *U.S. Energy Information Administration* Available: https://www.eia.gov/electricity/data/browser/ (Accessed: Jul. 13, 2024)

[5] Browne M W and Kansal P K 2002 Transient simulation of vapour-compression packaged liquid chillers *Int. J. Refrig.* 25 doi.org/10.1016/S0140-7007(01)00060-3, pp. 597–610

[6] Fu L, Ding G, and Zhang C 2003 Dynamic simulation of air-to-water dual-mode heat pump with screw compressor *Appl. Therm. Eng.* 23 no. 13 doi: https://doi.org/10.1016/S1359-4311(03)00109-1, pp. 1629–1645

[7] Ding G and Fu L 2005 Performance analysis and improvement of air-to-water chiller for application in wide ambient temperature range *Appl. Therm. Eng.* 25, no. 1 doi: https://doi.org/10.1016/j.applthermaleng.2004.03.007, pp. 135–145

[8] Sala-Cardoso E, Delgado-Prieto M, Kampouropoulos K and Romeral L 2020 Predictive chiller operation: A data-driven loading and scheduling approach *Energy Build.* 208 doi: https://doi.org/10.1016/j.enbuild.2019.109639.

[9] Zhu X, Zhang S, Jin X and Du Z 2020 Deep learning-based reference model for operational risk evaluation of screw chillers for energy efficiency *Energy* 213 doi: https://doi.org/10.1016/j.energy.2020.118833.

[10] Mtibaa A, Sessa V, Guerassimoff G, and Alajarin S 2024 Refrigerant leak detection in industrial vapor compression refrigeration systems using machine learning *Int. J. Refrig.* 161 doi: https://doi.org/10.1016/j.ijrefrig.2024.02.016, pp. 51–61

[11] Miller R and Wayne N 1983 Optimum simple step-stress plans for accelerated life testing *IEEE Trans. Reliab.* R-32 no. 1 doi: https://doi.org/10.1109/TR.1983.5221475, pp. 59–65

[12] Pruitt G R, Davis T M and Ross B A 2004 Methods for accelerated life evaluation of long-life cryocoolers *AIP Conf. Proc.* 710 doi: https://doi.org/10.1063/1.1774811, pp. 1239–1251

[13] Jayatilleka S 2018 Performance verification throughout the product life cycle using accelerated life testing *IEEE 2018 Annu. Symp. Reliab. Maintainab. RAMS* doi: https://doi.org/10.1109/RAM.2018.8462919.

[14] Comstock M C 1999 Development of analysis tools for the evaluation of fault detection and diagnostics in chillers Master's Thesis Purdue Univ. (West Lafayette, IN)

[15] McIntosh I B D, Mitchell J W and Beckmann W A 2000 Fault detection and diagnostics in chillers - part I: Model development and discussion *ASHRAE Trans. Technol. Collect.* 106, p. 268

[16] Hjortland A L 2018 Automated fault detection, diagnostics, impact evaluation, and service decision-making for direct expansion air conditioning *PhD Thesis Purdue Univ.* (West Lafayette, IN)

[17] Bendapudi S 2004 Development and evaluation of modeling approaches for transients in centrifugal chillers PhD *Thesis Purdue Univ.* (West Lafayette, IN)

[18] AHRI 2023 ANSI/AHRI Standard 550/590 (I-P) - Standard for Performance Rating of Water-chilling and Heat Pump Water-heating Packages Using the Vapor Compression Cycle *Air-Conditioning, Heating and Refrigeration Institute*

[19] Hoess A J 2022 Design and Commissioning of a Test Stand to Conduct Performance Degradation Studies and Accelerated Life Testing on Water-Cooled Variable-Speed Screw Compressor Chillers *Master's Thesis Purdue Univ.* (West Lafayette, IN)
[20] A J Hoess et al 2021 *IOP Conf. Ser.: Mater. Sci. Eng.* 1180 012033
[21] Hoess A J, Ziviani D, Groll E A and Braun J E 2022 Development and Application of Accelerated Life Test Cycles for Performance Degradation Study on Water-cooled Variable-speed Screw Compressor Chillers *Int. Compress. Eng. Conf. Purdue Univ.* (West Lafayette, IN)
[22] Dassault Systems 2024 Dymola - Thermodynamic Modeling Software *Dassault Systems* (France)
[23] Ma J, Kim D and Braun J E 2024 Reduced-dimension Bayesian Optimization for Calibrating Dynamic Models of Vapor Compression Cycles *20th Int. Refrig. Air Cond. Conf. Purdue* (West Lafayette, IN)

MoS2 Coatings in unsynchronized, dry-running Screw Compressors: Experimental Insights on Operational Efficiency and Durability

M Geissendorf[1*], D Aurich[1], B Spraffke[2], N F Lopes Dias[2], W Tillman[2], S Nicolai[3], J Debus[3], A Brümmer[1]

[1]TU Dortmund University, Chair of Fluidics, Dortmund, Germany
[2]TU Dortmund University, Institute of Materials Engineering, Dortmund, Germany
[3]TU Dortmund University, Experimental Physics 2, Dortmund, Germany

*E-mail: meik.geissendorf@tu-dortmund.de

Abstract. The discourse on unsynchronized, oil-free screw compressors versus their conventional, oil-lubricated counterparts encapsulates a pivotal shift towards more efficient and environmentally friendly industrial machinery. By forgoing gears and lubricants, these innovative compressors streamline manufacturing and assembly, while also eliminating the risk of lubricant contamination in the process gas. In this configuration, the torque generated on the gate rotor by the gas pressure is directly conveyed to the main rotor through the interfacing surfaces of the rotors. In the development of compressor technology, a significant improvement is the application of a molybdenum disulfide (MoS2) coating on the primary rotor. This application utilizes the properties of molybdenum disulfide, which is effective in reducing friction and wear under various operational conditions. By applying this coating, there is an anticipated increase in the wear resistance of the rotor, effectively reducing frictional losses during its operation. This reduction in friction is expected to improve the mechanical efficiency of the compressor and extend its operational lifespan. The investigation of contact interactions utilizes a custom-designed test rig, through which effective parameters are analyzed to facilitate conclusions regarding losses and the durability of the coating. Additionally, the geometry within the contact area is tactilely measured, enabling the quantitative determination of wear. Various layer thicknesses, surface roughness levels, and loads are examined in this context.

1. Introduction

Screw machines are classified as twin-shaft rotary displacement machines. The two counter-rotating rotors are enclosed in a housing. Working chambers are formed between the twisted rotors, which change volumetrically as they rotate. This results in a compression of the transported medium within the machine. This design has a comparatively recent technical history. The first prototype was tested in 1934 as a dry-running machine. Even then, the wear behavior caused particular problems, as the gearing seized when the rotors came into contact. Compared to piston compressors, screw compressors have considerable advantages, hence their widespread use in industrial applications. Typical uses range from air compression systems and compressor systems in chemical and process engineering to refrigerant compressors. Depending on the application, dry-running machines are used in combination with a synchronization gearbox or liquid-injected machines [1].

IOP Conf. Series: Materials Science and Engineering **1322** (2024) 012012 doi:10.1088/1757-899X/1322/1/012012

Compared to these conventional designs, dry-running, unsynchronized screw machines offer numerous advantages. These benefits stem from the elimination of synchronization gears, lubricants, and associated peripheral devices. Without synchronization gears, assembly is considerably simplified and there is no need to adjust the interlobe clearance [2]. The lower gap height results in higher delivery rates of the machine. To date, several theoretical and practical studies have been carried out on unsynchronized, dry-running screw machines. The operating times have so far been between 200 and 400 hours [2] [3]. Several publications highlight the importance of the tooth shape and of the roughness as fundamental influences on the stress and service life of the machine. Due to the many influencing parameters on the mechanics in contact, a systematic investigation of the roughness and the coating system is carried out in the course of this work. For this purpose, the main rotors are coated with molybdenum disulfite (MoS_2), whereby the coating thickness is varied. The base material (16MnCr5) is also examined in two different roughnesses. In addition to the kinematics of the rotors, the tactile measurement of the contact area provides information on wear resistance.

The following sections describe the test rig used to examine the rotor contact, the installed measurement technology and the tactile measurement. First the measured contours of the contact are pre-sorted with regard to the hertzian equivalent radius. Then the dynamic measurements are done. In addition to the methodology, the data analysis and error estimation are considered and the results of the dynamic measurement are classified according to the durability of the coatings. The necessary tactile measurements are verified using the manufactured pitch and evaluated with regard to the kinematics of the screw machine and the wear volume.

2. Experimental setup

The focus of this work is primarily limited to the contact zone of the profile. In order to keep the sliding speed as low as possible the contact should be close to the pitch circle. A corresponding modification of the profile is described in [4] and is also used in this work.

2.1. Methodology of data collection

The experimental investigation is divided into several steps. First, the rotors are pre-sorted by means of tactile measurement, as there are deviations from the theoretically designed profile due to manufacturing tolerances. The tactile measurement is carried out using the *Leitz Reference XI 543* measuring machine from *Hexagon*. Figure 1 shows the measurement of the contact area on the measuring machine, with

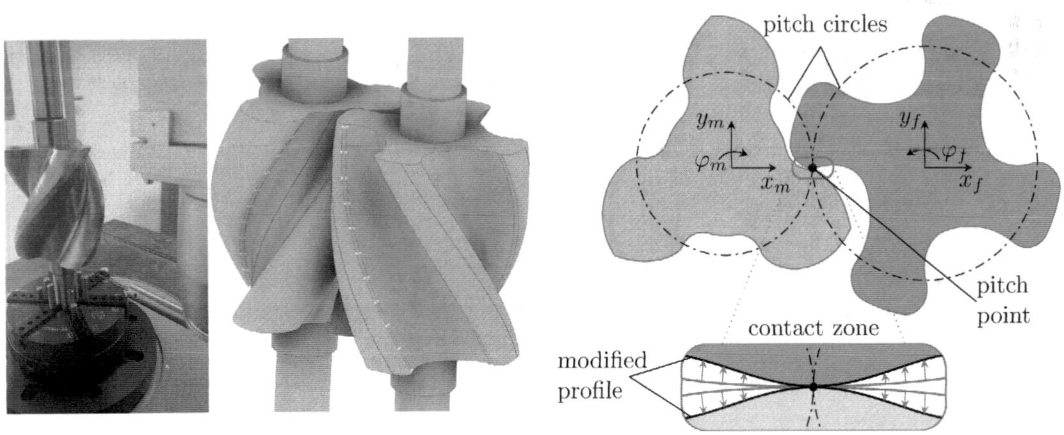

Figure 1. *Left*: Tactile measurement of an uncoated rotor on the Leitz Reference XI. The measured area of the contact is also visible in the adjacent 3D model. *Right*: Modified main (gray) and gate (darkgray) rotor profile and cut-out of the contact area in the cross section.

the 3D model of the rotors illustrating the scanned lines. The respective rotor is mounted in a rotary table so that the contact area can be measured section by section in conjunction with the axial degrees of freedom of the probe. This also achieves good repeat accuracy, which is described in more detail below. The profile modifications are also shown again using the cross section. The enlarged image illustrates that the contact is forced within the pitch circle, as the gap free ideal profile has been reduced in normal direction to the surface everywhere else.

The profile modification complicates the grinding process of the rotors, which is why manufacturing tolerances are to be expected. In order to carry out experiments that are as geometrically comparable as possible, the rotor pairs are therefore first assigned to each other. This is done using the Hertzian equivalent radius, which is defined by

$$r_{ij} = \frac{r_{m,i} \cdot r_{g,j}}{r_{m,i} + r_{g,j}}. \tag{1}$$

$r_{m,i}$ and $r_{g,j}$ are the respective curvature radii of the main and gate rotors contact area. The optimal rotor combinations are then determined using the Gaussian error squares, whereby the respective equivalent radius is considered in relation to the mean value $\overline{r}$ of all equivalent radii

$$\epsilon_{ij} = \sqrt{[r_{ij} - \overline{r}]}. \tag{2}$$

To compensate for the effect of the greater deviation from the mean equivalent radius $\overline{r}$, these rotors are provided with a higher MoS_2 layer.

2.2. Design of the rotor test rig

To investigate the fluid-free lubricant layer, application tests were conducted using a specially designed rotor test rig. The complete 3D model and data sheets for the measurement technology employed are available at [5]. The test rig setup comprises a drive and an output branch, each equipped with various components to ensure accurate measurements and analysis.

Figure 2 shows the actual test rig setup with a drive system visible on both sides. The servo motor on the right is used to prescribe the rotor speed and the one on the left side to modell the torque resulting from the gas forces. Centrally located, the air bearings are mounted on the test rig, with 6-axis force sensors positioned underneath. The air bearings are chosen to reduce friction in order that the loss mechanisms can be mainly attributed to the contact. In addition torque sensors are integrated before the flywheels into the setup. For visual monitoring, cameras are positioned to qualitatively record the behavior of the lubricant layer. These cameras are triggered by the torque measurement shafts to capture relevant points in time precisely. This setup ensures comprehensive measurements of the fluid-free lubricant layer's performance under various operational conditions.

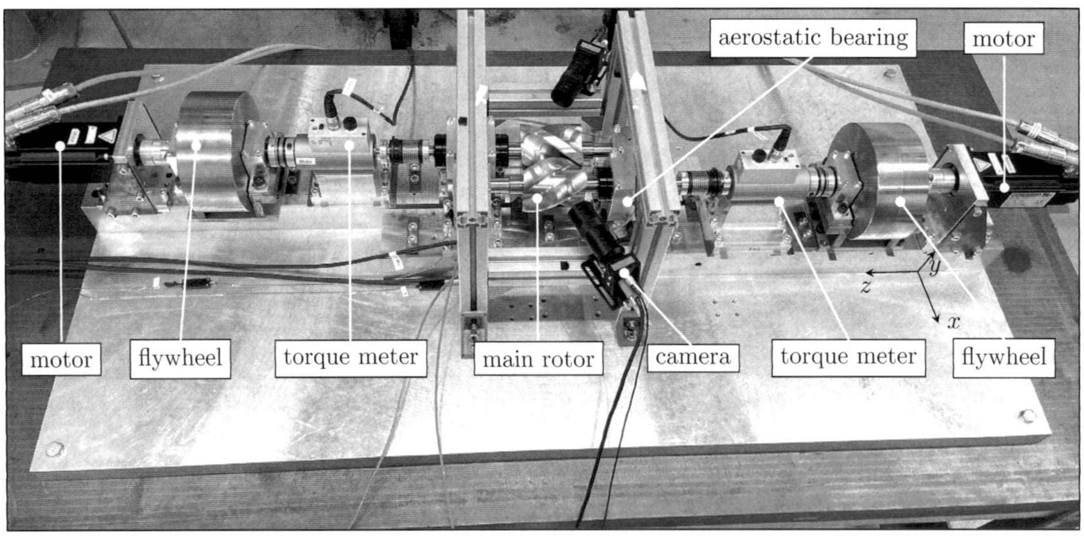

Figure 2. Test rig for dynamic measurement of the power loss and the force components resulting from the contact.

As described the drive system features a servo motor, specifically the *AKM43G*, known for its high performance and precision. This motor specifies the desired speed for the setup, operating with a rated speed of 6000 rpm. It delivers a rated torque of 3.01 Nm and a peak torque of 16.1 Nm. The brake system simulates the gas forces transferred from the gate rotor to the main rotor during contact. Flywheels serve as damping systems, significantly reducing mechanical vibrations during operation.

The measuring systems include torque measuring shafts, which are essential for analysing correlations and measuring the torques. The torque sensor model *4503B* from *Kistler* operates on the strain gauge principle, designed for measuring both torque and rotational speed, with a broad measurement range from 0.2 to 5000 Nm and speeds up to 50,000 rpm. It maintains an accuracy class of 0.05% full scale within the standard measurement range and 0.1% to 0.2% in the extended range. All measurement are done in the standard range which result in an accuracy class of 0.05%.

Furthermore, load cells are installed under the air bearings of the main rotor, which are used to measure forces and torques. The *K6D40* 6-axis force sensor is inserted for the measurement of the forces along three orthogonal axes. It has nominal force ranges of 500 N in the axes x and z, 2000 N in the axis x, in relation to the coordinate system in Figure 2. Calibration data, including individual calibration factors and error corrections, are determined and documented by the manufacturer using a calibration matrix to minimise measurement errors caused by crosstalk. After its calibration it maintains an accouracy of 0.2% full scale. For a more comprehensive insight into the technical specifications, the reader is referred to [5].

3. Data and error analysis

In the field of measurement, uncertainty is an inevitable aspect that affects all measurements, regardless of the precision of the equipment used. The primary terms include *measurand*, referring to the quantity being measured, and *true value* which is the actual value of the measurand but is never exactly known. Instead, measurements provide an estimate within an *uncertainty interval* where the true value is expected to lie with a certain probability [6].

In the following, some concepts of data analysis are described, which were applied in the course of data pre-processing. In particular, aspects of time series are addressed. Time series are a collection of observations that are carried out sequentially over time. The large amounts of data that can arise during recording are challenging. In addition, the subjective perception of the similarity between several time series depends on the user, the application and the type of analysis. Pre-processing of the data is therefore of great importance for the interpretation of the data [7].

3.1. Data Pre-Processing: Moving Average and Transformation

One problem of the measured data is the occuring noise component. In order to identify trends that may occur, it is necessary to smooth the data in advance. Therefore the dynamic time signals $X = [x_1, x_2, ..., x_n]$ are initially treated using the moving average in order to work out these trend developments. A window width k is defined for this purpose. The moving average μ_t is then calculated as follows

$$\mu_t = \frac{1}{k} \sum_{i=t-k+1}^{t} x_i. \tag{3}$$

t is the time at which μ_t is evaluated from the values x_i of the time series X with $t \geq k$. In the following, time signals that have been pre-processed using the moving average are referred to as trend developments.

In figure 3 the trend development of the radial force components is shown (left). The grey shaded areas visualise the areas of uncertainty. In the course of normalisation (right), the uncertainty ranges are also normalised in order to establish the relationship between the data series and the uncertainty. The exact interpretation and description will take place in the following chapter and the statistical properties of the time series will first be classified here. The trend developments of the force component for different layer thicknesses s are shown, whereby the equivalent radius r_{ij} of the pre-sorting is also indicated. The horizontal axis displays the number of contacts n_c resulting from the tooth contact frequency

$$n_c = z_m \cdot n_m \cdot t. \tag{4}$$

The number of contacts is determined by the number of teeth of the main rotor z_m, the associated speed n_m and the time t. Several problems arise from the progression that complicate the interpretation of the trend developments. The rotors differ not only in terms of the contact areas, but also in the layer thickness and the installation situation. The latter in particular depends on the user. In addition, the

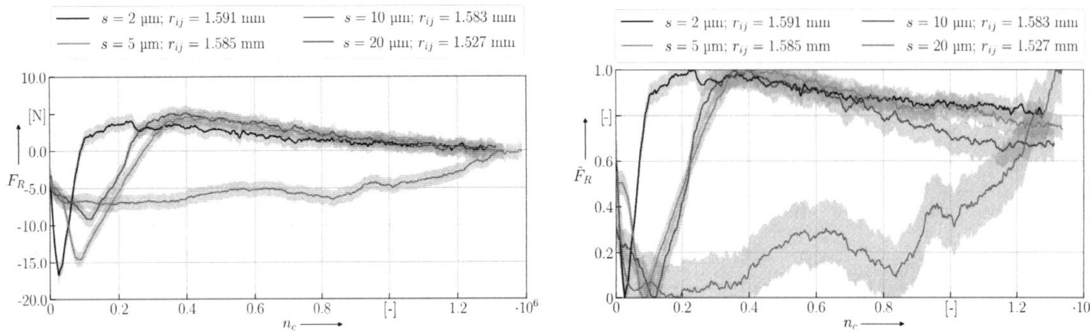

Figure 3. *Left*: Trend developments of a radial force component for different layer thicknesses s. *Right*: Min-max normalization of the trend developments.

outside temperature and humidity, the compressed air supply to the air bearings and the tolerance of the manufactured shaft seats also play a decisive role. Time series are therefore usually subjected to a transformation or normalization in order to ensure improved comparability and interpretability of the data [8][9]. One of the most commonly used methods is the z-transformation. The data is transformed in a way that it has a mean value μ of 0 and a standard deviation σ of 1. This results in the following transformation rule

$$z_i = \frac{x_i - \mu}{\sigma}. \tag{5}$$

The problem is that this method can only be applied to statistically stationary time series, i.e. those whose statistical properties such as mean value and standard deviation do not vary over time.

As the figures (Figure 3) illustrate, the statistical properties of the wear measurement change significant over time. An explanation for this is the change in the contact surface due to wear as the number of contacts increases. Another problem with this method is the influence of data outliers, which have a high impact on the standard deviation.

Due to the progression of the trend developments, min-max normalization is a suitable method. Here, the data is mapped into a predefined range [9]. For the trend curves of the force components and the moments, the extreme values can be easily determined. Consequently, a corresponding transformation according to equation 6 can be carried out, where $max(\overline{X})$ and $min(\overline{X})$ are the maximum and minimum values of the trend developments $\overline{X}$

$$\tilde{x}_i = \frac{x_i - min(\overline{X})}{max(\overline{X}) - min(\overline{X})}. \tag{6}$$

The normalised force curve $\tilde{F}_R$ can be seen on the right-hand side in Figure 3. The range of values is now mapped between 0 and 1. The normalisation allows a consideration invariant of amplitude and offsets, which means that the influences of the layer thickness can be evaluated, among other things.

A linear decrease in the force component is visible after a number of contacts of approximately $0.4 \cdot 10^6$. The observed drift does not seem to be cause of the wear, as the torque sensors do not confirm any progression. This is illustrated in Figure 5 (left), which shows the normalized power loss calculated from the measured torques. Compared to Figure 3, no linear drift can be recognised. From the specified accuracy data of the force sensors, there are several possible influences for the drift [5]. Temperature influences on the zero signal and the characteristic value can lead to a drift in the measurement signal, especially with large changes. To evaluate this, the temperatures at the load cells were therefore recorded. Figure 4 shows the recorded temperatures for two different coating systems. The mean value of the temperatures recorded on the two load cells is shown. A settling phase can be seen at $s = 20$ µm, which is not present for $s = 10$ µm. This is due to the fact that the sensors were deliberately switched on before the start of the measurement in order to eliminate the influence of pre-heating from the measurement. In order to minimise the influence of temperature, all further measurements are carried out without the pre-heating process. However, the linear drift only occurs after this heating process. Overall, the changes in temperature corresponding two the linear drift are less than one Kelvin. With the help of the data

sheet, the maximum changes in the measurement signal due to the temperature change can be estimated, which is therefore less than 0.5 N. However, Figure 3 shows that the changes in drift are up to 5 N. The influence of temperature therefore only seems to play a minor role.
Relative drift is another influencing factor. It involves a slow change in the measurement signal under constant load. A clear explanation can be given for this: Wear of the coating causes a significant change in the contact area and therefore the load. Due to the significantly higher wear rate of the coating compared to the base material, there are higher changes in relation to the load in the first period. This is also reflected in the trend developments (Figure 5). Subsequently, the load remains almost constant due to the small changes within the contact zone. The influence of relative creep increases here.

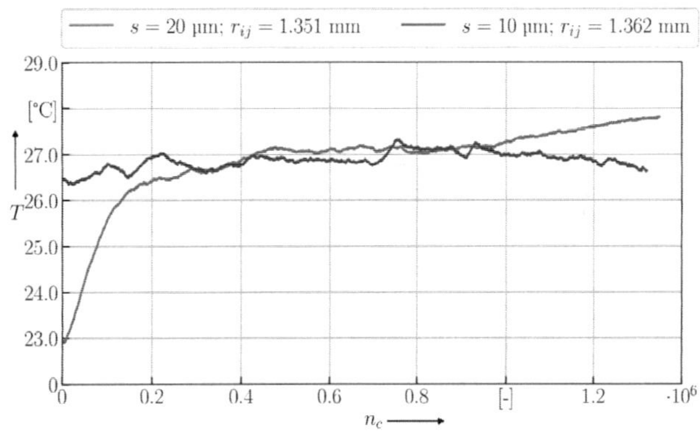

Figure 4. Trend development for two different layer thicknesses s. For $s = 20$ µm, the influence of the heating process on the measurement has been taken into account.

Subsequently, a regression analysis is performed for correction. Various approaches can be employed for this purpose. The LOESS method (Locally Estimated Scatterplot Smoothing) is chosen to take into account the non-stationary properties. LOESS is a non-parametric regression method used to model the relationship between a dependent (normalized) variable $\tilde{x}$ and one or more independent variables $\tilde{y}$. Unlike parametric models, such as linear regression, LOESS does not make assumptions about the form of the relationship between the variables. Instead, it estimates the relationship locally, making LOESS more flexible.

For each point $\tilde{x}_i$ in the dataset, a local model is estimated using a weighted average of neighboring points to get the value $\tilde{y}_i$. The weights w_{ij} are often cumputed using a tricube weighting function

$$w_{ij} = \left(1 - \left(\frac{|\tilde{x}_j - \tilde{x}_i|}{d_i}\right)^3\right)^3. \tag{7}$$

Here, d_i ist the distance to the nearest neighbor of $\tilde{x}_i$. A local linear regression is then applied to the weighted points. The coefficients of this regression are chosen to minimize the weighted sum of squared residuals

$$\min_{\beta_0,\beta_1} \sum_j w_{ij}(\tilde{y}_j - (\beta_0 + \beta_1\tilde{x}_j))^2. \tag{8}$$

This process is repeated for each point $\tilde{x}_i$ in the normalized dataset, creating a local model for each point. The LOESS estimate $\hat{y}$ for a point $\tilde{x}_i$ can be generally expressed by

$$\hat{y}(\tilde{x}_i) = \sum_j w_{ij}\tilde{y}_j. \tag{9}$$

With the LOESS estimate the linear drift in the dataset can be removed (Figure 5). The trends presented below in the analysis of the results include this correction without explicitly mentioning it again.

3.2. Sources of error and uncertainties
In order to be able to fully evaluate the measured values, the uncertainties that can be derived from the respective accuracy classes of the measurement sensors must be taken into account. This results in the measurement uncertainties listed in table 1. Besides the torques, which are evaluated via the torque measuring shaft, the forces are shown, whereby the different measuring ranges of the respective directions must be taken into account. The uncertainties of the measured values are also used to calculate the error propagation for determining the uncertainty of a calculated quantity from several measured quantities.

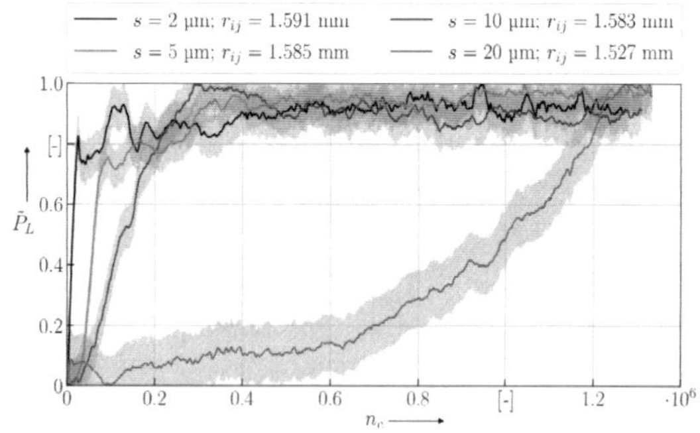

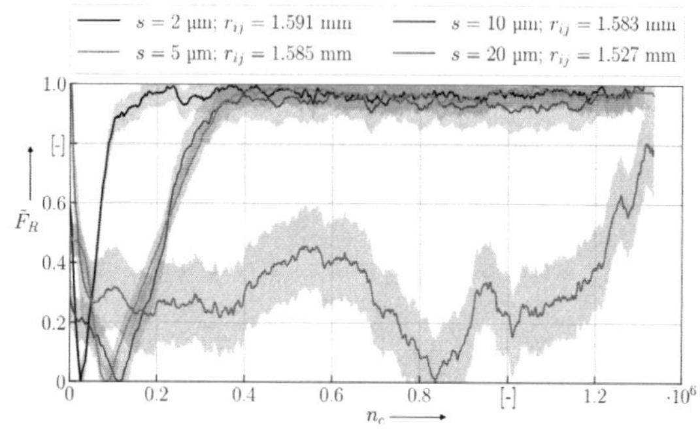

Figure 5. *Left*: Trend developments of a normalized power loss for different layer thicknesses. *Right*: Min-max normalization of the force component after the LOESS estimate correction.

Table 1. Specifications and uncertainties of the used sensors [5].

	Torque sensor	Force sensor (x)	Force sensor (y)	Force sensor (z)
Accuracy class	0.05	0.2	0.2	0.2
Measuring range	5 Nm	500 N	500 N	2000 N
Uncertainty u	±0.0025 Nm	±1 N	±1 N	±4 N

This is essential if a direct measurement of the desired variable is not possible and it has to be calculated from the measured values. For the power loss P_L this is the case.

To demonstrate the principle of error propagation, assume that the computed variable a is arbitrarily related to the measured variables $b_1, b_2, ..., b_n$ by $a = f(b_1, b_2, ..., b_n)$. For uncorrelated measurements, the uncertainty of the calculated quantity a is given by the root sum of the squares of the partial derivatives of the function with respect to each measured quantity. These partial derivatives are then multiplied by their respective uncertainties u_i [6]

$$u_a = \sqrt{\left(u_{b_1} \cdot \frac{\partial a}{\partial b_1}\right)^2 + \left(u_{b_2} \cdot \frac{\partial a}{\partial b_2}\right)^2 + ... + \left(u_{b_n} \cdot \frac{\partial a}{\partial b_n}\right)^2}. \tag{10}$$

In the course of the tactile measurements, random uncertainties caused by the user are also expected. The repeatability of the measurements is therefore assessed in advance. For this purpose, a rotor is installed and removed several times by several users in the setup of the measuring machine and the same routine is carried out. This involves measuring the lines within the contact area for each tooth of the rotor. To obtain the next section, the rotor is rotated by an angle $\Delta\Phi$ using the rotary table and moved by a height ΔZ. The scanning process then begins, whereby 200 points are recorded per millimetre. The mean radius of curvature is analysed to evaluate the repeatability, whereby each scan of each tooth is taken into account.

Figure 6 shows six repetitions of the analysed measurement. The error bars indicate twice the standard deviation of the measurements. Each error bar covers the range in which 95% of the measured values lie. The measured radii of curvature $r_{m,i}$ are all in a close range between 2.8 mm and 2.9 mm, which indicates a low variability in the measured values. The overlap of the error bars between the individual measurement points indicates that the differences between the measured values are within the measurement uncertainties. Therefore a high repeatability of the measurements can be expected, as the measured val-

ues are consistent and the uncertainties are relatively small compared to the measured values themselves. The low scatter of the measured values and the small uncertainties show that the measurements can be reliably repeated.
Furthermore, the measured contact sections can be used to make statements regarding the pitch errors. The relationship between pitch S and rotor length L can be determined using the wrap angle Φ via

$$S = \frac{2 \cdot \pi}{\Phi} \cdot L, \quad (11)$$

where the wrap angles of the main and gate rotor can be determined using the gear ratio $i = z_m/z_g$. Here z_g is the number of teeth of the gate rotor. To compare the measured contact lines the helical path of the rotor is untwisted. To do this, the point with the minimum distance to the pitch circle is determined for all measured points in the contact area

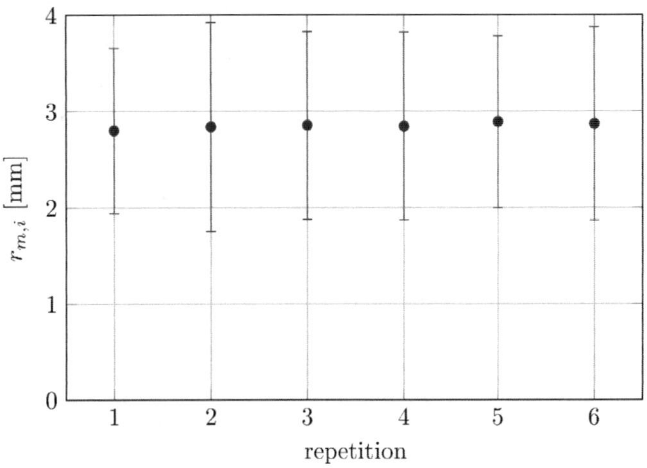

Figure 6. Repeatability of the curvature radii from one main rotor as a result of the tactile meassurement.

$$min(\Delta r_i) = min(r_{m,p} - r_i), \quad (12)$$

where $r_{m,p}$ is the radius of the pitch circle of the main rotor and r_i is the respective radii of the measured points. The same principle is used for the gate rotor.

The coordinate measuring machine already offers the option of converting the measured values into polar coordinates. The coordinates x_i and y_i can be used to determine the sectional wrap angle $\Delta\Phi$. Figure 7 shows these relationships graphically. Two different cross sections of a main rotor profile are illustrated. These are shifted by Δz and rotated by an sectional angle $\Delta\Phi$. This situation is illustrated at an arbitrary point. The pitch can thus be verified for each tooth by adding up the sectional angle. The results are listed exemplarily for a main rotor in Table 2.

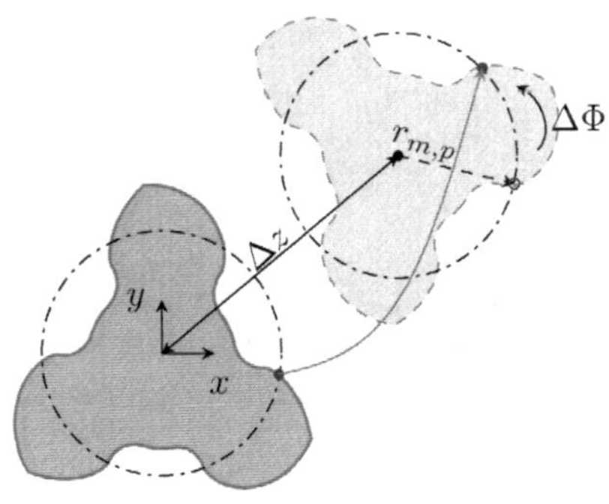

Figure 7. Illustration of the pitch deflection. The wrap angle can be determined section by section by rotating the measured point in the new rotor position.

4. Forces and moments in contact

In order to put the relationships into a better context, the force components and torques in the contact are first analysed in more detail. For better evaluation of the forces, the profiles were designed that for the most part only one pair of teeth is in contact. The contact force F_C transmitted by the secondary rotor is divided into three orthogonal components: axial force F_A, radial force F_R and tangential force

Table 2. Calculated wrap angles and corresponding pitches listed as an example for the teeth of a measured main rotor ($\Phi_{\texttt{ideal}} = 133\,°$, $S_{\texttt{ideal}} = 270$ mm).

Tooth	Wrap angle Φ	Pitch S	Deviation $S/S_{\texttt{ideal}}$
1	132.88 °	270.92 mm	0.34 %
2	133.27 °	270.11 mm	0.04 %
3	132.69 °	269.25 mm	0.27 %

F_T [10]. Assuming that the tangential force acts on the pitch circle, there is a direct relationship between the measured torque M_m and tangential force via the pitch circle radius via

$$F_T = \frac{M_m}{r_{m,p}}. \tag{13}$$

The axial force is only dependent on the helix angle β

$$F_A = F_T \cdot \tan(\beta). \tag{14}$$

The geometric relationship between the tangential and radial force becomes clear with the help of the normal section. The radial force can therefore be calculated using the angle α

$$F_R = \frac{F_T}{\cos(\beta)} \cdot \tan(\alpha). \tag{15}$$

Due to the geometric change in the contact area, the angle cannot be converted directly using the pitch angle. Accordingly, this must be determined using the measured forces. Assuming small radial forces ($\cos(\alpha) \approx 1$), the contact force F_C can be determined via

$$F_C = \frac{M_m}{r_{m,p} \cdot \cos(\beta)}. \tag{16}$$

The contact force can also be used to estimate the maximum pressure so that the applied moment can be better interpreted. According to [11], the maximum normal pressure p_0 between two cylindrical bodies can be calculated via

$$p_0 \approx \left(\frac{E^* \cdot F_C}{\pi \cdot l \cdot r}\right)^{0.5}. \tag{17}$$

E^* is the equivalent modulus of elasticity of the two bodies, l the length of the contact and r the Hertzian equivalent radius. The interested reader is referred to [11] for more detailed information on the calculation. The applied moment $M_g = 0.5$ Nm thus corresponds to a maximum normal pressure in the contact of $p_0 = 0.805$ MPa with a contact force of approx. $F_C = 25$ N. For $M_g = 1.5$ Nm, pressures of up to 2.0 MPa occur in the contact, which corresponds to a contact force of approx. 55 N.

Figure 8 shows the force equilibrium in the case of a driving main rotor using a section of the top view and the cross and normal section of the rotor teeth. The coordinate systems of the test rig (Figure 2)is also displayed for better categorisation in the overall context of the test rig. The tangential component points in the direction of the y-axis, the axial component in the direction of the z-axis. The radial force always acts towards the centre of the rotor and points in the direction of the x-axis. While the axial and tangential forces can be clearly assigned, a change to the relative system of the main rotor is advantageous for interpreting the radial component.

During the operation the main rotor rotates with an angle φ_m around its origin. The gate rotor rotates counterwise around its origin. Investigations by [4] show that in the relative system of the main rotor, the gate rotor first moves towards the main rotor and then moves away from it again. For ideal

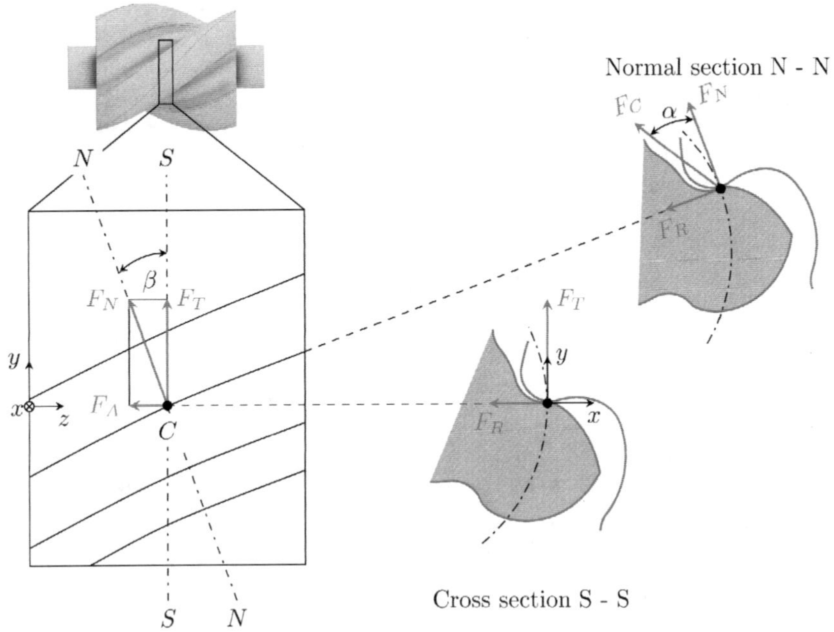

Figure 8. Forces on the main rotor are illustrated in the section of the top view and the cross and normal section at the contact point C.

contact between two rigid bodies, there is only a single contact point C and the sliding speed is zero. Figure 9 shows the described relative movement of the gate rotor in the two-dimensional relative system located on the main rotor. The interested reader is referred to [4] for a more detailed description. The contact takes place in the pitch point with a main rotor rotation of $\varphi_m = 0$. If the elastic deformation of the rotors is also included in the considerations, the rotors will slide next to the pitch point in addition to the pure rolling contact. Accordingly, a significant proportion of the frictional force is characterised in the radial direction. To assess the durability of the coating, the radial force component is analysed under this assumption.

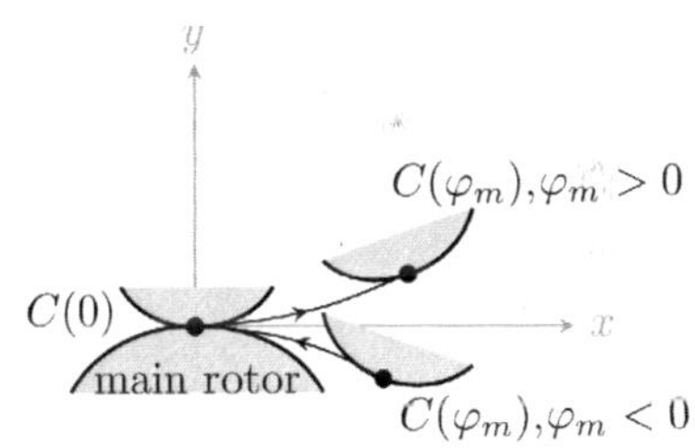

Figure 9. Relative movement of the gate rotor according to [4].

In addition to the measured forces, the torques are also analysed. The functional relationship between the power P and the torque M as well as the angular velocity ω is

$$P = M \cdot \omega = M \cdot 2 \cdot \pi \cdot n, \tag{18}$$

where n is the speed of the respective rotor. As a result of the friction losses, the torque at the main rotor is correspondingly lower than at the gate rotor. The power loss P_L can therefore be determined via the power at the gate and main rotors

$$P_L = P_g - P_m. \tag{19}$$

5. Measurement results

Using the pre-processing methods described above, the trend developments for the measured forces and torques are worked out. These are assessed in terms of wear resistance. Decisive influencing factors are the specified braking torque, the roughness of the rotors and the layer thickness of the coating.

5.1. Variation of the layer thickness and contact force

The durability of the MoS_2 layer is analysed under different loads. For this purpose, the braking torque on the gate rotor is varied and the rotors are analysed once at $M_g = 0.5$ Nm and once at $M_g = 1.5$ Nm.

Figure 10 shows the normalised trend developments of the power loss and the radial force for the lower load case. The trend developments are shown over the number of contacts (NoC) n_c. The normalised power loss $\tilde{P}_L$ and the normalised radial force $\tilde{F}_R$, which is mainly responsible for the loss mechanisms in the contact, are evaluated. The results of the different load cases for the different layer thicknesses are analysed in more detail below.

In the case of power loss, there is a clear tendency for higher layer thicknesses to be accompanied by longer durability. While the coating with $s = 2$ µm is already worn out after a few contacts, the coatings with $s = 5$ and $s = 10$ µm show a failure of the coating at approx. $0.1 \cdot 10^6$ number of contacts and $0.2 \cdot 10^6$ respectively. In addition, the highest coating thickness appears to exhibit increased durability. This may be due to a transfer of the coating to the gate rotor. A similar wear behaviour can be seen in the radial force trends. Here too, the coating thickness of $s = 20$ µm appears to be particularly suitable for longer operating times. In addition, a renewed drop can be seen after approx. $0.6 \cdot 10^6$ NoC, which again speaks in favour of a transfer and new formation of the lubricating layer.

Figure 11 shows the curves at higher loads, where the power losses increase continuously. In addition, the higher layer thicknesses do not show any advantages. The force components tends to a very chaotic behaviour after $0.4 \cdot 10^6$ NoC. This suggests that the lubricating layer is completely worn out. The resulting steel-to-steel contact causes micro-welds to form, which subsequently tear open again. Very high temperatures also occur at times during contact, so that iron oxide can be found in the contact area, among other things. The pure MoS_2 coatings therefore do not appear to be a suitable choice for higher loads.

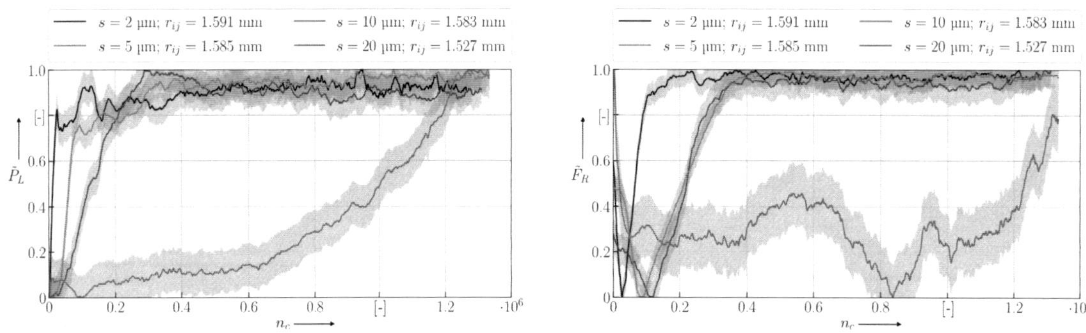

Figure 10. *Left*: Trend developments of the normalized power loss for different layer thicknesses s and a torque of $M_g = 0.5$ Nm. *Right*: Normalization of the radial force component for a torque of $M_g = 0.5$ Nm.

5.2. *Variation of roughness*

In addition to the layer thickness and the loads, variations in roughness are also analysed. In comparison to the measurements shown above, rotors with a roughness of $R_z = 8 - 10$ are used instead of $R_z = 3 - 4$. Ideally, the higher roughness should ensure a longer residence time of the coating. The roughness specifications here always refer to the rotors before coating.

Figure 12 again shows results of the experiments with the load of $M_g = 0.5$ Nm. Similar tendencies are recognisable as with rotors with a finer surface, whereby the loss mechanisms improve with increasing layer thickness. A comparison of the gradients highlights this correlation. Both the power losses and the radial force show a flatter increase in the trend developments compared to the rotors with a roughness of $R_z = 3 - 4$. It can therefore be assumed that the higher roughness improves the transfer behaviour of the lubricant layer between the main and gate rotor. The roughness valleys serve as a reservoir for the MoS_2, which increases the residence time.

However, for higher loads $M_g = 1.5$ Nm no improvements can be recognised due to the higher roughness. It is noticeable that even a higher layer thickness does not seem to improve with regard to the loss. The chaotic progression of the force component also indicates a corresponding failure of the lubricating layer.

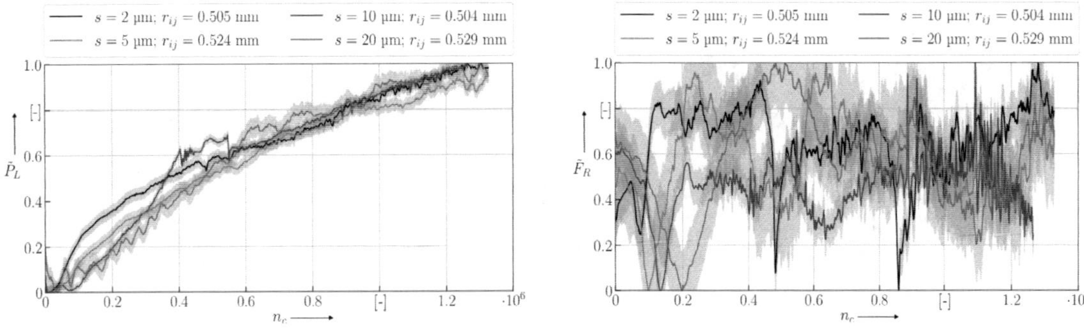

Figure 11. *Left*: Trend developments of the normalized power loss for different layer thicknesses s and a torque of $M_g = 1.5$ Nm. *Right*: Normalization of the radial force component for a torque of $M_g = 1.5$ Nm.

Overall, the MoS_2 coatings appear to achieve a longer service life with increasing coating thickness. Among other things, a higher roughness also has a positive effect because of the higher residence time of the coating in the contact area. The MoS_2 coating only seems suitable for a smaller range of loads.

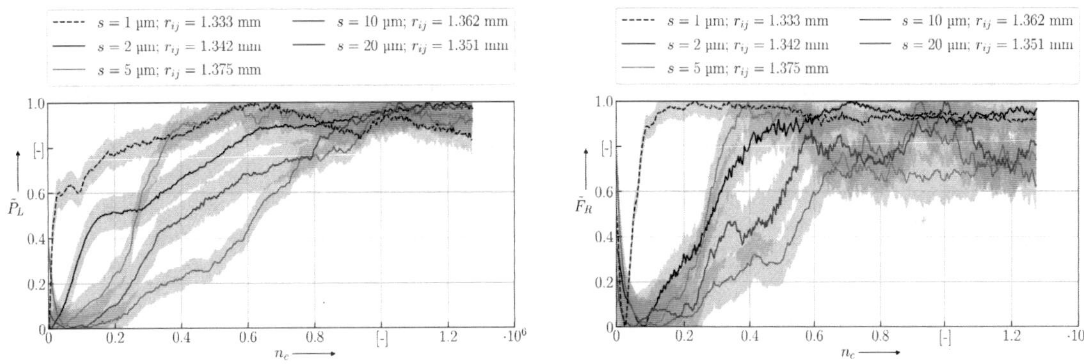

Figure 12. *Left*: Trend developments of the normalized power loss for different layer thicknesses s and a torque of $M_g = 0.5$ Nm and $R_z = 8 - 10$. *Right*: Normalization of the radial force component for a torque of $M_g = 0.5$ Nm and $R_z = 8 - 10$.

5.3. *Wear*

To obtain a qualitative statement on wear, the rotors are measured tactilely as described above. Once the measurements on the test rig have been completed ($n_c \approx 1.3 \cdot 10^6$), the rotors are subjected to another tactile measurement. Due to the sensitivity of the rotors with regard to their alignment to each other, the rotors are not removed during a wear measurement. This means that only integral statements can be made about the overall wear, but not at different times during the measurement. The results of the tactile measurements are shown below. The measured lines of the contact area are projected in such a way that the rotor is untwisted. The original contours before the wear measurement are then used as a reference. The resulting difference Δz, the width of the contact and the projected length can then be used to calculate the wear volumes using numerical integration.

Figure 14 shows the evaluated tactile measurements for a layer thickness of $s = 2$ µm, a torque of $M_g = 0.5$ Nm and a roughness of $R_z = 3 - 4$. Here x is in the direction of the rotor axis, where the measured contact lines are plotted. y is in the radial direction and z perpendicular to x and y. The

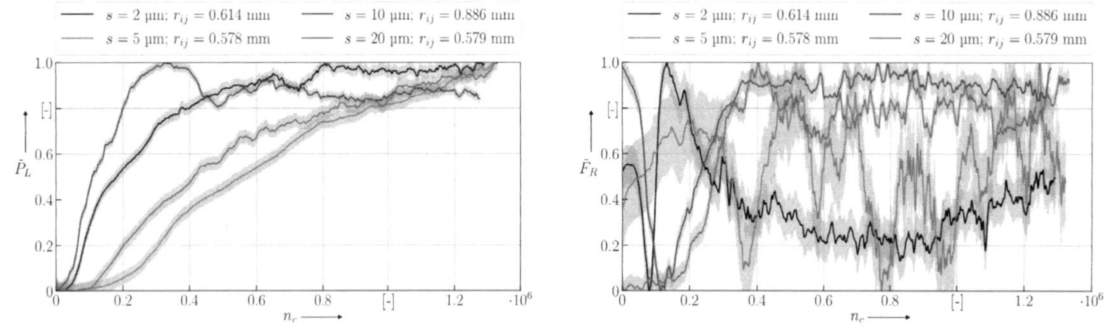

Figure 13. *Left*: Trend developments of the normalized power loss for different layer thicknesses s and a torque of $M_g = 1.5$ Nm and $R_z = 8 - 10$. *Right*: Normalization of the radial force component for a torque of $M_g = 1.5$ Nm and $R_z = 8 - 10$.

various contact lines are shown in a surface plot, with interpolation between the individual measurement contours. Δz indicates the difference between the original and the wear contours.

The pitch point of the main rotor is $y = r_{m,p} = 30$ mm and that of the gate rotor is $y = r_{g,p} = 40$. The figure clearly reveals that the material accumulates to the left of the main rotor and to the right of the gate rotor pitch point. A positiv Δz indicates here an accumulation of material, a negative Δz represents a removal of material. The relative movement of the gate rotor to the main rotor described above can be seen here. The gate rotor enters the contact area of the main rotor on the right-hand side (see figure 9). Due to non-ideal contours and elastic deformation, the highest wear occurs outside the pitch point. The main rotor shows high signs of wear on the entrance side of the gate rotor due to occuring sliding. The same can also be observed on the gate rotor. There, the material is worn away on the left-hand side. The figure also shows that the material is accumulated primarily at $y \geq 30$ next to the pitch point of the main rotor and at $y \leq 40$ of the gate rotor.

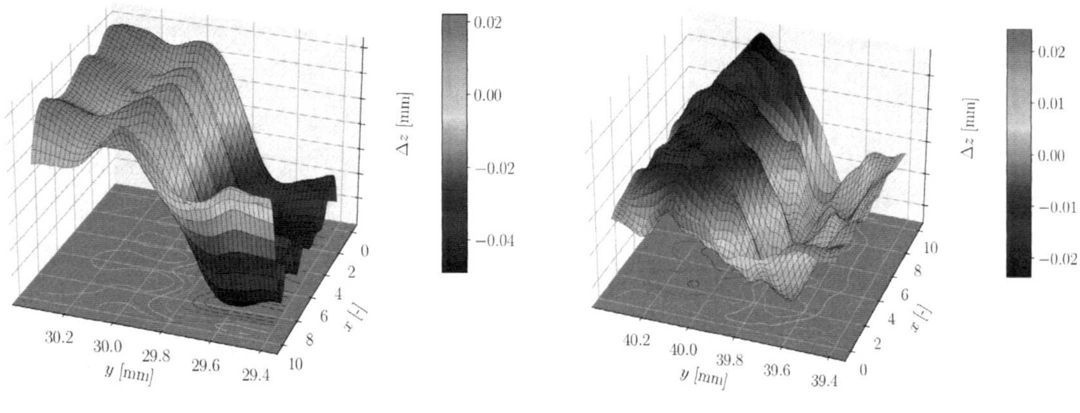

Figure 14. *Left*: Wear of the main rotor for $s = 2$ µm, a torque of $M_g = 0.5$ Nm and $R_z = 3 - 4$. *Right*: Wear of the gate rotor for $s = 2$ µm, a torque of $M_g = 0.5$ Nm and $R_z = 3 - 4$.

In figure 15 the tactile measurements for the same layer thickness ($s = 2$ µm) and roughness ($R_z = 3 - 4$), but for the higher torque ($M_g = 1.5$ N) is shown. The same rotor pair is involved here, whereby only the direction of rotation has been changed. This means that wear measurements for higher loads can be carried out on the other tooth flank of the main and gate rotor. Overall, the effects are similar to those of the low loads. However, the wear that occurs is up to 4 times higher than before, which is directly related to the higher load.

It is also noticeable that parts of the main rotor tooth are not engaged ($x = 0$). This can be explained by the coverage. It was selected in such a way that, if possible, only one pair of teeth is in contact at the same time. As the alignment of the rotors to each other is never ideal, no contact can occur in some sections, whereby other areas are subjected to correspondingly higher wear.

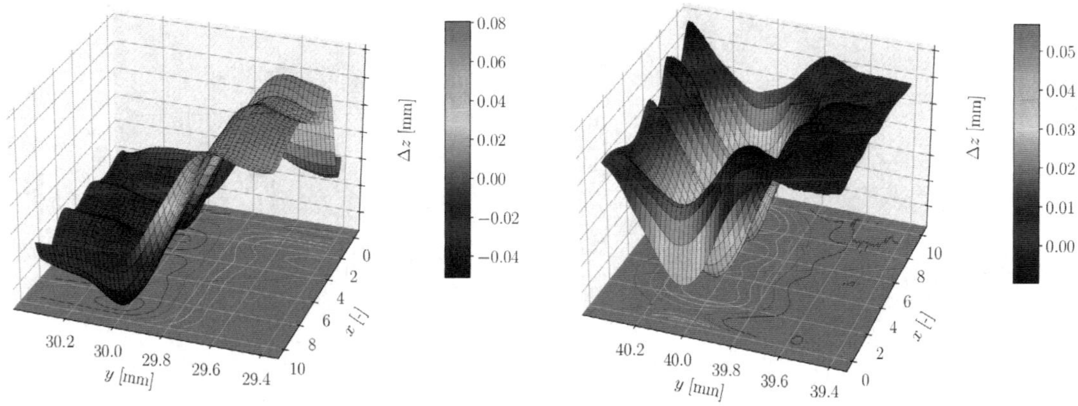

Figure 15. *Left*: Wear of the main rotor for $s = 2$ µm, a torque of $M_g = 1.5$ Nm and $R_z = 3 - 4$. *Right*: Wear of the gate rotor for $s = 2$ µm, a torque of $M_g = 1.5$ Nm and $R_z = 3 - 4$.

Based on the previous results, the layer thickness $s = 20$ µm is also compared with the extreme case of the thinnest. Figure 16 shows the lower load case for a roughness of $R_z = 3 - 4$. It can be seen that the wear is generally lower compared to the thin layer system (Figure 14). This agrees with the results from the trend developments. The residence time of the coating in the contact appears to be significantly increased. Transfer mechanisms and new layer formation are also a possible explanation. It can also be seen that the accumulation of material on both the main rotor and the gate rotor is lower. The lubricating layer appears to enable continuous favouring of the frictional properties in the contact.

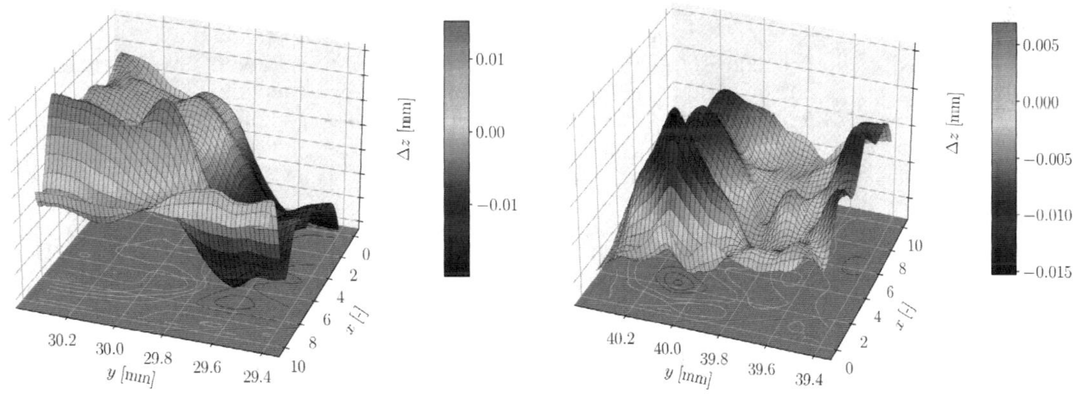

Figure 16. *Left*: Wear of the main rotor for $s = 20$ µm, a torque of $M_g = 0.5$ Nm and $R_z = 3 - 4$. *Right*: Wear of the gate rotor for $s = 20$ µm, a torque of $M_g = 0.5$ Nm and $R_z = 3 - 4$.

Finally, the influence of roughness is also taken into account in figure 17. It appears that again, several flank sections of the main rotor have little to no engagement. On the other hand, the load has a much greater effect in other parts. In the centre area, high signs of wear appear compared to the edge area ($\Delta z_{max} = 0.02$). Both the worn material and the accumulated portion are of a similar order of magnitude to the rotors with $R_z = 3 - 4$. The effect of roughness appears to have a favourable effect, particularly in the running-in phase. However, as the tactile measurements only allow a comparison of the original contour with the contour after the application tests, these cannot be recognised here. These results reinforce the impression that although the higher roughness has a positive effect for running-in processes, this has only a minor effect on real operating times.

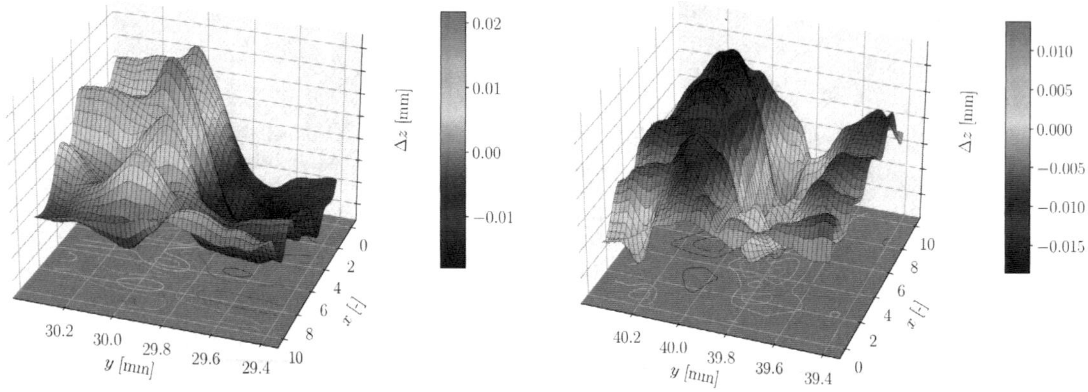

Figure 17. *Left*: Wear of the main rotor for $s = 20$ µm, a torque of $M_g = 0.5$ Nm and $R_z = 8 - 10$. *Right*: Wear of the gate rotor for $s = 20$ µm, a torque of $M_g = 0.5$ Nm and $R_z = 8 - 10$.

6. Summary and outlook

As part of this work, fluid-free lubricant coatings were investigated in dry-running, unsynchronised screw rotors. These combine the advantages of wet-running and synchronised, dry-running machines. The problems of the lubricant layers are mainly due to wear phenomena in contact. To date, only a few studies are available for screw rotors. This publication is intended to provide a systematic approach to specifically investigate the contact of screw rotors and to make conclusions about the wear mechanisms. For this purpose, a test rig was set up to analyse the loss mechanisms. In addition to the coating thickness and the roughness of the rotors, the load in the contact was also varied. Statistical normalisation allows the individual measurements to be better compared with each other in order to prevent offsets due to the installation situation or external influences such as humidity and temperature.

The analysis of the durability of the MoS_2 layer under different loads and layer thicknesses has provided several key insights. At lower loads, higher layer thicknesses, particularly 5 µm and 10 µm, exhibit significantly longer durability of the coating. The 20 µm layer shows the highest durability, possibly due to the transfer and reformation of the lubricating layer on the gate rotor. The normalized power loss and the radial force show a flatter increase, indicating more effective lubrication and longer operational times.

Under higher loads, the power losses continuously increase, and higher layer thicknesses do not demonstrate significant advantages. The force components exhibit chaotic behavior, suggesting complete wear of the lubricating layer, leading to direct steel-to-steel contact, micro-welding, and high temperatures. Thus, pure MoS_2 coatings do not seem suitable for higher loads.

Additionally, variations in rotor roughness were analyzed. Higher roughness ($R_z = 8 - 10$) compared to finer surfaces ($R_z = 3 - 4$) ideally ensures a longer residence time of the coating. At lower loads, higher roughness improves the lubrication transfer behavior between the main and gate rotors, with the roughness valleys serving as reservoirs for MoS_2, leading to a longer residence time. However, at higher loads, no improvements due to higher roughness were observed, and the chaotic force component progression indicates the failure of the lubricating layer.

The tactile measurements reveal insights into the wear behavior of rotors under various conditions. The results show that material accumulates predominantly to the sides of the pitch points of the main and the gate rotor. The side of the material accumulation depends on the respective direction of rotation. This pattern indicates significant wear due to sliding, particularly on the entrance sides. When comparing different torques, the higher loads result in up to four times more wear, highlighting a direct correlation between load and wear severity. For different layer thicknesses, thicker layers ($s = 20$ µm) exhibit generally lower wear compared to thinner layers, suggesting a prolonged coating residence time and improved frictional properties due to the lubricating layer. Additionally, higher surface roughness appears to facilitate better performance during the running-in phase, although its long-term impact on wear is minimal.

Overall, MoS_2 coatings achieve longer service life at low loads with increasing coating thickness, and higher roughness has a positive effect on the lubricant layer's residence time in the rotor contact. Tests with doped molybdenum disulphite are planned for the near future. Materials such as silver, tungsten and carbon are to be incorporated into the coating system. It is expected that the wear and friction behaviour will be improved compared to pure MoS_2. In addition, investigations using Ramann spectroscopy are also planned, which can be used to analyse the transfer mechanisms of the coating.

These findings underline the potential of optimized MoS_2 coatings for enhancing the durability and performance of screw rotors under specific conditions. By further refining the coating compositions and surface treatments, it is anticipated that advancements in wear resistance and lubrication efficiency can be achieved, paving the way for more reliable and efficient machinery in industrial applications.

Acknowledgement

Funded by the Deutsche Forschungsgemeinschaft (DFG, German Research Foundation). Gefördert durch die Deutsche Forschungsgemeinschaft (DFG) – Projektnummer 357505886.

References

[1] Rinder L 1979 *Schraubenverdichter* (Wien, New York: Springer-Verlag) DOI 10.1007/978-3-7091-8554-4

[2] Temming J 2007 *Stationärer und instationärer Betrieb eines unsynchronisierten Schraubenladers* (Dissertation Universität Dortmund)

[3] Dämgen U 1998 *Beanspruchung von verschleißbeständigen Werkstoffverbunden in Schraubenkompressoren* (Dissertation Universität Dortmund)

[4] Aurich D, Brümmer A, Wittig A, Stangier D, Tillmann W, Thomann C A, Wittrock A and Debus J 2022 *IOP Conference Series: Materials Science and Engineering* **1267** 012005 DOI 10.1088/1757-899X/1267/1/012005

[5] M Geissendorf D A 2024 *Testrig MoS2 Coatings in unsynchronized, dry-running Screw Compressors: Experimental Insights on Operational Efficiency and Durability* (Zenodo) DOI 10.5281/zenodo.11452550

[6] C Ratcliff B R 2015 *Doubt-Free Uncertainty In Measurement: An Introduction for Engineers and Students* (Springer International Publishing Switzerland) DOI 10.1007/978-3-319-12063-8

[7] J D Cryer K C 2008 *Time Series Analysis* (Springer New York) DOI 10.1007/978-0-387-75959-3

[8] F T Lima V S 2023 *Big Data Research* **34** 100407 DOI https://doi.org/10.1016/j.bdr.2023.100407

[9] E Ogasawara M L 2010 *The 2010 International Joint Conference on Neural Networks (IJCNN)* 1–8 DOI 10.1109/IJCNN.2010.5596746

[10] Engelmann J 2019 *Maschinenelemente kompakt: Auswahl, Gestaltung und Dimensionierung in Theorie und Praxis* (Springer Berlin) DOI 10.1007/978-3-662-57955-8

[11] Popov V 2010 *Kontaktmechanik Und Reibung* (Springer Berlin Heidelberg) ISBN 9783642133022 DOI 10.1007/978-3-662-45975-1

IOP Conf. Series: Materials Science and Engineering **1322** (2024) 012013 doi:10.1088/1757-899X/1322/1/012013

Test rig setup for particle wear analysis in screw pumps

P Moor*, M Kuhr and P Pelz

Fluidsystemtechnik, Technische Universita t Darmstadt, Otto-Berndt-Str. 2,
64287 Darmstadt, Germany

*E-mail: pascal.moor@fst.tu-darmstadt.de

Abstract. For many applications multi-screw pumps can be operated with minimal wear rates, thus offering long service life. Pump manufacturers are however met with increasing demand for pumps in applications with particle-laden fluids, which often lead to high wear rates. Thus, the need for design guidelines and models for the estimation of the life expectancy arises. The scientific literature does not offer such models or published systematic studies. In this work the authors present a test rig setup for the systematic analysis of wear in operation with particle-laden fluids in multi-screw pumps. The setup is designed with repeatability of the test environment and reliability aspects in mind and draws inspiration from food and chemical processing industries.

1. Introduction

All technical systems, including screw pumps, are operated in a conflict between effort, availability and acceptability. In order to reach new markets and remain relevant in existing markets, these dimensions of quality need to be met to the customers satisfaction. The development of new markets and applications in particular can present with new and challenging requirements. For many positive displacement pumps, especially for multi-screw pumps, abrasive particles suspended in the pumped fluids cause high wear rates. Historically the standard literature recommends the use of multi-screw pumps only for non-abrasive applications and either with large particles or with filtered fluids. These pumps are the focus of the presented research and will furthermore simply be called *screw pumps*. In order to reduce the wear rates in such critical applications, a reduction in operating speed is often recommended [1]. In order to respond to increasing customer demand for screw pumps in applications with abrasive particles, the wear behaviour of different types of screw pumps needs to be investigated in a controlled manner. Thereby, critical operating parameters are to be identified and a model for the estimation of an expected lifetime for the pumps at given parameters is to be developed.

For the analysis of wear in screw pumps, the respective pump construction needs to be further specified. The way in which the spindles are supported, as well as the differentiation between timed and untimed construction significantly impact the behaviour of the sealing gaps during operation. These are the most critical areas for wear in screw pumps, since increasing the gap heights directly reduces the volumetric efficiency for otherwise unchanged parameters. For untimed three-spindle screw pumps, the achievable tolerances allow for smaller sealing gap heights compared to timed two-spindle screw pumps, see **Figure 1**. Therefore, the three-spindle pumps can be expected to be more susceptible to abrasive wear, in particular three-body abrasion [2].

IOP Conf. Series: Materials Science and Engineering **1322** (2024) 012013 doi:10.1088/1757-899X/1322/1/012013

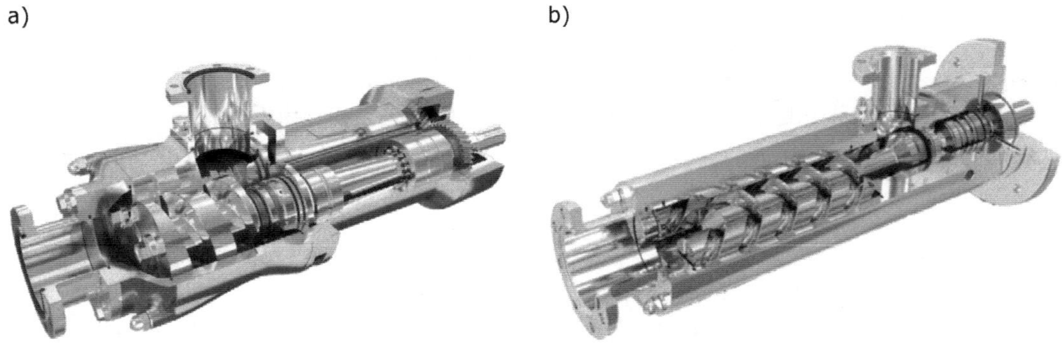

Figure 1. Multi-spindle screw pumps are available in multiple configurations. In two-spindle pumps, see a) [3], the spindles are often timed by spur gears, the shafts are often supported by external bearings. Three-spindle pumps often realise an untimed design and are often do not feature dedicated bearings to support the shafts, see b) [4].

Wear can generally not be analysed in an isolated generic experiment, since the wear of components is specific to the system and the application. Even the applicability of application specific wear tests needs to be carefully evaluated. Preliminary works to the ISO/DIS 9632 test method shows, that gear pumps with similar performance in practice may still show different wear behaviour in a controlled test environment [5].

For two-spindle screw pumps the work of Schmidt shows, that reliable models for the wear in timed two-spindle screw pumps cannot be derived from the tribometer experiments performed [6]. Schmidt recommends to focus future wear analysis on particle concentration, the flow conditions within the sealing gaps and the enclosed cavities, as well as the particle size distributions. Within the context of multi-phase pumping in the petroleum industry, Dorenbos et al. [7] review the experiences from multiple applications of two-screw pumps. These field studies do not allow for general design guidelines, concluding, that the best available solution often is barely good enough.

For three-spindle screw pumps, the scientific literature shows no systematic investigations of particle wear. The failure analysis of a three-spindle oil pump in the aftermath of an accident in a hydroelectric power plant highlights the damaging potential of wear debris accumulation in a closed loop system which led to a cascading failure mechanism on a system level [8]. The wear behaviour of the pump itself was not analysed.

In order to perform systematic studies on the wear behaviour of timed two-spindle and untimed three-spindle screw pumps, the presenting authors of this research paper have developed a new test rig setup aiming at a repeatable test environment for the wear experiments. Similar to the test rig presented by Schmidt [6], the suspension with abrasive particles will be prepared in a tank with a mixer, while the operating pressure of the pump is regulated by throttling the pumps outlet with a pinch valve. The setup differs however in the choice of the test fluid, the sensors as well as additional efforts regarding the temperature control and the filtering of worn abrasive particles. The test parameters are extended to cover multiple viscosities and particle properties in order to better resolve the impact on the wear behaviour of the pumps. Preliminary measurements to support the theses presented in this research paper are not yet available, since the test rig is currently being constructed.

2. Development of a test rig setup for particle wear analysis

In order to ensure reliable and repeatable operation of the wear test rig for displacement pumps, the requirements for the test rig need to be defined and characterised. At the same time, in order to maintain a reasonable level of effort, suitable boundary conditions and limitations to the achievable quality of the representation of real-world open-loop systems are to be defined. Whenever possible, the boundary conditions are chosen in accordance with the real-world application of the respective pumps in mind. The requirements to be met by the test rig setup and necessary restrictions in order to achieve good scientific results with feasible efforts are covered by the following sections. Furthermore, the choice of the operating medium and the components will be elaborated on, before discussing the composition of the test rig setup and the methodology of the wear tests.

2.1. Goal of the experimental works

The focus of the experimental work by the presenting author is the analysis of the wear behaviour of the types of multi-spindle screw pumps specified before. The wear of the pump components is to be characterised by gravimetric measurement of the spindles and characterisation of the deviation of the spindle profile from the original state by means of a coordinate measuring machine. The physical wear of the components is then to be linked to the progressing degradation of the hydraulic properties of the pump characterised by pressure and volume flow at fixed pump speed and state of the throttle.

The experimental work focuses on abrasive and erosive wear in pumps due to the presence of abrasive particles in the pumped media. Further wear characteristics, such as run-in behaviour, which has been researched i.e. for external gear pumps [9] or the wear due to cavitation are not considered in the work. Therefore, care must be taken to exclude the impact of such wear processes on the wear within the tests.

2.2. General requirements

Previous works [10] with abrasive particle suspensions in the context of erosion wear at the institute for fluid systems have shown, that the abrasive particles themselves are subject to wear. The sharp edges are worn down, thus reducing the erosive effects. For the wear tests in screw pumps, Schmidt [6] shows that the particles additionally are ground up during the wear experiments, thus shifting the median to smaller particle sizes. Since these effects also impact the way, the particles are drawn into the gaps, these effects are to be mitigated and monitored. The test pumps provided by Leistritz Pumpen GmbH provide a volumetric flow rate of up to 200 l/min. It is expected that the wear tests may take multiple hours, some parameter combinations may even require multi-day tests. With concentrations of up to 3 % by mass, filtering the entire suspension flow directly after the pump is therefore not feasible. The filtering circuit is required to provide a repeatable test environment, while keeping the use of abrasive particles to a feasible level. Therefore, the requirement for the filtering circuit is set to allow for continuously filtration of the suspension in the tank with a flow rate that allows the total fluid volume of the test rig to be filtered once per test day.

Unlike in the work of Schmidt, the tests will be performed with variable particle and fluid properties. The tests are to be performed with multiple viscosities, particle types and concentrations as well as particle sizes at different pump speeds and pressure levels. Ultimately parameter variations of the spindle and liner material are to be performed. Therefore, the scope of the experimental work is large, which highlights the need for well placed restrictions in order to reduce experimental complexity and allow for a robust test rig setup at feasible costs. Similar to the research of Schmidt [6], the wear behaviour of the pumps is characterised by measurements of the temperature, pressure and volumetric flow rate.

For the design of the test rig, that needs to be capable of handling abrasive suspensions, conceptual inspiration can be drawn from multiple industries e.g., food processing plants with regards to dead-volume free designs and from mining applications with regards to handling abrasive suspensions, which are often called slurries.

2.3. Choice of the test fluid

The choice of the test fluid is based on ecological aspects as well as the requirements defined by the test pumps. The untimed three-spindle screw pumps to be studied, relies on the hydrodynamic journal bearing effect of the spindles in the housing in order to run most efficiently and with minimal wear. In order to ensure that the wear observed in the studies is only to be attributed to the presence of abrasive particles in the fluid, a contact free operation in all operating points needs to be ensured. Therefore, the fluid needs to offer lubricating properties and the viscosity should not fall below a critical threshold. Within the parameter variations for the spindle and liner material, steel components will be tested. Therefore, the test fluid also needs to offer corrosion inhibiting properties.

Worn-down abrasive particles will have to be filtered out and disposed of. Therefore, the choice of fluid is also subject to ecological considerations. After literature research with a focus on the food, cosmetics and mining industries, no suitable test fluid was found that combines adjustable viscosity, lubricating and corrosion inhibiting properties with ease of disposal for both the test fluid and contaminated particles while offering consistent fluid properties without ageing effects.

With a desired viscosity range of 5 – 200 cSt, mineral hydraulic oils were chosen as the test fluid, since a repeatable and robust operation can be expected, while the disposal of the fluid, filters and particles can be performed within a well-established and regulated recycling system.

2.4. Choice of the abrasive particles

In order to achieve high wear rates, the pumps are to be operated with abrasive suspensions. The influences of the particle properties, such as particle size, size distribution, hardness and shape are not yet known for the wear system within the screw pumps. Therefore, three different abrasive media with different properties are to be analysed. For the tests the following abrasive particle types are chosen:

- Silicon Dioxide - SiO_2
- Aluminium Oxide - Al_2O_3
- Silicon Carbide - SiC

These abrasive media are characterised by different hardness levels and characteristic particle shape and will give new insights into the wear of the abrasive particles within the different screw pumps. The most relevant particle sizes for the wear of the pumps are those close to and slightly above the sealing gap heights [11]. The particle size distribution will be documented for the respective abrasive media type, however no structured parameter variation with regards to the particle size distribution is planned.

2.5. Choice of hardware components

The test rig is designed with a reduction of dead-volumes in mind, wherever possible. Special focus is hereby put on the suction side, since dead-volumes allow for particles to settle out and for agglomeration. Periodic introduction of agglomerated particles into the pump is expected to significantly impact the wear behaviour of the pump compared to continuous particle concentration. The test rig is modular in construction in order to allow for this hypothesis to be tested later-on in dedicated studies with discontinuous particle concentration. This is of relevance, since applications with particle suspensions may not necessarily offer continuous particle concentrations or dead-volume free designs. Modularity and ease of maintenance require for connections to be placed conveniently throughout the piping. These connectors need to satisfy dead-volume free design criteria. Therefore, clamp-type couplings based on DIN 32676 and ISO 2852 which are common in the food and chemical process industries have been chosen for all suction and low-pressure connections. Due to constraints regarding the maximum test pressure and the pump flanges, the modular pressure plenum has been designed with hydraulic flanges based on ISO 6162. Minor dead-volumes that arise from these flanges do not directly affect the wear measurement in case of the flanges on the pressure side and are representative of real pump applications in the case of the suction side flange of the pump.

In order to reduce wear to the components of the test rig, the piping and components need to be designed according to the following principles. Disruptions to the flow are to be minimised. Thus, changes to the cross-sectional area of the pipelines are to be avoided, where they are required, they are to be designed with smooth transitions. The bend radius of hoses and pipes should be as large as possible in order to reduce wear in these areas. Elastic hoses are considered to be beneficial over (stainless) steel pipes, since they are easily replaced and often offer longer service life in abrasive applications. While for many abrasive applications in the agricultural, construction and mining industries, reinforced specialty hoses are available, these hoses are generally not suitable for oil-based fluids. Only hoses which cater to tar and asphalt products have been found to satisfy the requirements within the test rig. Such hoses are available with a maximum working pressure of up to 1.4 MPa which is not sufficient to cover the desired operating parameters for the tests. Therefore, hoses will only be used for the return circuit and the suction side. Since these are low pressure applications, readily available hydraulic suction and pressure hoses are used. In order to minimise hose wear, the maximum flow velocity is limited to ~ 2 m/s.

In order to allow for reconfigurations of the modular test rig and allow access to the test pump for service and analysis without draining the lines, the lines are equipped with valves. These valves need to be suitable for applications with oils and particle suspensions, dead-volumes are not desired. The valves need to offer a soft sealing mechanism in order to ensure reliable closing of the valve when particles are present in the sealing zone and with progressing wear of the valve. Within the relevant industries, two valve types are reported to be suitable for such applications. Membrane valves allow to throttle or close-off flows, while offering long service life and good repairability. Even simpler and therefore more common in mining or single-use pharmaceutical applications are pinch valves. Both types are designed with minimum dead-volume and therefore well suited for the application in the test rig. Since for these valves the test rig concept does not require throttling, pneumatic pinch valves are chosen for their minimal impact on the flow and the availability with flanges according to DIN 32676.

For the tests, the pump needs to be operated at different pressure levels, which is achieved by throttling. Throttles experience high wear rates in abrasive applications. Therefore, a continuously adjustable throttle with good wear properties is needed. These requirements are met by membrane and pinch valves. The pressure ratings for common membrane valves however do not cover the required pressure range for the tests. Manual pinch valves from the mining industry are available with higher pressure ratings up to and beyond 5 MPa and are compatible with oil. The only component that is exposed to the abrasive medium and thus subject to wear, is the sleeve. Therefore, a manual pinch valve has been selected for the throttling of the flow.

For the pressure relief system, the critical requirement is the suitability for particle suspensions. Since conventional, spring-based pressure relief systems often rely on critical sealing surfaces, reliable sealing when subjected to particle suspensions cannot be guaranteed. Therefore, a burst disc-type relief system is chosen for the test rig.

2.6. In-line measuring equipment

For the characterisation of pump wear behaviour, pressure, temperature and volumetric flow rate measurements as well as friction torque are considered. The monitoring of the test parameters requires measurement of the particle concentration. The necessary considerations for the application in the test rig are summarised below. **Table 1** gives an overview of the types of sensors used in the test rig.

Table 1. Sensors used in the wear test rig, including the special considerations for the handling of abrasive particles.

Measurement	type	reference	special considerations
pressure	capacitive ceramic sensor element	Suchy Messtechnik SD-45/SD-47	front mounted ceramic reduces dead-volume
temperature	thermocouples	standard Type J stainless steel immersion sleeves	immersion sleeves reduce dead-volume
flow rate	Coriolis-type flow meter	Endress+Hauser Proline Promass I 300	straight, single-tube design reduces dead-volume and wear to the sensor
friction torque	torque measuring shaft	standard measuring shaft KTR-SI overload coupling	overload coupling protects test rig in the event of a jam

For the pressure sensors, the abrasive particles may lead to wear to the sensor and depending on the construction, the particles may clog the sensing cavity, thus distorting the measurements. Therefore, sensors with front mounted membranes are chosen, see **Figure 2**, which also offer minimal dead-volume and are common in food and chemical industries. The membrane however is thereby exposed to the flow. Sensors with a front mounted aluminium oxide ceramic membrane are recommended for abrasive applications and are expected to offer reliable pressure measurement for the long measuring campaigns.

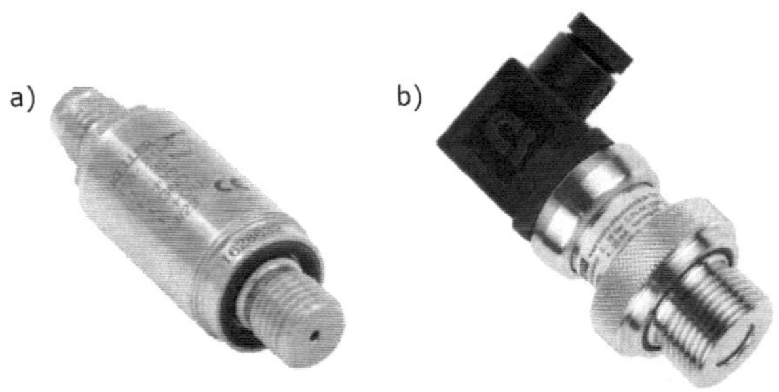

Figure 2. a) Pressure sensors with small sensing cavities are prone to clogging [12], b) Pressure sensors with front-mounted ceramic membrane minimise dead-volume and increase wear resistance [13].

The temperature sensors are isolated from the abrasive suspension flow by immersion sleeves in order to prevent wear to the sensor and also minimise the effects of dead-volumes. Furthermore, surface probes are placed on the pumps.

For precise flow measurement, test rigs from the hydraulic oil industries often rely on screw- or gear-type flow meters, which are not suitable for particle suspensions. For abrasive applications, magnetic-inductive flow meters are often chosen, since they offer full cross-sectional area without dead-volume. The minimum required electric conductivity of the fluid however does not allow for these sensors to be used with mineral oils. A Coriolis-type flow meter with a straight, single-tube design has been chosen. This type is suitable for oil-based particle suspensions and offers a dead-volume free installation for the inline measurement of mass flow.

As stated above, considerable efforts are made in order to give little chance for particle agglomeration by minimising dead-volumes. However, in order to verify whether the pump has been subject to consistent particle concentration or fluctuating particle concentrations, the particle concentration needs to be continually recorded. This measurement is even more important, since the suspension in the tank cannot be expected to be homogenous, see **Figure 3**. Therefore, the concentration within the suction line is expected to differ from the global concentration that is set by the addition of a set mass of particles to the tank. Due to the abrasive nature of the particles, optical measurement systems are not suitable, while sensors based on ultrasonic extinction are best suited for high particle concentrations and are, according to the manufacturers, limited with regards to the minimum particle concentration that can be resolved. Therefore, a different approach to the measurement of the concentration is chosen. With known fluid and particle density, the particle concentration of the suspension can be determined by measuring the density of the suspension. This can be achieved via the Coriolis-type flow meter.

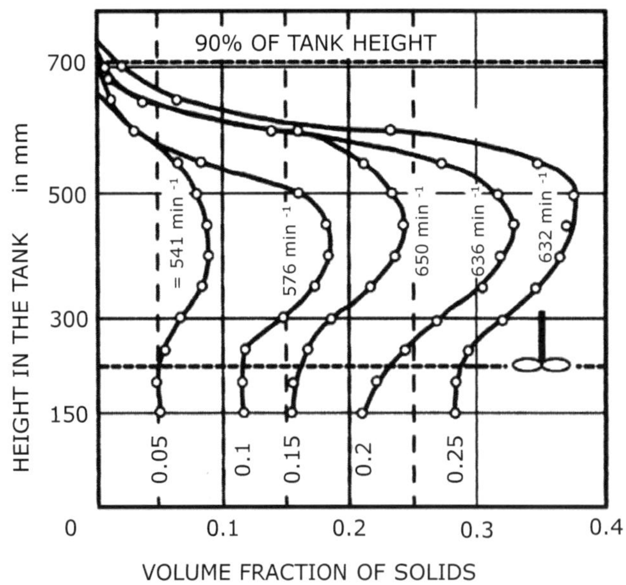

Figure 3. The concentration of solid particles in a suspension mixing tank is generally not homogenous along the tank height. The plot shows a typical distribution – based on [14].

The torque required to drive the pump at a known operating point indicates changes in the friction which can be attributed to the ongoing wear in the pump which increases surface roughness, hinders the formation of lubricating films and eventually leads to increased gap heights at the sealing gaps. Therefore, the torque is recorded by means of a torque measuring shaft. With the introduction of particles into the flow, the chance of a catastrophic failure of the pump increases. Particles with sizes greater than the gap height may become jammed between the spindles, thus suddenly stopping the rotation of the spindles. In order to prevent further damage beyond the measurement pump, especially to the torque measuring shaft, a torque limiting safety coupling is introduced.

2.7. Composition of the test rig

The concept shown in **Figure 4** has been developed at the institute for fluid systems with the requirements, restrictions and availability of components in mind. The suspension is prepared in the suspension tank by mixing via a pitched blade propeller. The fluid temperature is controlled via a heating element and a tube coil heat exchanger. Due to higher sinking velocities, larger particles are statistically more likely to be present in the lower part of the tank. Therefore, the main circuit extracts the suspension form the lower half of the tank, with sufficient distance from the walls and bottom. A cylindrical tank with a dished bottom and total drain has been chosen, which is equipped with baffles in order to aid the mixing and view ports in order to validate the quality of the suspension. The suction side is designed with minimal interference to the flow in mind and offers measurement of the pressure and fluid temperature as well as mass flow rate and suspension density via the Coriolis-type flow meter. All couplings within the suction line, up to the pump flanges are based on clamp connection according to DIN 32676 and ISO 2852, which are compliant with the strict requirements in the food processing industry. Thus, the suction line is considered dead-volume free. In order to reduce the amount of piping that is subject to the full operating pressure, the pressure plenum with pressure and temperature measurement is flanged directly to the pump. The pressure plenum has been designed in accordance with the German standard AD 2000, while offering considerable wear volume in order to ensure safe and reliable operation. Critical components of the test rig will be checked for wear periodically.

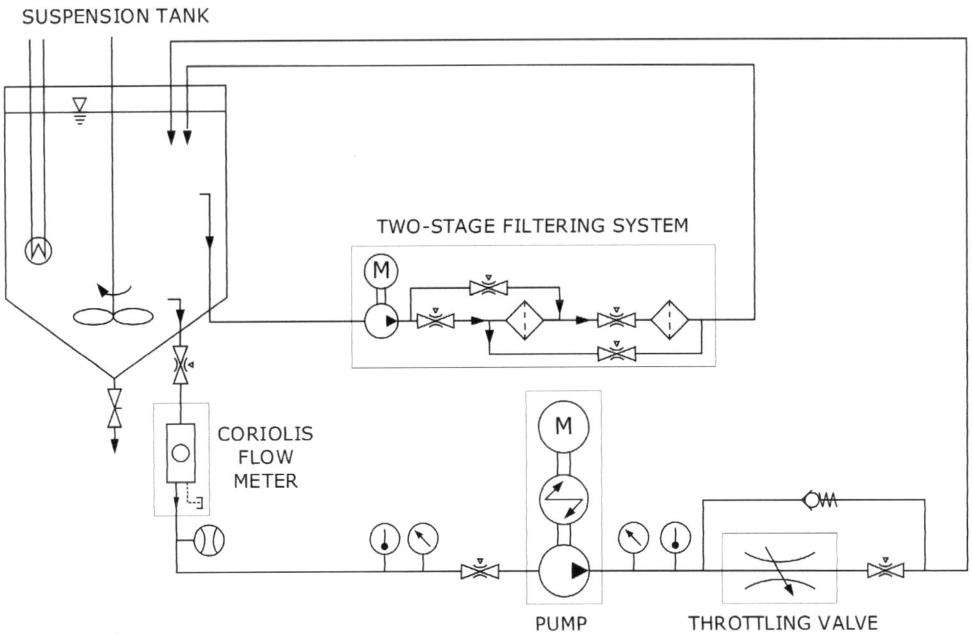

Figure 4. Concept schematic of the test rig for particle wear analysis in screw pumps.

The manual pinch valve that is used for the throttling does not offer dead-volume optimised flanges, therefore flanges according to DIN 1092 are used. DIN 1092 flanges are also used for the burst disc-type relief system. Since these components are located downstream of the pump, these dead-volumes may be accepted. In order to aid service and reconfiguration of the test rig, the pneumatic pinch valves are positioned according to the schematic in **Figure 4**. The fluid is finally reintroduced into the suspension tank, well below the fluid level in order to aid the mixing of the suspension while minimising the chance of forming gas bubbles that may lead to additional wear based on mechanisms not to be considered in this research project.

As per the requirements defined above, filtering in the main circuit at the total volumetric flow rate is not feasible. Therefore, a secondary, low pressure filtering circuit is introduced, which allows for the filtering rate to be set independent of the flow rate in the test section. The filtering circuit is operated by a secondary pump at a reduced flow rate. In order to reduce the mass of particles used by the test rig, a two-stage filtering system has been developed, see **Figure 4**, that allows the reintroduction of large particles into the tank, while fine particles are permanently removed from the circuit. A course filter removes the particle fraction with a size larger than a predefined threshold. These particles are expected to still contribute to the abrasive wear within the pumps, since the particles are similar in size to the gap heights. The prefiltered fluid will then be continuously filtered by a standard 10 µm filter. The prefilter is periodically backflushed back into the suspension tank. Thus, a continuous shift of the particle size distribution towards smaller particles can be countered. The mass of particles removed by the fine filter can be periodically determined and replaced with fresh particles. The particle size distribution in the tank can be expected to be shifted towards smaller particles in the upper parts of the suspension tank, therefore the filtering circuit extracts the suspension from the upper regions.

2.8. Wear test methodology

As has been shown in 2.2, the influence of the ageing particles on the wear behaviour in a closed-loop test setup is unknown. Therefore, the commissioning phase of the test rig is to be used for the analysis of this effect. After the test rig setup has shown, that it allows for wear-free operation when no particles are introduced into the fluid, the fluid will be laden with particles of a known particle size distribution. While operating the test rig without the use of the filtering section, the degradation of the particles may be analysed. The suspension will be sampled periodically in order to characterise the shift in the particle size distribution. Thereby, the requirements for the filtering section, i.e. the requirement to allow filtering of the total fluid volume once per day, can be tested for validity. The parameters of interest for the analysis of the wear to the particles are the size distribution as well as the sharpness of the particles, characterised by their sphericity [15]. With knowledge about these effects, the filtering rate may be adapted to the flow rate in the measuring section for the different particle types in order to allow for a consistent and comparable test environment across the range of parameters.

Once the test setup has been validated, the wear tests can be executed with high confidence in the repeatability of the test environment, without requiring continuous monitoring of the particle size distribution. Currently, no qualified estimation of the mean time to failure can be given, therefore the tests will at first be stopped at predefined intervals in order to check for wear. Once these preliminary tests have been performed, an estimation for the mean time to failure can be made and the main wear tests can be run to failure.

For the analysis of wear, a new set of spindles and liners will be installed for each test. Before the test, the properties, i.e. the mass and profile of the components will be documented. After the operation at constant parameters, within the aforementioned restrictions, the pump will be disassembled and the components are to be analysed. The most significant wear rates are expected in the meshing zones of the spindles; however, the pumps will also be analysed for further signs of wear. This is especially important in order to avoid mistakes when drawing conclusions from the results of the frictional torque measurement.

3. Outlook and applicability of the test results

The test rig setup presented in this research paper is currently being built up at the institute for fluid systems. By running the test rig without the filtering unit for preliminary tests, the severity of wear to the particles is to be characterised. Thus, the applicability of the restriction to the filtering task discussed above may be tested and considered for the final setup. With the finalised test rig, extensive parameter studies may be performed with the goal of developing a model for the prediction of the expected service life of the pumps in abrasive applications. Further, by analysis of the wear-affected zones and variations to the materials and geometric shape, guidelines for the design of screw pumps for abrasive applications may to be derived.

The research project is focused on direct transfer of the knowledge generated with the experimental work to the industry partner involved in the project. Since the operating conditions to be provided by the test setup are closely related to the real-world applications faced by the pumps in many industries, the findings are expected to give new insight into the wear mechanisms relevant for screw pumps. Thereby, first models for the prediction of the life expectancy of the pumps are expected to arise from the experimental works.

As of now, the manufacturers have two known approaches to handle abrasive applications, one being the rejection of such applications, the other being the recommendation of reduced pump speed in order to reduce the wear rates. The experiences gathered by purposely running the pumps in inhospitable conditions are expected to offer some new insight into the processes and might lead to additional instructions for the reduction of wear in critical applications. Furthermore, it is expected, that the boundaries of known uncritical applications might be shifted, thus allowing to operate pumps wear-free in slightly harsher conditions as is possible today. Thereby, efforts expended in the industrial application may be reduces.

References

[1] Karassik I J 2001 *Pump handbook* 3rd edn (New York: McGraw-Hill)
[2] Rabinowicz E and Mutis A 1965 Effect of abrasive particle size on wear *Wear* **8** 381–90
[3] 2023 *Leistritz Hyper-Twin* https://hygienic.leistritz.com/de-de/hygienic/products/image-thumb_2261_image-full/hyper-twin.webp (accessed 12 Jun 2024)
[4] 2023 *Leistritz Hyper3* https://hygienic.leistritz.com/de-de/hygienic/products/image-thumb_2260_image-full/hyper3.webp (accessed 12 Jun 2024)
[5] Crook A 1991 Contaminant Sensitivity of Hydraulic Pumps – Development of the ISO/DIS 9632 Test Method *SAE Transactions* **100** 50–3
[6] Schmidt S 1999 Verschleiß von Schraubenspindelpumpen beim Betrieb mit abrasiven Fluiden *Dissertation* Technische Fakultät, Universität Erlangen-Nürnberg
[7] Dorenbos C, Müeller-Link D and Jäeschke A 2001 Sand Handling During Multiphase Operations With Twin-Screw Pumps *All Days SPE International Thermal Operations and Heavy Oil Symposium (Porlamar, Margarita Island, Venezuela, 12.03.2001 - 14.03.2001)* (SPE)
[8] Milović L and Zrilić M (eds) 2009 *FAILURE ANALYSIS OF ROTARY SCREW PUMP STRUCTURAL INTEGRITY AND LIFE (Ankaran, Slovenia, October 23-24, 2009)* 9th edn
[9] Mucchi E, D'Elia G and Dalpiaz G 2012 Simulation of the running in process in external gear pumps and experimental verification *Meccanica* **47** 621–37

[10] Budde I, Ludwig G and Pelz P F 2016 *Pump Users International Forum 2016: 3rd International Rotating Equipment Conference : 14 and 15 September 2016, Congress Center Düsseldorf, Germany ; proceedings (Pumps + Systems. Compressors, Compressed Air and Vacuum Technology)* (Frankfurt am Main: Pumps + Systems Association within VDMA e.V)
[11] Vetter G, Wirth W, Korner H and Pregler S 2000 *Multiphase Pumping With Twin-Screw Pumps - Understand And Model Hydrodynamics And Hydroabrasive Wear*
[12] KELLER Druckmesstechnik AG 2024 *retina_preview_65152867f210c.jpg* (accessed 10 Jun 2024)
[13] Suchy Messtechnik 2023 *sd-47-drucksensor-mit-frontbuendiger-keramik-g-3-4-kl-0-5.jpg* (accessed 10 Jun 2024)
[14] Kraume M 2002 *Mischen und Rühren* (Wiley)
[15] Bahadur S and Badruddin R 1990 Erodent particle characterization and the effect of particle size and shape on erosion *Wear* **138** 189–208

Acknowledgement

We would like to thank Leistritz Pumpen GmbH for funding of and the substantial contributions to the project.

Yet another structured mesh generator for screw machines simulation

Y Ji and M Möller

Delft University of Technology, Delft Institute of Applied Mathematics, Delft, the Netherlands

E-mail: y.ji-1@tudelft.nl

Abstract. High-quality structured mesh generation is essential for the numerical simulation and design optimization of screw machines, particularly rotary twin-screw compressors used in high-pressure gas production. Existing meshing techniques often struggle to accurately capture their intricate geometries, limiting their application potential. This paper proposes a novel Isogeometric Analysis (IGA)-based mesh generation approach specifically tailored for twin-screw compressors. Our method leverages high-order B-spline parameterizations derived from boundary representations to enable efficient and precise mesh generation. A novel boundary correspondence technique using Schwarz-Christoffel mapping is introduced to further enhance mesh quality and preserve critical profile features. Additionally, we integrate elliptic grid generation techniques within the IGA framework, combined with a block-diagonal Jacobian-preconditioned Anderson acceleration algorithm, to efficiently solve the associated nonlinear systems. This approach, built upon the open source Geometry + Simulation Modules (G+Smo) library, demonstrates its effectiveness in generating high-quality structured meshes that meet stringent requirements for screw machine simulations, providing a viable alternative for design optimization.

1. Introduction

Rotary positive displacement machines (PDMs), such as twin-screw compressors, play a vital role in industrial applications by compressing air and gases to high pressures. These machines operate by harnessing the mechanical power of two counter-rotating rotors, which draw fluid through an inlet, trap it within chambers formed between the rotors, and reduce the chamber volume through rotation to compress the fluid. Given their extensive use, optimizing the efficiency of these machines is crucial for improving energy efficiency and reducing carbon emissions. Even minor improvements in performance can lead to significant global energy savings [1].

Numerical simulation is indispensable for the analysis and optimization of twin-screw compressors [1, 2]. Effective simulations and structural design optimization require an accurate and time-dependent geometric model of the fluid flow passages between the rotors and the casing. This requirement underscores the importance of generating a structured high-quality hexahedral computational mesh, which is challenging due to the inherent complex geometry of the compressors, narrow clearances, and variable gap sizes. Additionally, the need for precise geometry representation demands high mesh resolution, particularly around the clearances and gaps, presenting significant computational challenges.

Specialized tools such as SCORG™[3] and TwinMesh™[4] are commonly employed to address these challenges. These tools are specifically designed for rotary twin-screw compressors and generate sequences of computational meshes for fluid and solid domains at predetermined time intervals. These meshes are then processed using commercial simulation packages such as ANSYS CFX™[5]. However, despite their widespread use, these mesh generators typically produce only linear meshes with straight-sided cells, which may limit their ability to accurately capture complex geometries and result in less precise simulations. In response, tools like CFX meshing in ANSYS (from V2019 onwards) offer capabilities that support high-order curved boundaries, particularly around turbomachinery profiles.

Our research addresses this limitation by leveraging high-order B-spline parameterizations instead of linear meshes. Starting from the input point cloud boundary representations, we first convert them into high-order B-spline representations. The quality of the resulting parameterization is critically dependent on the matching of boundary parameters. To this end, we introduce a new boundary correspondence method that employs Schwarz-Christoffel mapping while preserving the exact geometry of the spline-represented boundary. With the matched boundary geometry, we then incorporate elliptic grid generation within the isogeometric analysis (IGA) framework. Additionally, we employ a block-diagonal Jacobian-preconditioned Anderson acceleration (AA) algorithm to expedite the solution of the involved nonlinear equations. This approach, built upon the open source Geometry + Simulation Modules (G+Smo) library [6], provides a viable alternative for design optimization in screw machine simulations.

2. Complete workflow of our mesh generator

In this section, we briefly introduce the workflow of our mesh generator. The input is a point cloud representing the geometry of one lobe of the main and gate rotor, and the output is a structured hexahedral mesh for the twin-screw machine. The primary workflow is illustrated in Figure 1.

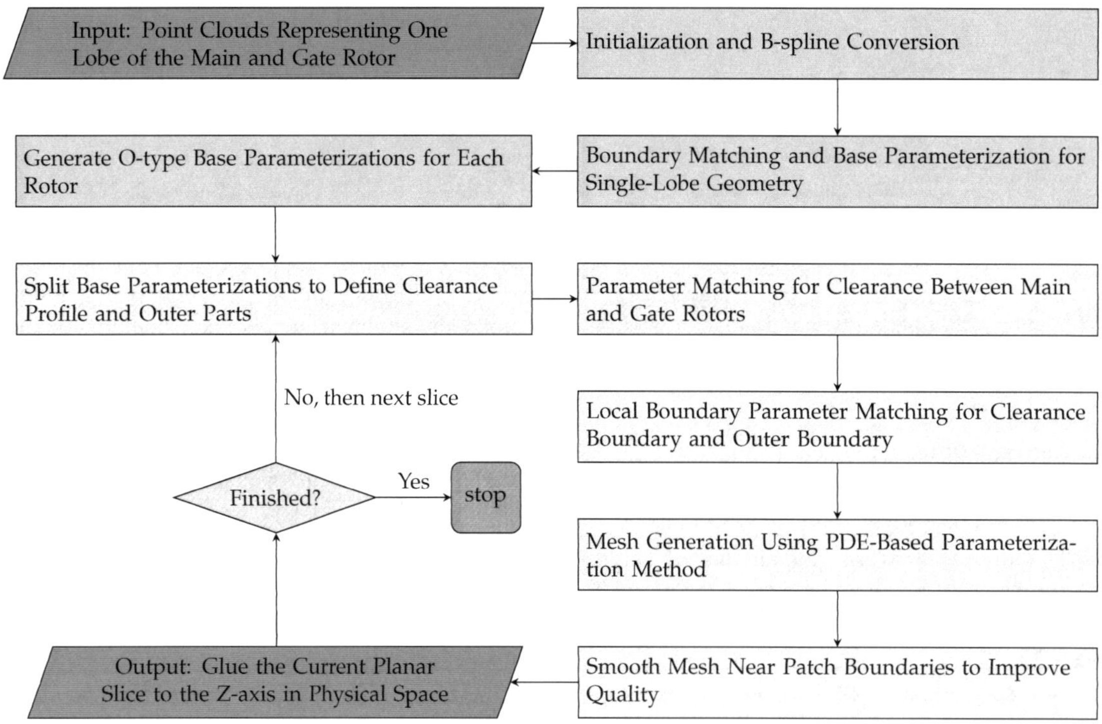

Figure 1. Primary workflow of our structured hexahedral mesh generator for twin-screw machines.

The detailed workflow is as follows:

Step 1 **Initialization and B-spline Conversion:**

- Generate the B-spline representation of the casing curve for each rotor. ◁ Section 3.1

- Convert the input point cloud into a B-spline representation using a constrained least squares B-spline fitting method. ◁ Section 3.2

Step 2 **Mesh Generation for Base Parameterization**

- Perform boundary parameter matching using Schwarz-Christoffel mapping to ensure high-quality mesh generation. ◁ Section 3.3
  Figure 2 shows the resulting mesh generated before (left) and after (right) employing the boundary parameter matching technique. It is evident that the quality of the mesh is greatly improved by using the boundary parameter matching technique.

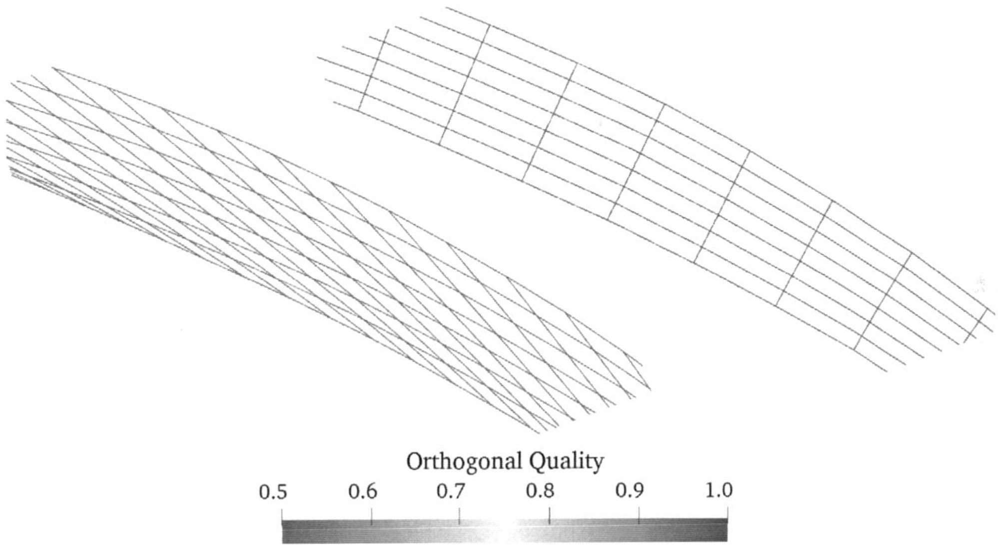

Figure 2. Effect of boundary parameter matching: mesh without boundary parameter matching (left) and mesh with boundary parameter matching (right).

- Employ the PDE-based elliptic parameterization method to generate a high-quality mesh using the matched boundary curves. ◁ Section 3.4
  Figure 3 illustrates the role of different parameterization methods on the geometry of a single lobe of the main rotor. The four subfigures present a comparison of linear parameterization without boundary correspondence (Figure 3(a)), linear parameterization with boundary correspondence (Figure 3(b)), PDE-based parameterization with boundary correspondence (Figure 3(c)), and enhanced PDE-based parameterization with boundary correspondence (Figure 3(d)). As shown in Figure 3(a), the linear parameterization without boundary correspondence leads to a mesh with poor orthogonality and misaligned boundaries, which can negatively impact the accuracy of the simulation results. The introduction of boundary correspondence, as seen in Figure 3(b), significantly improves the alignment of the mesh boundaries but still leaves room for improvement in overall mesh quality. The application of the PDE-based parameterization method with boundary correspondence (Figure 3(c)) further enhances the mesh by ensuring better control over the element distribution, resulting in improved orthogonality and element quality. Finally, the enhanced PDE-based parameterization (Figure 3(d)) provides the highest mesh quality, as indicated by the color map that represents the orthogonality of the mesh. Higher values in the color map correspond to better orthogonality, which is crucial for ensuring accurate and reliable simulation results.

IOP Conf. Series: Materials Science and Engineering **1322** (2024) 012014 doi:10.1088/1757-899X/1322/1/012014

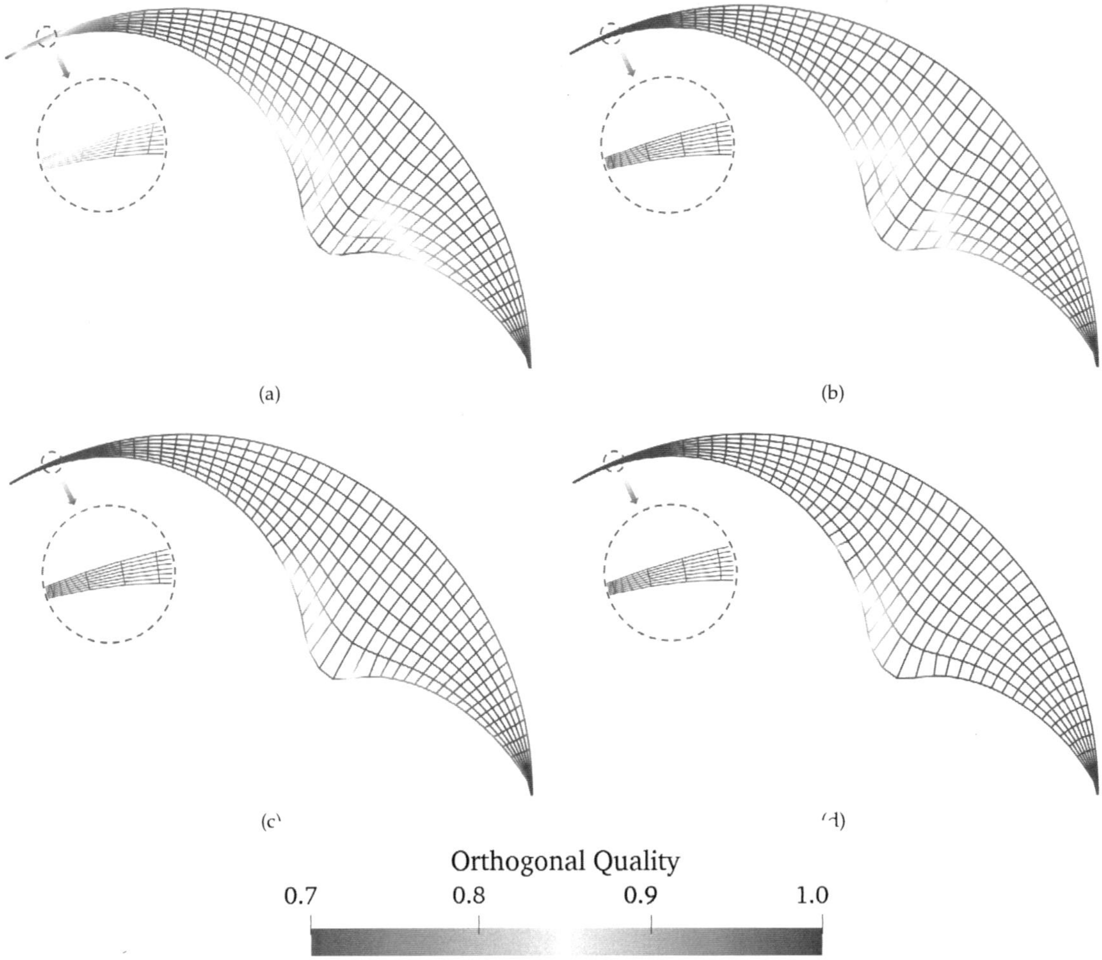

Figure 3: Comparison of different parameterization methods for a single lobe geometry of a main rotor: (a) linear parameterization without boundary correspondence, (b) linear parameterization with boundary correspondence, (c) PDE-based parameterization with boundary correspondence, and (d) enhanced PDE-based parameterization with boundary correspondence. The color map represents the orthogonality of the resulting mesh, with higher values indicating better orthogonality.

- Obtain the complete base parameterization for each rotor by rotating the parameterization of one lobe, as shown in Figure 4. The algorithm leverages the inherent symmetry of the screw compressor geometry to achieve this consistent alignment. Because the endpoints of each lobe align precisely with the corresponding casing boundaries, the meshing algorithm automatically ensures node matching between consecutive lobes, particularly at the lobe tips, where the radial clearance with the casing is critical. This alignment is vital for maintaining high mesh quality and ensuring accurate numerical simulations.

IOP Conf. Series: Materials Science and Engineering **1322** (2024) 012014 doi:10.1088/1757-899X/1322/1/012014

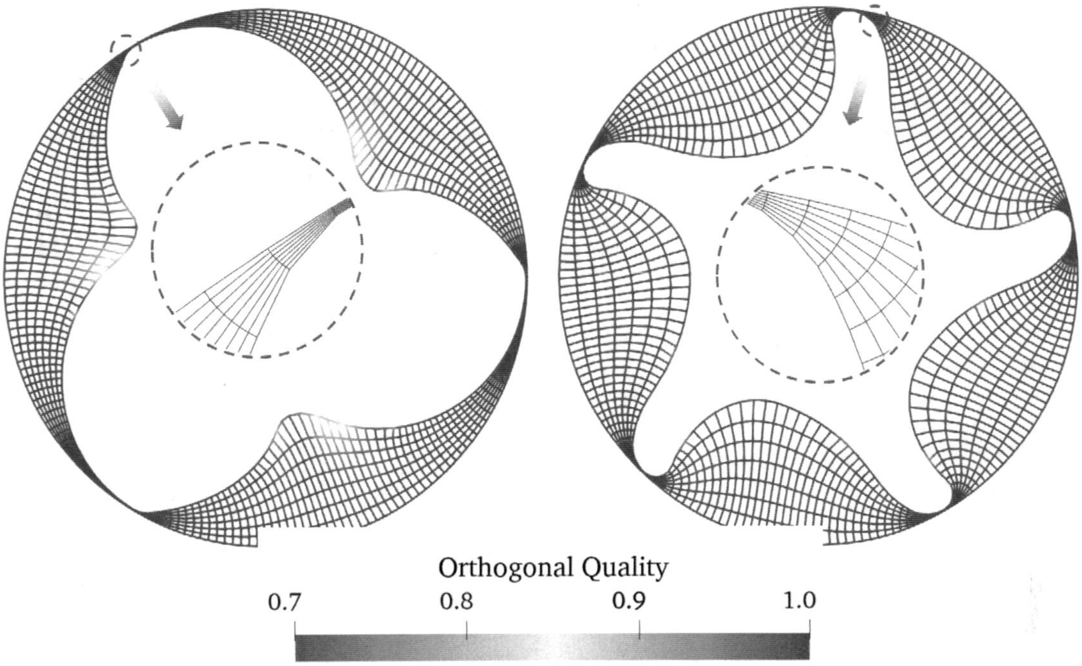

Figure 4: Base parameterization for the main rotor and gate rotor. The color map represents the orthogonality of the resulting mesh, with higher values indicating better orthogonality.

Step 3 **Time Step Rotation and Parameter Computation:**

- For each time step, rotate the base parameterization around the center of each rotor.
- Compute the parameter values of the top and bottom CUSP points using the B-spline point inversion algorithm.

Step 4 **Mesh generation:**

- Split the base parameterization at the parameter values of the top and bottom CUSP points to form the four boundaries of the clearance, as shown in Figure 5(a).

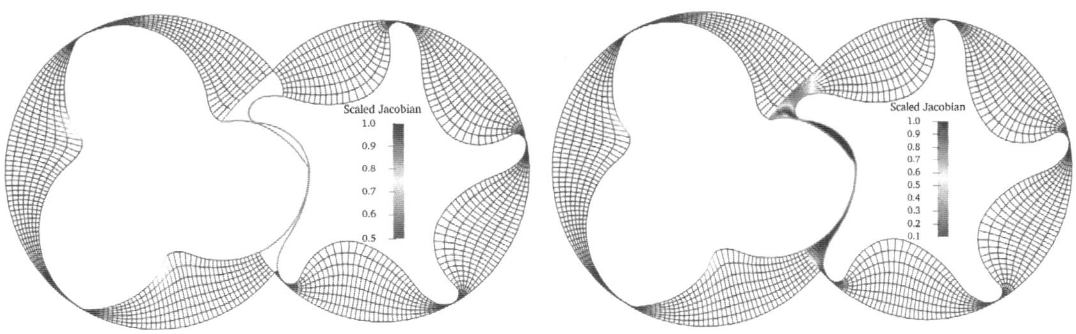

(a) Split at CUSP points (b) Resulting mesh without smoothing

Figure 5: Splitting of the base parameterization at CUSP points and the initial mesh without smoothing.

- Apply the boundary parameter matching technique to the current clearance and generate the mesh, as shown in Figure 5(b). ◁ Section 3.3 & Section 3.4

- Select a small segment of the clearance boundary and the outer boundary of the top and bottom clearance parts for local boundary parameter matching. ◁ Section 3.3
  As shown in Figure 5, the quality of the generated mesh is poor near the top CUSP region due to the limited space available for the clearance mesh. By employing local boundary parameter matching, the resulting mesh, illustrated in Figure 6, shows a significant improvement in quality. This enhancement highlights the effectiveness of local boundary parameter matching in addressing mesh quality issues in constrained regions.
- Generate the mesh for each rotor using the PDE-based parameterization method, as shown in Figure 6. ◁ Section 3.4

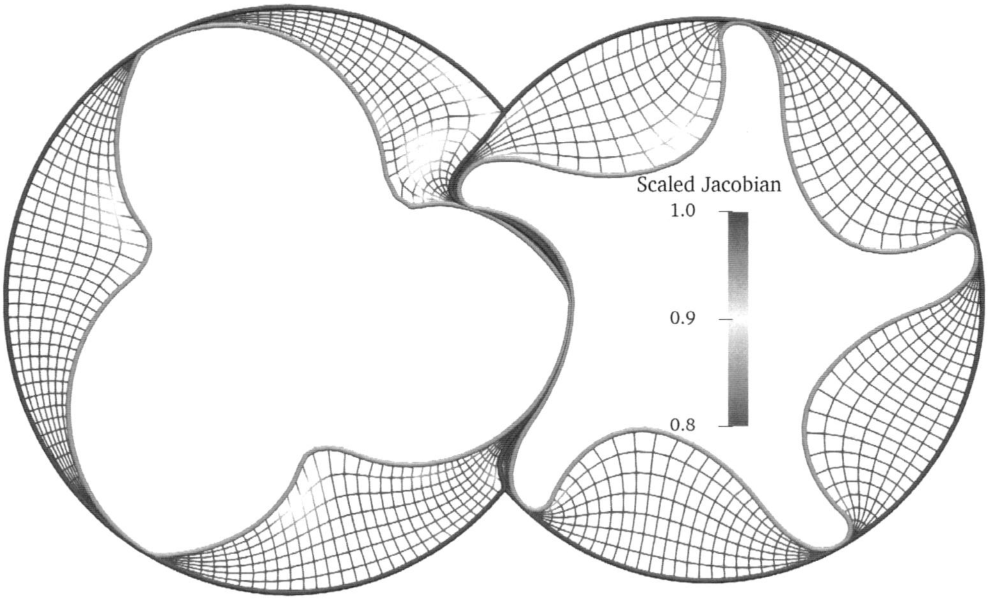

Figure 6. Resulting mesh for a representative slice *before* smoothness improvement.

- To enhance the smoothness of the clearance part, split the clearance at the CUSP points and apply the PDE-based parameterization method to the clearance again. ◁ Section 3.4
  As shown in Figure 6, the smoothness near the top CUSP point is insufficient. To address this issue, we applied additional smoothness improvement processing to improve the mesh quality. Figure 7 shows a representative slice of the resulting mesh after this step. The color encoding in the figure represents the Scaled Jacobian, a commonly used metric for assessing the mesh orthogonality. Higher values indicate better quality, with values close to 1 representing near-perfect elements, while lower values indicate elements that may be distorted or poorly shaped. Additionally, the presence of a negative value indicates a fold or an invalid element in the resulting mesh. Despite a localized drop near the CUSP, the positive and relatively high minimum value across the mesh confirms that the mesh remains valid and that the overall quality, particularly smoothness, is well maintained.

Step 5 **Output:**

- Write the grid points and connectivity data to files.
- If it is not the last slice, return to Step 3; otherwise, finish.

Using the aforementioned methods, the resulting structured hexahedral mesh is shown in Figure 8. This mesh demonstrates the effectiveness of our method in generating high-quality, structured hexahedral meshes capable of accurately capturing the intricate geometries characteristic of twin-screw compressors. The mesh maintains high levels of orthogonality and element quality throughout the domain, as indicated by the Scaled Jacobian metrics. Such attributes are crucial for ensuring the reliability and accuracy of subsequent simulations. In the next section, we will explain each substep in detail.

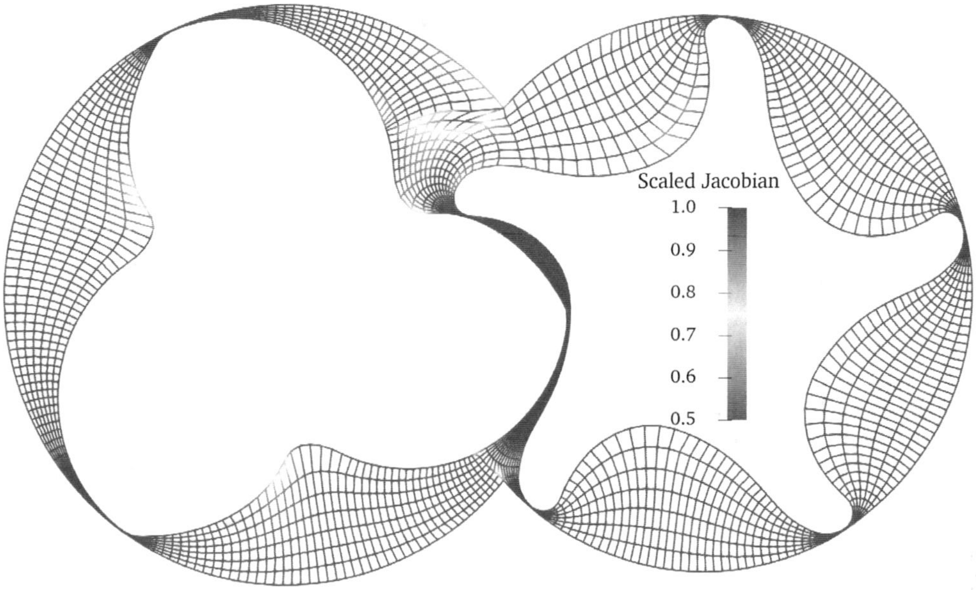

Figure 7. Resulting mesh for a representative slice *after* smoothness improvement.

Figure 8. Structured hexahedral mesh generated using the proposed methods.

3. Ingredients used in our mesh generator

This section provides a detailed introduction to the ingredients used in our mesh generator.

3.1. *Brief introduction to B-splines*

Given a non-decreasing sequence of real numbers

$$\Xi = \{\xi_0, \xi_1, \ldots, \xi_{n+p+1}\}, \quad \xi_{i+1} \geq \xi_i,$$

where p is the degree, then a set of degree-p B-spline basis functions over the knot vector Ξ can be established using the well-known Cox-de Boor recursion formula [7]

$$N_{i,p}(\xi) = \frac{\xi - \xi_i}{\xi_{i+p} - \xi_i} N_{i,p-1}(\xi) + \frac{\xi_{i+p+1} - \xi}{\xi_{i+p+1} - \xi_{i+1}} N_{i+1,p-1}(\xi), \tag{1}$$

starting from

$$N_{i,0}(\xi) = \begin{cases} 1, & \text{if } \xi \in [\xi_i, \xi_{i+1}), \\ 0, & \text{otherwise}, \end{cases}$$

where the ratios of the form $0/0$ are defined as zero.

Denote the spline function space spanned by these B-spline basis functions as

$$\mathbb{S}_p^{\Xi} = \left\{ \psi \in L^2([\xi_p, \xi_{n+1}]) : \psi|_{\xi=\xi_i} \in C^{k_i},\ \psi|_{\xi \in (\xi_i, \xi_{i+1})} \in \mathbb{P}^p \right\}, \tag{2}$$

where $\mathbb{P}^p$ represents the function space of degree-p polynomials, $k_i = p - m_i$ is the smoothness degree, and m_i is the multiplicity of the i-th knot. Then a degree-p B-spline curve is defined as a vector-valued function in the spline function space $\mathbb{S}_p^{\Xi}$

$$\mathcal{C}(\xi) = \sum_{i=0}^{n} \mathbf{P}_i N_{i,p}(\xi),\ \xi \in \hat{\Omega} = [0, 1], \tag{3}$$

where $\mathbf{P}_i$ is the i-th control point, and $\hat{\Omega}$ is referred to as the parameter domain. In other words, a B-spline curve is defined as a linear combination of B-spline basis functions with control points serving as coefficients. Figure 9(a) shows quadratic B-spline basis functions over an knot vector $\Xi = \{0, 0, 0, 0.2, 0.4, 0.6, 0.8, 0.8, 1, 1, 1\}$, while Figure 9(b) displays a B-spline curve defined over Ξ.

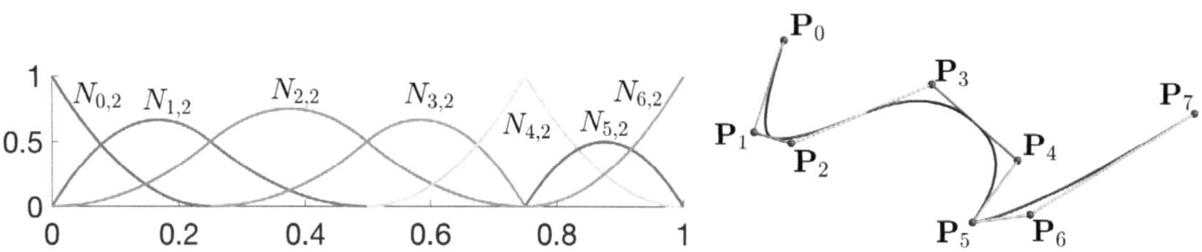

(a) B-spline basis functions (b) A B-spline curve

Figure 9. (a) Quadratic B-spline basis functions defined over the knot vector $\Xi = \{0, 0, 0, 0.2, 0.4, 0.6, 0.8, 0.8, 1, 1, 1\}$; (b) A B-spline curve constructed using these basis functions, with the control points marked as filled red circles.

The extension to bi-variate tensor-product B-spline basis functions is straightforward. Given two knot vectors Ξ and $\mathcal{H}$ in ξ- and η-direction, respectively, bi-variate B-spline basis functions are computed by the tensor product of univariate B-spline basis functions $\{N_{i_1}^p(\xi)\}_{i_1=0}^{n_1}$ and $\{N_{i_2}^q(\eta)\}_{i_2=0}^{n_2}$, i.e.,

$$N_{i_1,i_2}^{p,q}(\xi, \eta) = N_{i_1,p}(\xi) N_{i_2,q}(\eta),\ i_1 = 0, 1, \ldots, n_1,\ i_2 = 0, 1, \ldots, n_2. \tag{4}$$

Similar to B-spline curves, a tensor-product B-spline surface is defined by the linear combination of the above bi-variate basis functions (4) and control points:

$$\mathcal{S}(\xi, \eta) = \sum_{i_1=0}^{n_1} \sum_{i_2=0}^{n_2} \mathbf{P}_{i_1,i_2} N_{i_1,i_2}^{p,q}(\xi, \eta),\ (\xi, \eta) \in \hat{\Omega} = [0, 1]^2 \tag{5}$$

where $\mathbf{P}_{i_1,i_2}$ are the control points, and $\hat{\Omega}$ is the parameter domain.

3.2. *Constrained B-spline fitting*

In our workflow, we leverage the periodic nature and geometric features of the rotor profile. We begin with an input point cloud representing a single-lobe geometry of the rotor profile. The initial step is to convert this point cloud into a B-spline curve.

To achieve this, we first identify the point with the maximum distance from the center of the rotor. We then reorder the point cloud so that this point serves as the starting point. This reordering improves the boundary correspondence between the rotor profile points and its casing circle.

With the reordered points, we apply a constrained least-squares B-spline fitting algorithm using chord-length parameterization, as the two end points are crucial for geometry accuracy. Specifically, the control points $\{\mathbf{P}_i\}_{i=0}^n$ are determined by solving the following linear system:

$$\left[\begin{array}{c|c}\mathbf{A}^{\mathrm{T}}\mathbf{A} & \mathbf{I}_{2\times 2}\\ \hline \mathbf{I}_{2\times 2} & \mathbf{O}\end{array}\right]\left[\begin{array}{c}\mathbf{P}_{\mathrm{f}}\\ \hline \mathbf{P}_{\mathrm{c}}\end{array}\right] = \left[\begin{array}{c}\mathbf{A}^{\mathrm{T}}\mathbf{r}_{\mathrm{f}}\\ \hline \mathbf{r}_{\mathrm{c}}\end{array}\right], \tag{6}$$

where $\mathbf{P}_{\mathrm{f}} = [\mathbf{P}_1, \mathbf{P}_2, \ldots, \mathbf{P}_{n-1}]^{\mathrm{T}}$ and $\mathbf{P}_{\mathrm{c}} = [\mathbf{P}_0, \mathbf{P}_n]^{\mathrm{T}}$ denote the free and the constrained first and last control points, $\mathbf{r}_{\mathrm{f}} = [\mathbf{r}_1, \mathbf{r}_2, \ldots, \mathbf{r}_{m-1}]^{\mathrm{T}}$ and $\mathbf{r}_{\mathrm{c}} = [\mathbf{r}_0, \mathbf{r}_m]^{\mathrm{T}}$ denote the corresponding free and the constrained points on the rotor or the casing circle, $\mathbf{I}_{2\times 2}$ is the identity matrix, and

$$\mathbf{A} = \begin{bmatrix} N_1(t_1) & N_2(t_1) & \cdots & N_n(t_1)\\ N_1(t_2) & N_2(t_2) & \cdots & N_n(t_2)\\ \cdots & \cdots & \ddots & \cdots\\ N_1(t_{m-1}) & N_2(t_{m-1}) & \cdots & N_n(t_{m-1}) \end{bmatrix}$$

is the configuration matrix of $m-1$ rows and $n-1$ columns. The coefficient matrix in the linear system (6) is sparse and symmetric, making it efficient to solve using a standard Cholesky decomposition solver.

3.3. *Boundary parameter matching using Schwarz-Christoffel map*

With the two long boundary B-spline curves defined, our next step is to generate a spline-based parameterization from these curves. The parameter speed of these boundary curves significantly affects the quality of the resulting mesh. As shown in Figure 3(a), mismatched boundary parameters result in poor-quality mesh. To address this issue, we employ the Schwarz-Christoffel (SC) map [8, 9], a specialized conformal mapping technique extensively used in computational geometry and complex analysis.

Consider an open and simply connected polygon $\mathcal{P}$ on the complex plane $\mathbb{C}$, as shown in Figure 10. The vertices of $\mathcal{P}$, labeled $\omega_1, \omega_2, \ldots, \omega_n$, are arranged in a counterclockwise direction, with $\omega_{n+1} = \omega_1$ and $\omega_0 = \omega_n$ for indexing purposes. The interior angles at these vertices are denoted by $\alpha_1, \alpha_2, \ldots, \alpha_n$. For each vertex j, we define $\beta_j = \alpha_j/\pi - 1$, which lies in the interval $(-1, 1]$. According to the interior angle sum theorem for polygons, the sum of the interior angles is $(n-2)\pi$, thus $\sum_{k=1}^{n} \alpha_k = (n-2)\pi$.

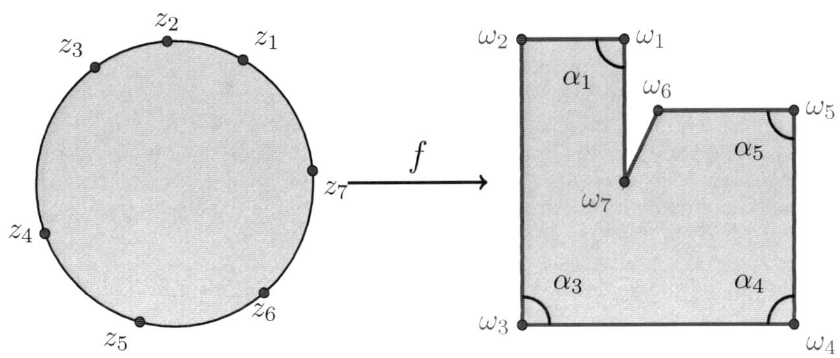

Figure 10. Notational conventions for the Schwarz–Christoffel mapping.

Given these conditions, a conformal mapping f from the unit disk $\mathcal{D}$ to the polygon $\mathcal{P}$, as illustrated in Figure 10, can be described by the Schwarz-Christoffel formula [8]:

$$f(z) = f(z_0) + C \int_{z_0}^{z} \prod_{j=1}^{n} \left(1 - \frac{\zeta}{z_j}\right)^{\beta_j} d\zeta, \tag{7}$$

where $f(z_0)$ and C are complex constants with $C \neq 0$. The points $z_1, z_2, \ldots, z_n$ are prevertices located on the boundary of $\mathcal{D}$ in counterclockwise order, such that $\omega_k = f(z_k)$ for each $k = 1, 2, \ldots, n$. The well-known Riemann mapping theorem ensures the existence of an analytic function f with a nonzero derivative that maps the open unit disk $\mathcal{D}$ onto $\mathcal{P}$, ensuring $f(\mathcal{D}) = \mathcal{P}$ and providing a bijective correspondence over the domain $\mathcal{D}$.

In the calculation of the SC formula (7), the number of vertices in the polygon domain $\mathcal{P}$ significantly affects the computational cost. To enhance computational efficiency, we employ a polygon simplification subroutine to reduce the number of points required to represent a polyline. Specifically, we utilize the Douglas-Peucker algorithm [10], which implements a point-to-edge distance tolerance strategy. In our implementation, we set the tolerance to half the gap between the rotor and its casing. The simplified result is shown in Figure 11. Initially, we densely sampled the rotor profile and its casing with 1002 points to accurately capture the geometries. The Douglas-Peucker algorithm then simplifies the polygon, reducing the number of points to 46, while maintaining the essential shape of the profile.

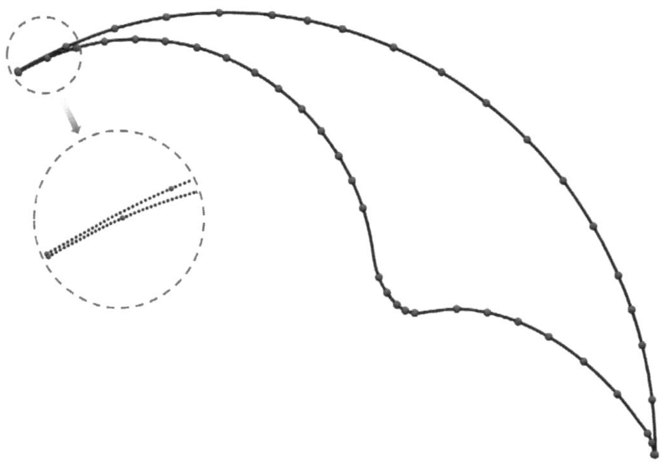

Figure 11: Polygon simplification using the Douglas-Peucker algorithm. The original dense sampling (1002 points) is reduced to 46 points, maintaining the key features of the geometry.

Figure 12 illustrates the markers used for boundary parameter matching using the SC map. Using these markers, we first compute the parameter values on the rotor profile using the B-spline point inversion algorithm. These parameter values are then assigned to the corresponding markers on the casing profile. Subsequently, we reparameterize the casing curve using the B-spline reparameterization technique [9]. Figure 3 presents the resulting meshes obtained from different parameterization methods for a single lobe geometry of a main rotor. Figures 3(a) and 3(b) show the meshes generated by simple linear interpolation along the two long boundaries. It is clear that the boundary parameter matching step significantly enhances the mesh quality, as evidenced by Figure 3(b) compared to Figure 3(a). Specifically, with good parameter matching, even a simple linear interpolation can produce a high-quality mesh.

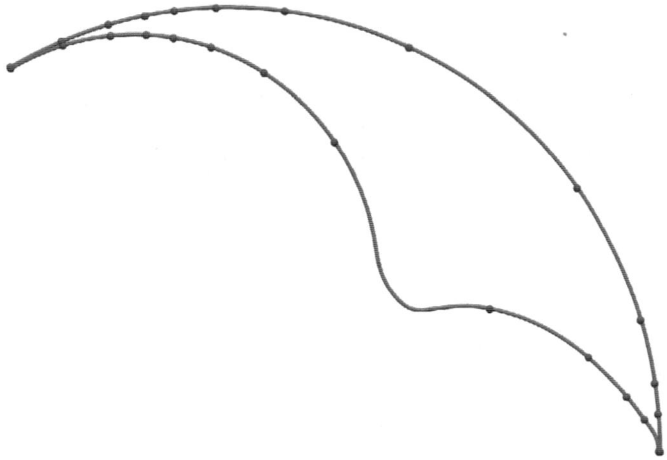

Figure 12: Boundary parameter matching using the SC map.

3.4. *PDE-based parameterization method*

Although various analysis-suitable parameterization methods have been developed over the past decade within the IGA community, many existing methods struggle to address the parameterization construction problem for twin-screw geometries with extreme aspect ratios. To the best of our knowledge, the only effective solution for this challenging geometry is the PDE-based parameterization method [11, 12]. Therefore, we adopt this approach to generate an analysis-suitable parameterization for the complex twin-screw geometry.

The cornerstone of the PDE-based parameterization method involves solving the following nonlinear elliptic PDE system:

$$\begin{cases} \nabla \cdot \mathbb{A}\nabla\xi(x) &= 0, \\ \nabla \cdot \mathbb{A}\nabla\eta(x) &= 0, \end{cases} \quad \text{s.t. } \boldsymbol{x}^{-1}|_{\partial\Omega} = \partial\hat{\Omega}, \tag{8}$$

where $\mathbb{A}$ denotes the tensor field. One typical choice for $\mathbb{A}$ is the identity matrix $\mathbf{I}_{2\times 2}$, as adopted in [11]. Alternatively, it can be set to $\mathtt{DIAG}(1/|\mathcal{J}|, 1/|\mathcal{J}|)$, as specified in [12]. In this paper, we refer to the former as the standard PDE-based parameterization method and the latter as the enhanced PDE-based parameterization method.

We denote by $\mathbb{S}_{p,q}^{\boldsymbol{\Xi},\boldsymbol{\mathcal{H}}}$ the spline space spanned by bivariate B-spline basis functions of degrees p and q, with knot vectors $\boldsymbol{\Xi}$ and $\boldsymbol{\mathcal{H}}$, respectively. The subspace $(\mathbb{S}_{p,q}^{\boldsymbol{\Xi},\boldsymbol{\mathcal{H}}})^0 = \{N_i \in \mathbb{S}_{p,q}^{\boldsymbol{\Xi},\boldsymbol{\mathcal{H}}} : N_i|_{\partial\hat{\Omega}} = 0\}$ consists of basis functions that vanish on $\partial\Omega$. The variational principle reduces the non-linear PDE system (8) to an equivalent discretization scheme:

$$\forall N_i \in (\mathbb{S}_{p,q}^{\boldsymbol{\Xi},\boldsymbol{\mathcal{H}}})^0 : \begin{cases} \int_\Omega \nabla\mathbf{N} \cdot \mathbb{A}\nabla\xi(x)\,\mathrm{d}\Omega &= 0, \\ \int_\Omega \nabla\mathbf{N} \cdot \mathbb{A}\nabla\eta(x)\,\mathrm{d}\Omega &= 0, \end{cases} \quad \text{s.t. } \boldsymbol{x}|_{\partial\hat{\Omega}} = \partial\Omega, \tag{9}$$

where $\mathbf{N}$ denotes the collection of B-spline basis functions in the spline space $(\mathbb{S}_{p,q}^{\boldsymbol{\xi},\boldsymbol{\mathcal{H}}})^0$. In this paper, we employ a preconditioned Anderson acceleration algorithm using a block-diagonal Jacobian as the preconditioner for solving the nonlinear system (9). Figures 3(c) and 3(d) display the meshes produced by more advanced PDE-based parameterization methods. The enhanced parameterization, shown in Figure 3(d), demonstrates a significant improvement in mesh quality. This advancement highlights the effectiveness of the proposed approach in refining the parameterization process.

By incorporating the aforementioned ingredients into the complete workflow described in Section 2, we complete the entire process of our mesh generator. The resulting structured hexahedral mesh is shown in Figure 8.

4. Conclusions and outlook

In this paper, we have presented a novel spline-based mesh generation approach specifically tailored for twin-screw compressors. The high-order B-spline parameterizations derived from boundary representations enable efficient and precise mesh generation, addressing the limitations of existing meshing techniques in capturing the intricate geometries of these machines. Our method incorporates a novel boundary correspondence technique using Schwarz-Christoffel mapping, which significantly enhances the resulting mesh quality and preserves critical profile features. Additionally, by employing IGA-based elliptic grid generation techniques and a block-diagonal Jacobian-preconditioned Anderson acceleration algorithm, we achieve efficient solutions to nonlinear systems. This combination ensures that our approach can generate high-quality structured meshes that meet the stringent requirements for twin-screw compressors.

There are some limitations in our mesh generator. First, the boundary parameter matching step, while effective and robust, is relatively time-consuming. Finding an alternative efficient solution for boundary parameter matching would be beneficial. Second, there are several global parameterization computation steps in our workflow, which may reduce the computational efficiency of our method. Additionally, the presented method does not sufficiently utilize information from previous slices when generating the mesh for the current slice. Developing more efficient mesh generation methods using local smoothing techniques would be helpful. Overall, future work will focus on further refining the boundary parameter matching step and exploring more efficient local smoothing techniques to enhance the overall computational performance of the mesh generation process.

References

[1] Kovacevic A, Stosic N and Smith I 2007 *Screw compressors: three dimensional computational fluid dynamics and solid fluid interaction* Vol. 46 Springer Science & Business Media

[2] Stosic N, Smith I and Kovacevic A 2005 *Screw compressors: mathematical modelling and performance calculation* Springer Science & Business Media

[3] SCORG 2023 *Screw compressor rotor grid generator* `http://pdmanalysis.co.uk/scorg`

[4] TwinMesh 2023 *Reliable CFD analysis of rotary positive displacement machines* `https://www.twinmesh.com`

[5] ANSYS 2023 *ANSYS CFX: Turbomachinery CFD Software* `https://www.ansys.com/products/fluids/ansys-cfx`

[6] Mantzaflaris A 2019 *An overview of geometry plus simulation modules* International Conference on Mathematical Aspects of Computer and Information Sciences Springer pp. 453–456

[7] Piegl L and Tiller W 1996 *The NURBS book* Springer Science & Business Media

[8] Driscoll T A and Trefethen L N 2002 *Schwarz-Christoffel mapping* Vol. 8 Cambridge university press

[9] Ji Y et al 2024 *Boundary parameter matching for isogeometric analysis using Schwarz–Christoffel mapping* Engineering with Computers 1–19

[10] Douglas D H and Peucker T K 1973 *Algorithms for the reduction of the number of points required to represent a digitized line or its caricature* Cartographica: the international journal for geographic information and geovisualization 10 (2) 112–122

[11] Hinz J, Möller M and Vuik C 2018 *Elliptic grid generation techniques in the framework of isogeometric analysis applications* Computer Aided Geometric Design 65 48–75

[12] Ji Y et al 2023 *On an improved PDE-based elliptic parameterization method for isogeometric analysis using preconditioned Anderson acceleration* Computer Aided Geometric Design 102 102191

IOP Conf. Series: Materials Science and Engineering **1322** (2024) 012015 doi:10.1088/1757-899X/1322/1/012015

Simulation analysis of the internal flow field in single screw compressor using local re-meshing method

W Wu*, P Zhang, Q Wang, L Huang, Y Sun, T Feng

School of Energy and Power Engineering, Xi'an Jiaotong University, Xi'an, Shaanxi, PR China

E-mail: weifengwu@xjtu.edu.cn

Abstract. The single screw compressor (SSC) is considered an excellent competitor to the twin screw compressor due to its symmetrical structure and balanced gas forces on the screw rotor. To enhance the performance of SSC, it is crucial to have a thorough understanding of the internal flow field, particularly in the leakage channels. Nevertheless, the vertical relationship between the star wheel axis and the screw axis leads to mesh distortion, which presents technical challenges for mesh division in CFD simulations. In this paper, the method of local re-meshing is employed to simulate the comprehensive flow field using the Forte software. The mesh is regenerated near the motion boundary, while the mesh in the remaining areas remains unchanged. The pressure and temperature distribution of the internal flow field of the compressor are obtained, as well as its variation during operation. The results indicate a noticeable pressure gradient change in the leakage gap. The pressure and temperature of gas gradually increase along the axis direction of the screw, reaching the maximum value in the exhaust channel. The research can provide support for the future optimization design of the SSC.

1. Introduction

The single screw compressor was invented by French engineer Zimmern in 1962[1]. As shown in Figure 1, its structural feature is that a screw drives a symmetrically arranged star wheel to rotate, and the working chamber formed by groove, groove flank, and the cylinder surface changes periodically. The SSC exhibits superior mechanical balance performance in comparison to the twin screw compressor. This characteristic enables it to effectively accommodate higher pressure ratios and rotating speeds[2], making it particularly suitable for oil-free air compression, two-phase expansion, and vapour compression applications. Investigating the SSC's internal flow field is an important way to clearly state the loss during the working process and enhance its efficiency.

Theoretical numerical analysis of its working process could reveal much more detailed information than experiments, which mainly refer to the p-V diagram of the SSC.

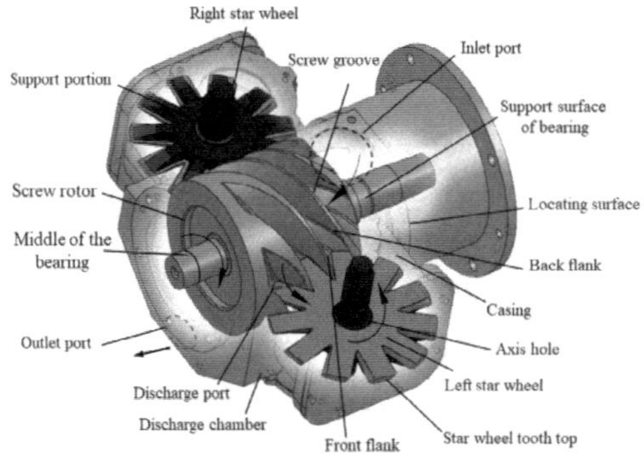

Figure 1. Structure of the SSC[2]

In recent years, Computational Fluid Dynamics (CFD) technology has garnered significant attention in the realm of volumetric compressors. Twinmesh stands out with its robust meshing capabilities tailored for positive displacement machines, including screw machines[3-5], scroll machines, and gear pumps. Presently, CFD methodologies have attained maturity in addressing scroll, screw, Roots, vane compressors, and expanders, offering sophisticated software solutions[6, 7]. However, scant literature exists regarding the flow simulation of single screw compressor machines, with only a sparse collection of articles focusing on single-screw expanders. This scarcity can be attributed to the unique meshing configuration of single screw compressors, wherein the screw and star wheel engage in vertical meshing, inducing complex gas flow patterns influenced by rotational stirring effects. Attempts to employ mesh deformation methods encounter challenges such as excessive stretching and twisting, leading to negative volume issues that impede solution continuity. Moreover, the evolving complexity of meshing profiles, transitioning from straight line envelope to multi-column envelope profile and surface envelope profile, further compounds mesh generation difficulties, particularly concerning intricate structures on the tooth flank. The quality of the grid remains a crucial determinant in three-dimensional simulations of single screw compressors, imposing constraints on their broader application.

Several researchers have endeavoured to explain the gas flow dynamics within single screw compressors through various methodologies. The operational sequence of a single screw compressor unfolds as a continuous process: following the completion of suction, gas volume diminishes gradually within the meshing interface of the screw and star wheel until it connects with the exhaust port, subsequently entering the exhaust piping. To capture the entirety of this compression process, an analytical approach focusing on a specific working chamber post-suction has been adopted. Ning delineated a computational domain formed by the groove, groove flank and the cylinder surface[8]. The screw rotor and star wheel execute rotary motion at a consistent speed ratio, governing gas compression via boundary motion settings. Moreover, recent work by Suman et al. analysed the feasibility of diverse computational methodologies in simulating the single screw flow field. Among the approaches investigated, the Fluent dynamic mesh technique was explored, wherein screw and star wheel motion induced fluid mesh deformation, subsequently addressed through mesh smoothing and local reconstruction[9]. Another method involved STAR-ccm+'s overlapping grid technique, which entailed generating two mesh regions occupying the same space, with the part mesh dynamically coupled to the static background mesh, enabling adaptation to various motion forms. Additionally, the Pumplinx mesh morphing method deformed the mesh via boundary motion.

In other studies, Casari et al. analysed the problems that occurred in the result calculated in OpenFOAM[10, 11]. The result is the grid was partially missing. Additionally, the team's simulations of a single-screw expander using the overlapping grid method showed large pressure gradients, presumably due to gas entering the working chamber at high pressure, spreading across the screw outer diameter, and potentially causing local pressure gradient issues[12, 13]. Scholars from Beijing University of Technology endeavored to simulate the flow field of a single-screw expander, revealing notable pressure fluctuations[14, 15]. However, the pressure changes in the middle position groove and the exhaust side position groove are not obvious, and the lowest pressure occurs in the position of the outer diameter of the screw that flows from the first groove to the second groove, which is not described in detail in the thesis, and it is presumed that the low quality of the mesh at this position in the calculation has caused the local pressure gradient to be larger. In another paper, it was found that there were vacancies on the surface of the star wheel in the results, and there may have been a missing mesh problem during the solution process, which would have caused an error in the data transfer at that location and reflected in the results as a void on the surface of the cloud diagram. Existing studies have shown that some methods can be used to simulate the flow field in a single screw expander, but some problems need to be solved and it needs to be verified whether they can be applied to the simulation solution of a single screw compressor.

In this study, local re-meshing methods were employed in conjunction with Forte software under the ANSYS platform to simulate the flow field of a SSC. Forte software offers distinct advantages, notably its capacity to automatically monitor surface component gaps and augment mesh density at these regions, thereby enhancing mesh quality. Additionally, using empirical gap models facilitates the optimization of mesh quality, particularly in addressing flow issues at gap regions. Consequently, this approach holds promise for simulating internal flow fields within SSC.

2. Numerical calculation model

The machine considered in this study is a 37 kW air compressor. The single screw compressor (SSC) discussed is an oil-injected type with an elliptical surface envelope, which demonstrates significant potential for applications in high-temperature gas compression fields, such as high-temperature heat pumps and water vapour compression.

For the simulation, the internal flow domain of the SSC was simplified to some extent. The calculation area excludes the screw shaft, star wheel support, and bearings to streamline the numerical simulation. The closed spiral was not included in this study, and its surface was approximated as a cylindrical surface. Other surfaces were modelled to accurately reflect the actual compressor, with the star wheel surface rotating around its axis and the screw rotating around its own axis.

Lubricating oil, which plays a crucial role in lubrication and cooling within the compressor, was not considered in this simulation. This omission may impact the simulation results to some degree.

2.1 geometric model

The 3D geometric model and physical object of the compressor are shown in Figure 2, computational domain is shown in Figure 3. Air enters the suction end of the single screw compressor (SSC) through the air inlet and the star wheel cavity, then is drawn into the unsealed screw groove. The star wheel meshes with the screw rotor to compress and pressurize the air after suction. Ultimately, the high-pressure gas exits the working chamber through the upper and lower triangular holes. The post-exhaust process, including high-pressure gas mixing and exhaust, is not considered in this model. Nonetheless, this computational model is capable of simulating the entire thermodynamic working process of the SSC.

The medium used in the simulation is modelled as an ideal gas, primarily composed of nitrogen and oxygen.

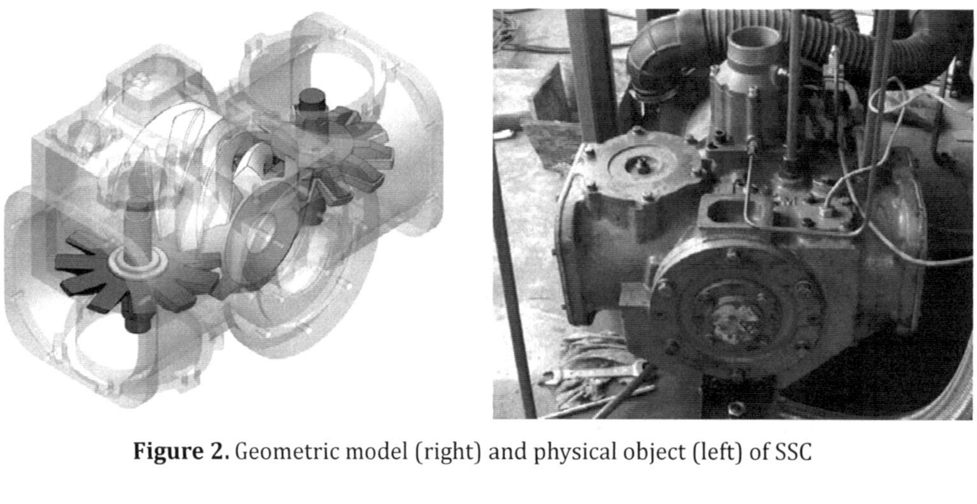

Figure 2. Geometric model (right) and physical object (left) of SSC

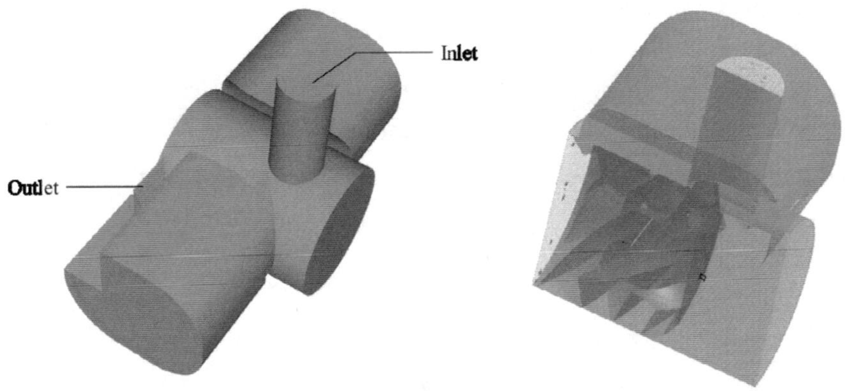

Figure 3. Computational domain of the SSC

Table 1. Primary structural parameters of the SSC

Property	Symbols	Value
Screw lobe	Z_1	6
Star wheel teeth	Z_2	11
Screw diameter	D_1	182mm
Star wheel diameter	D_2	194mm
Center distance	A	145.5mm
Star wheel tooth width	b	28mm
Inlet pressure	P_{in}	100kPa
Outlet pressure	P_{out}	700kPa
Rotation speed	ω	2960r/min

The specific design dimensions are shown in Table 1. In addition, all gaps have been modelled in SpaceClaim software, and the specific parameters are shown in Table 2.

Table 2. Important clearances

Clearance	Real	Model
clearance between the teeth tip of star wheel and the screw groove	0.04-0.06mm	0.1mm
clearance between the flank of star wheel teeth and the screw groove	0.02-0.04mm	0.05mm
clearance between the surface of star wheel teeth and the shell	0.05mm	0.05mm
clearance between the surface of screw rotor and the shell	0.05-0.06mm	0.05mm

2.2 dynamic mesh

The local re-meshing method was employed to reconstruct the grid in regions near the moving walls while maintaining the original grid in other areas. Numerical calculations were performed using Forte software, which has demonstrated accuracy and reliability in twin-screw simulations. This approach allows for the effective meshing of complex areas.

ANSYS Forte software uses 3D transient compressible flow simulations for CFD analysis. The software automatically generates the grid, eliminating the need for pre-defined grids. It is suitable for calculating various types of compressors, including reciprocating pistons, screw, and scroll compressors. During simulations, Forte can automatically detect small gaps between surfaces and perform mesh refinement as needed. Additionally, it utilizes an empirical gap model to improve the resolution of meshes in these gap regions. Flow within the gaps is calculated using the Poiseuille flow model based on pressure differences. During setup, it is necessary to configure global size and refinement settings. For gap meshes, the gap size should be defined in the modelling software according to actual conditions and then imported into Forte as patches for meshing.

The overall grid size is 5mm, while the gap density grid size is 0.625mm, 1/8 of the overall grid size. The divided mesh is shown in Figure 4 below at one transient time.

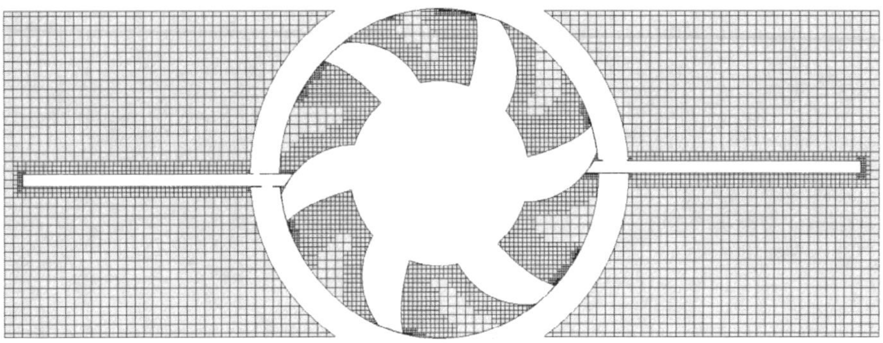

Figure 4. Mesh of computational domain in the SSC

The single crew compressor is a rotating machine characterized by complex internal flow, high speed, and high flow Reynolds numbers. To achieve more accurate calculations, the RNG k-ε model was employed. This model adjusts the turbulent viscosity, accounts for rotational motion, and enhances the accuracy of solutions for strong swirling flows. It also addresses the viscosity of low Reynolds number flows, thereby improving overall solution precision. To accurately describe the flow near the wall, multi-layer grids in the boundary layer would be ideal; however, this approach demands substantial memory and computational resources. Therefore, this study uses standard wall functions, which are well-suited for resolving flow conditions with high Reynolds numbers.

3. Result and discussion

The compressor performance is determined by the interaction between the fluid flow conditions and the geometry of the SSC. The transient simulation provides the possibility of a detailed analysis of the internal flow of the SSC.

To extract flow field data such as pressure changing with time in the working chamber, several monitoring points were created in the flow field. The location of the monitoring points is shown in Figure 5. The -20 point was located at the entrance of the primitive volume, and the detected data is the pressure at the moment when the screw groove is closed. The number -20 means that the point is 20mm away from the vertical plane where the centre line of the star wheel is located. The 0 point is located on the vertical plane where the centre line of the star wheel is, monitoring the pressure during the compression process; the 60 point is located at the position of the exhaust orifice, monitoring data at the end of the compression process and the exhaust process. The right side of Figure 5 is the pressure data at point 0. Due to the structural characteristics and working principle of SSC, the physical parameters in the working process such as pressure, velocity, density, torque and flow rate and so on are periodic. The characteristics can be a criterion to judge whether the simulation is converged or not. Therefore, iterative can be considered converged when these parameters present relatively stable periodicity.

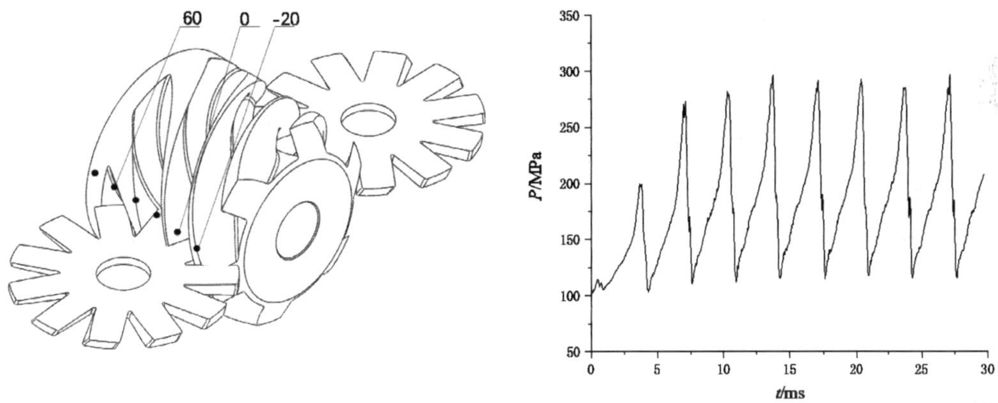

Figure 5. Monitor point location(left) and pressure variation of monitor point with time(right)

3.1 Unsteady properties of major fluid regions

The moment when the star wheel teeth mesh into the screw groove is defined as the beginning of the fluid work in this screw groove, and the moment when the star wheel teeth mesh out of the screw groove is defined as the end of the fluid work. Figure 6 shows the pressure contour plots on the screw rotor around the meshing pair of the SSC, where the pressure distribution in each screw groove at different rotation angles can be observed. When the star wheel rotation angle is -29.1°, the screw groove is just closed, and the gas in the screw groove begins to be compressed. The corresponding working chamber at this time is represented by the letter S in the figure, surrounded by white lines. At this time, due to the influence of the pressure difference, gas from the high pressure working chamber leaks into the low pressure working chamber.

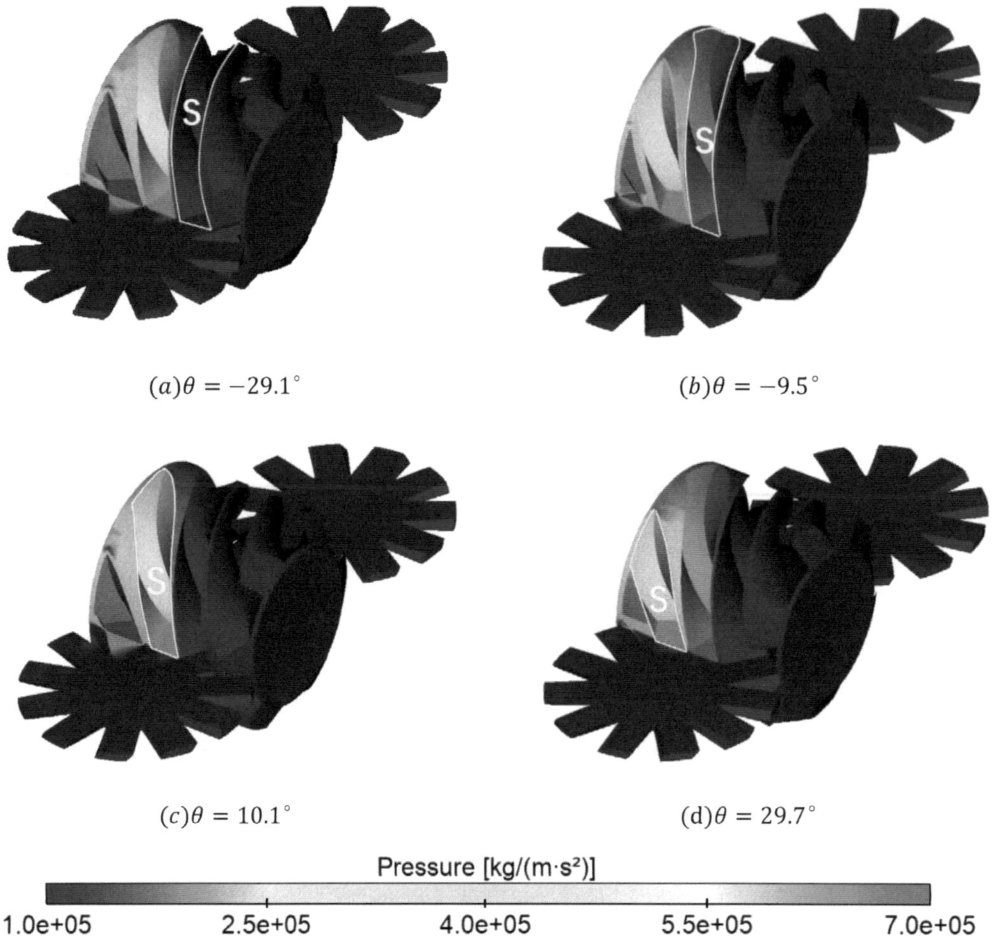

Figure 6. Pressure contour plots of the rotors' pressure field in one cycle

As the star wheel's rotation angle increases, the screw pushes the gas toward the exhaust end, causing a decrease in the screw groove volume. The screw and star wheel work together to compress the gas, converting mechanical energy into pressure energy. When the star wheel rotation angle reaches 27.8°, the screw groove connects to the exhaust pipe, resulting in a low pressure within the screw groove. This causes some of the gas in the exhaust pipe to flow back into the screw groove. As illustrated in Figure (d), when the working chamber first connects to the exhaust pipe, the pressure in the working chamber has not yet reached the exhaust pressure and remains significantly lower. This discrepancy is attributed both to practical mechanical factors and to substantial gas leakage observed in the simulation. Figure 6 illustrates the pressure magnitude and distribution in the gap between adjacent working chambers. The pressure difference between these chambers is the primary driving force behind leakage.

Monitoring points are established in the three-dimensional model, with overlapping detection ranges. To capture the complete working process of the chamber during compression, data from these monitoring points are collected, synchronized based on temporal continuity, and combined to represent the overall compression process. Figure 7 illustrates the pressure changes in the SSC's compression chamber, from suction closure through compression to discharge, and compares these results with experimental data. It can be seen from Figure 7 that the pressure in the screw channel gradually increases. Once compression begins, the pressure in the chamber rises steadily for a period. Exhaust starts when the star wheel rotation angle reaches 27.8°.

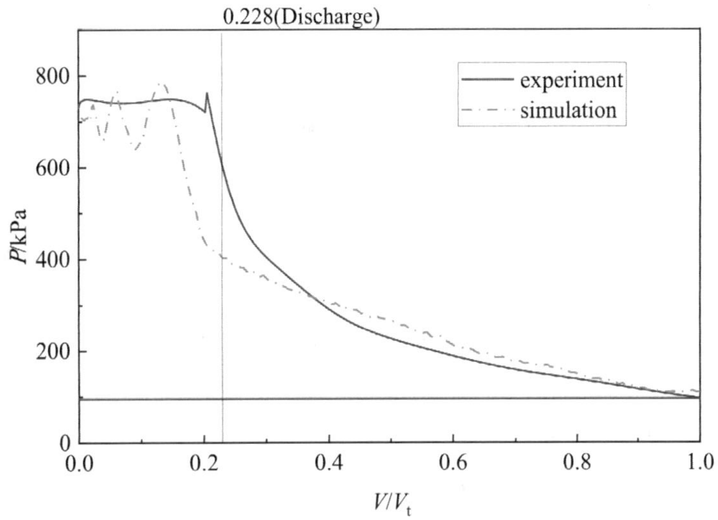

Figure 7. Diagram of pressure changes with a specific volume of the working chamber

There is a significant discrepancy between the pressure curves obtained from numerical simulations and experimental results. This difference arises from the sealing effect of lubricating oil. In experiments, lubricating oil is sprayed into the working chamber to seal it, resulting in minimal leakage. In contrast, the numerical simulations assume greater leakage.

During the early stage of compression, the pressure difference across the leakage channel is small, leading to minimal leakage. Because the leakage amount from the high-pressure side is smaller than the amount flowing out to the low-pressure side, the pressure in the cavity remains higher. In the later stages of compression, the length of the high-pressure side leakage path decreases, resulting in more gas escaping from the working chamber and a lower pressure. During the exhaust stage, when the working chamber connects to the exhaust pipe, high-pressure gas can flow back into the chamber, causing a sharp increase in pressure. Additionally, intermittent suction and exhaust lead to airflow pulsations in the exhaust pipe, causing pressure to fluctuate around the set 700 kPa and gradually diminish.

During the experiment, lubricating oil was sprayed into the compressor. The presence of lubricating oil during the exhaust process creates some resistance. When exhaust begins, the gas cannot flow into the exhaust chamber quickly, leading to increased resistance and higher losses during the exhaust process. Consequently, the final compression pressure is higher. The pressure at the exhaust end is influenced by both the pressure in the gas tank and the pressure in the exhaust pipe.

3.2 leakage flow

The SSC have no sealing elements and use gap seals. Due to the complex shape of the meshing pair, there are many leakage channels. Leakage is the main factor affecting the efficiency, input power, gas transmission volume and other performance of SSC. The velocity field in the leakage channel will be analysed below.

Figure 8 and Figure 9 show the flow at the gap between the star wheel teeth and the screw. The gap size at the tooth top is relatively stable, but the length of the gap at the tooth flank varies widely. The plane where the velocity vector diagram is located is 2mm above the horizontal plane.

It can be seen from Figure 9 that the high-pressure side is below the star wheel teeth on the right side. Affected by the pressure difference, the air flows from high-pressure to low-pressure in the leakage channel. It can be seen from the figure that the gas flow velocity in the leakage channel on the tooth top side is larger, and the gas flow velocity on the tooth flank is smaller. This is because the tooth side is an elliptical envelope shape and the gap is smaller; while the tooth tip is a parallel plate type leak channel.

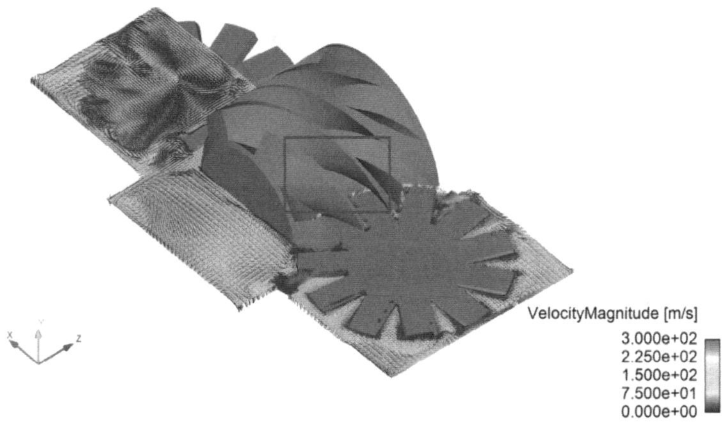

Figure 8. Horizontal plane speed vector illustration

The pressure difference between different screw grooves and the pressure gradient between the leakage channels are shown in Figure 10. Affected by the rotation of the screw and star wheel, the gas in the screw groove has a moving speed similar to that of the side of the screw groove. It can be seen in Figure 11. When the primitive volume is connected to the exhaust chamber and begins to exhaust, it has a horizontal leftward velocity component, and this is the main velocity direction. Compared with the tooth top, the flow rate of the leakage channel between the screw and the casing is smaller, but the leakage channel is wider, and its leakage occupies a dominant position.

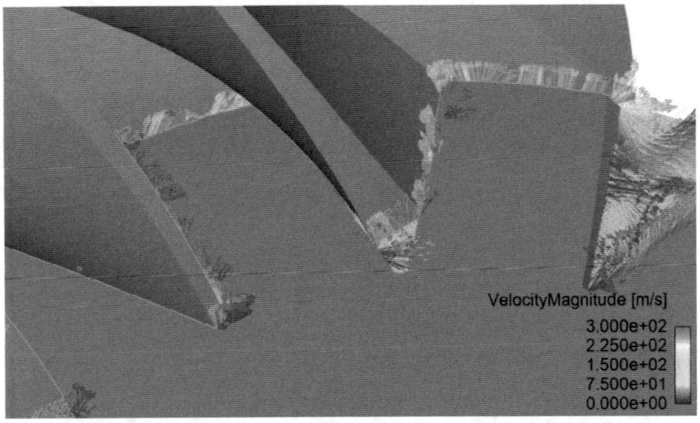

Figure 9. Velocity vectors of the gap between the star wheel and screw rotor

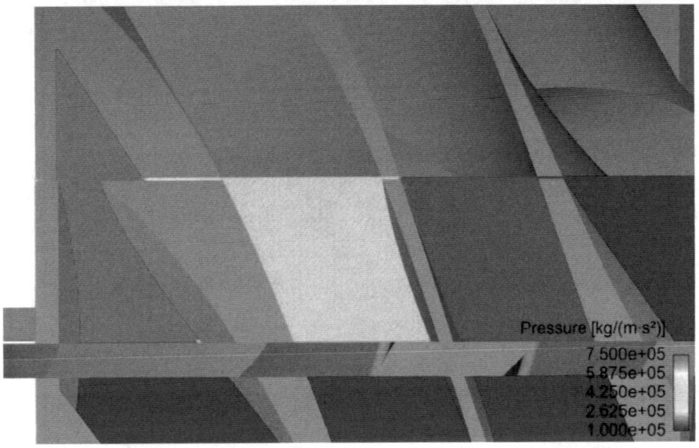

Figure 10. Diagram of vertical plane pressure distribution

Figure 11. Velocity vector distribution diagram on a vertical plane

3.3 Gas flow state in the working chamber

The Figure 12. (a) shows the pressure distribution in the working chambers, including the pressure distribution at the exhaust orifice. It can be seen from the figure(a) that as the screw rotates, the pressure in the working chamber is also increasing, and there is a leakage channel between the screw and the case. The pressure gradually decreases from the high-pressure side working chamber to the low-pressure side working chamber in the gap, and the gas passes through this channel, leaking from the high-pressure side to the low-pressure side. Figure (c) and (d) show the velocity distribution in the screw channel. Pushed by the star wheel and screw, the gas has a velocity component from the low-pressure end of the screw to the high-pressure end and has the same angular velocity as the screw rotation direction at the same time. Due to the stirring effect of the star wheel screw, the specific gas velocity varies.

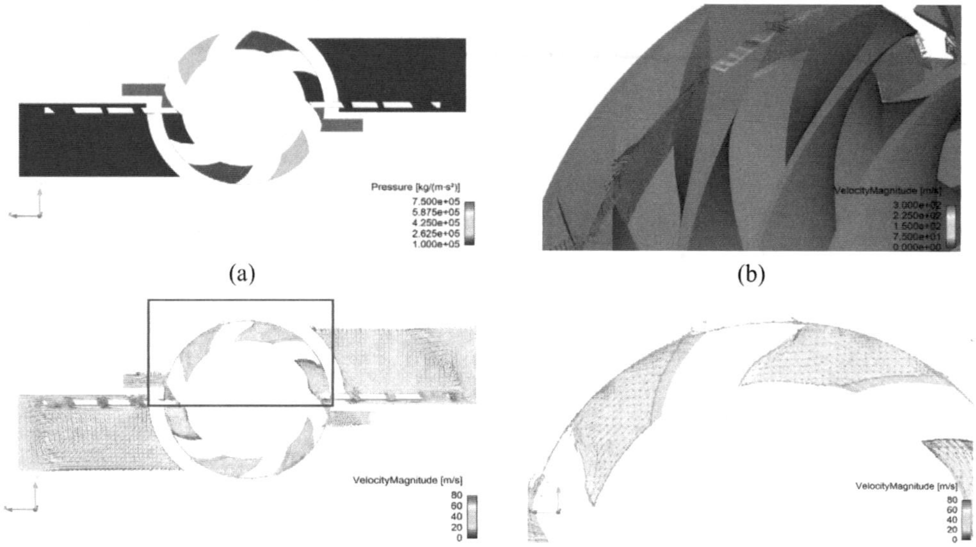

(a) (b)

(c) (d)

Figure 12. Diagram of velocity and pressure distribution at the vertical plane

Figure (b) shows the components of the three-dimensional velocity vector on the vertical plane. It can be seen that because the screw rotates counter clockwise, the gas is pushed by the side of the screw groove and also has a counter clockwise velocity component. The closer to the exhaust orifice, the smaller the velocity component of the center of mass of the working chamber in the screw axial direction, and the closer its speed is to the screw rotation speed, which is about 20m/s, and the distribution is more uniform; the unsealed working chamber is affected by the air intake, its velocity distribution is uneven.

For the gap between the screw and the casing, driven by the pressure difference between the adjacent working chambers, gas leaks from the exhaust end to the intake end. The gas flow velocity in the gap has a smaller velocity component in the screw circumferential direction and a larger velocity component in the axial direction, with a velocity of about 75m/s.

4. Conclusion

This paper primarily discusses the internal fluid simulation and calculation process of the single screw compressor (SSC), with a particular focus on its internal flow field. The flow field was analyzed using the local re-meshing method with Forte software, which allowed for identifying gaps and performing mesh refinement. An empirical gap model was employed to address the issues associated with the poor resolution of the mesh in these gaps. The flow within the gap was solved using the Poiseuille flow model, which depends on the pressure difference between the inlet and outlet of the leakage channel.

The results of the calculations provided insights into the pressure changes and velocity field during the working process and their distribution from the inlet to the outlet. The flow field analysis detailed the gas flow from the intake pipe to the star wheel chambers and working chambers. As the gas moves through the working chamber, it is progressively compressed due to the rotation of the screw and star wheel, leading to a decrease in volume and an increase in pressure. This validated the effectiveness of the method for calculating the internal flow field of the single screw compressor. Generally, the simulation results were consistent with experimental findings. The discrepancies between the simulation and experimental results are likely due to the exclusion of lubricating oil in the simulation. Future work should focus on simulating the working process of the SSC using a three-dimensional approach that incorporates the effects of oil.

References

[1] Bernard Z 1962 *Worm rotary compressors with liquid joints*

[2] Zhang Z and Wu W 2021 *Numerical investigation of thermal deformation of meshing pairs in single screw compressor* Applied Thermal Engineering 188 p. 116614

[3] Papes I, Degroote J and Vierendeels J 2015 *New insights in twin screw expander performance for small scale ORC systems from 3D CFD analysis* Applied Thermal Engineering 91 p. 535-546

[4] Xueming H et al 2015 *A twin-screw rotor profile design method based on computational fluid dynamics* Advanced Multimedia and Ubiquitous Engineering: Future Information Technology Springer

[5] Gu H et al 2021 *Analysis, modeling and simulations of an innovative sliding vane rotary compressor with a rotating cylinder* Energy conversion and management 230 p. 113822

[6] Montenegro G et al 2014 *CFD simulation of a sliding vane expander operating inside a small scale ORC for low temperature waste heat recovery* SAE Technical Paper

[7] Spille-Kohoff A Hesse J and El Shorbagy 2015 A *CFD simulation of a screw compressor including leakage flows and rotor heating* IOP Conference Series: Materials Science and Engineering IOP Publishing

[8] Ning X 2012 *Key technology research and CFD-based simulation of the single screw compressor*

[9] Suman A et al 2016 *Different Numerical Approaches for the Analysis of a Single Screw Expander* Energy Procedia 101 p. 750-757

[10] Casari N et al 2020 *CFD Simulations of Single-and Twin-Screw Machines with OpenFOAM* Designs 4(1) p. 2

[11] Casari N et al 2017 *Computational Models for the Analysis of positive displacement machines: Real Gas and Dynamic Mesh* Energy Procedia 129 p. 411-418

[12] Randi S et al 2018 *Numerical analysis of oil injection effects in a single screw expander* IOP Conference Series: Materials Science and Engineering IOP Publishing

[13] Casari N et al 2020 *Investigation of flow characteristics in a single screw expander: A numerical approach* Energy 213 p. 118730

[14] Shen L et al 2021 *Numerical Study On The Internal Irreversible Flow Loss of Single Screw Expanders*

[15] Shen L et al 2020 *Dynamic mesh technology for the simulation study of single screw expander* Computational and Experimental Simulations in Engineering: Proceedings of ICCES2019 24 Springer

Acknowledgements

This work was partly supported by the National Natural Science Foundation of China (No. 52076162), Youth Innovation Team of Shaanxi Universities, and Scientists and Engineers Team Construction of Qin Chuangyuan Platform (No. 2024QCY-KXJ-004).

A Comparative Study of Structured and Cut-Cell Grids Applied to an Oil-Injected Screw Compressor

M S Barrubeeah[1], A Saravana[1], S Bhaduri[1], D Low[2], E A Groll[1] and D Ziviani[1*]

[1] Ray W. Herrick Laboratories, School of Mechanical Engineering, Purdue University, West Lafayette, USA.
[2] Hitachi Global Air Power (HGAP), Michigan City, USA.

*E-mail: dziviani@purdue.edu

Abstract. Discretized Partial Differential Equations (PDEs) for numerical modeling of fluid mechanics systems require the use of various iterative linear solvers. In this study, two different CFD solvers, Ansys CFX and Converge, were employed to develop a transient 3D CFD model of an oil-injected screw compressor. Each solver utilized distinct meshing techniques—Ansys with a structured grid and Converge with a cut-cell grid. The numerical models have been validated utilizing experimental measurements. Presence of turbulent boundary layer and shear flow topology in the rotating fluid domains required implementing k-ω Shear Stress Transport (SST) turbulence model along with the physics needed to capture multifluid interfaces. In CONVERGE CFD, Semi Implicit Pressure Linked Equation (SIMPLE) algorithm was used for initial model development and pressure-velocity coupling in a collocated grid model, eliminating checkerboard numerical oscillations with the Rhie-Chow interpolation scheme. Ansys CFX handles pressure-velocity interactions through its solver framework, which operates without requiring a separately defined pressure-velocity coupling algorithm. These model parameters and pressure velocity coupling algorithms are kept unchanged while comparing preconditioned Successive Over Relaxation (SOR) and Incomplete Lower and Upper triangular (ILU) decomposition. These two solvers yielded to different numerical instabilities and rate of convergence, affecting the simulation clock time.

1. Introduction

Computational Fluid Dynamics (CFD) modeling plays a critical role in understanding and optimizing the performance of screw compressors, which are widely used in various industrial applications. Discretized partial differential equations (PDEs) for numerical modelling of fluid mechanics processes involve several different iterative solvers that rely on spatial and temporal discretization methods. Developing accurate and efficient CFD models for the internal flow of screw compressors is essential for improving their performance and investigate novel design features. The presence of lubricant oil complicates the modelling approach as multi-phase modelling approaches need to be included. In recent studies, various CFD solvers and methods have been employed to simulate and analyze the complex fluid dynamics within screw compressors. For instance, *Yang et al. [1]* utilized ANSYS CFX with an inhomogeneous Euler-Euler multiphase flow model and the $k-\omega$ Shear Stress Transport (SST) turbulence model for their analysis of an oil-injected twin-screw refrigeration compressor, successfully capturing detailed pressure pulsations and their attenuation throughout the

IOP Conf. Series: Materials Science and Engineering **1322** (2024) 012016 doi:10.1088/1757-899X/1322/1/012016

compressor. Similarly, *Andres et al.* [2] used ANSYS CFX for simulations of a two-stage twin screw compressor, demonstrating good agreement between CFD predictions and experimental data for volumetric flow rates and power consumption. *Ding et al.* [3] adopted the Simerics–MP+ CFD package with a Mixed Timescale Coupling method to model conjugate heat transfer in a twin-screw compressor, accurately predicting temperature distribution and thermal expansion of compressor components. Moreover, Rane et al. [4] employed a finite volume method-based solver with key-frame re-meshing for CFD analysis, providing detailed insights into pressure, temperature, and flow rates in screw compressors. However, their study faced challenges with mesh complexity, particularly for twin screw compressors. Recent advances include the work of *Wang et al. [5]*, who employed an Arbitrary Lagrangian-Eulerian (ALE) method using the finite volume method (FVM) to capture pressure fluctuations and trends with different speeds and wrap angles in twin-screw compressors. Another study by *Wang et al.* [6] demonstrated the use of the SCORG solver with Key-Frame re-meshing for CFD analysis, highlighting its effectiveness in handling complex geometries despite challenges in time consumption and accuracy. While existing investigations effectively addressed numerous aspects of fluid dynamics within screw compressors, they encountered notable limitations. Specifically, challenges related to mesh complexity and discretization adversely impacted both the accuracy and computational time. Additionally, the comprehensive assessment of iterative linear solvers, such as Incomplete Lower-Upper (ILU) and Successive Over Relaxation (SOR), was not fully realized, leaving unresolved issues pertaining to numerical stability and convergence rates.

In this study, a transient 3D CFD model of an oil-injected air screw compressor has been developed utilizing two different numerical approaches: a structured grid and a cut-cell grid. The simulation setup and main boundary conditions are illustrated in Fig. 1. This screw compressor design featured a single oil injection port.

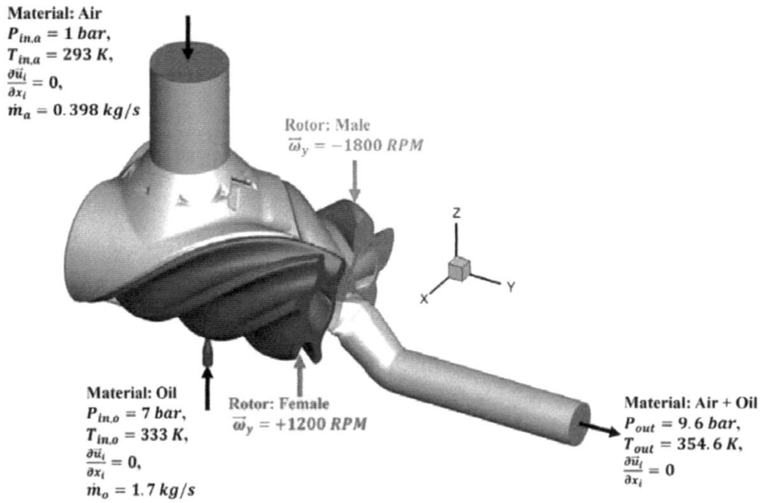

Figure 1. Description of imposed boundary conditions

The presence of turbulent boundary layers and shear flow topology in the rotating fluid domains required the implementation of the $k - \omega$ SST turbulence model. The Semi-Implicit Pressure Linked Equation (SIMPLE) method was utilized for pressure-velocity coupling in a collocated grid model, eliminating checkerboard numerical oscillations with the Rhie-Chow interpolation scheme. These model parameters and pressure-velocity coupling algorithms were kept unchanged while comparing preconditioned SOR and incomplete lower and upper triangular (ILU) decomposition. These solvers yielded different numerical instabilities and rates of convergence, affecting the simulation clock time. By addressing the limitations and leveraging the strengths identified in these prior studies, this research aims to further refine and validate CFD models for screw compressors, thereby enhancing their design optimizations and operational efficiency.

2. Numerical modelling

Unsteady three-dimensional compressible Navier Stokes equation is utilized for simulating the flow inside the screw compressor. The governing equations implemented in the numerical code are discussed as follows, represented in Einstein's index notation,

Continuity equation:

$$\frac{\partial \rho_m}{\partial t} + \frac{\partial u_i}{\partial x_i} = 0 \tag{1}$$

where, ρ_m is mixture density, u_i is mass averaged velocity vector.

Momentum equation:

$$\frac{\partial}{\partial t}(\rho_m u_i) + u_j \frac{\partial}{\partial x_j}(\rho_m u_i) = -\frac{\partial p}{\partial x_i} + \mu \frac{\partial^2 u_i}{x_{i,j,k}^2} \tag{2}$$

where, p is pressure and μ is dynamic viscosity of fluid.

Species transport equation:

$$\frac{\partial}{\partial t}(\rho_m Y_k) + u_j \frac{\partial}{\partial x_j}(\rho_m Y_k u_i) = -\frac{\partial}{\partial x_i}(\rho_m D_k \frac{\partial Y_k}{\partial x_i}) \tag{3}$$

where, Y_k is mass fraction of species k existing in a single phase and D_k is the coefficient of diffusion of the same species. Diffusion is a molecular phenomenon which only depends on molecular structure and size for a single species at a constant pressure and temperature condition [7], which allows implementing single species diffusion model.

Energy equation:

$$\frac{\partial}{\partial t}(\rho_m h_m) + \frac{\partial}{\partial x_i}(\rho_m h_m u_i) = \frac{\partial}{\partial x_j}\left((\mu)\frac{\partial T}{\partial x_j}\right) + \rho_m g_j u_i h_{fg} \tag{4}$$

where, h_m is mixture averaged sensible enthalpy, μ_t is eddy viscosity, T is temperature, g_j is component of gravitational acceleration along the velocity vector u_i, and h_{fg} is latent heat at pressure p and temperature T.

Void fraction transport equation:

$$\frac{\partial \alpha}{\partial t} + \frac{\partial}{\partial x_i}(u_i \alpha) = 0 \tag{5}$$

where, α is the void fraction, which is the ratio of volume of gas phase to the total volume of a finite volume cell present in the discretized domain. The code calculates mixture density as, $\rho_m = \alpha \rho_g + (1-\alpha)\rho_l$. The subscripts g and l represents gas and liquid phases respectively. However, the void fraction calculated from equation (5) is not transported to adjacent cells without solving the species transport equation (3) and equation (6) as follows,

$$\alpha = \frac{Y_g}{\rho_g}\left(\frac{Y_g}{\rho_g} + \frac{1-Y_g}{\rho_l}\right)^{-1} \tag{6}$$

The transient turbulence statistics of this problem is modelled utilizing Reynolds Averaged Navier-Stokes (RANS) approach. Turbulence is a chaotic rotational flow, in which RANS represents the velocity and scalar parameters as a summation of mean and fluctuating quantities.

$$u_i(x,t) = \bar{u}_i(x,t) + u_i'(x,t) \tag{7a}$$

$$\phi_i(x,t) = \bar{\phi}_i(x,t) + \phi_i'(x,t) \tag{7b}$$

where, ϕ is any scalar parameter, over bar denotes ensembled averaged quantity and prime denotes fluctuating quantity. This representation of turbulent quantities develops an issue with closure of mean momentum (equation 8a), mean scalar transport (equation 8b) and pressure Poisson (equation 9) equations by introducing extra covariance terms, which are known as Reynold's stress term $(\overline{u_i'u_j'})$ and scalar flux vector $(\overline{u_i'\phi_i'})$.

$$\frac{\partial}{\partial t}(\rho\bar{u}_i) + \bar{u}_j\frac{\partial}{\partial x_i}(\rho\bar{u}_i) = -\frac{\partial \bar{p}_i}{\partial x_i} + \mu\frac{\partial^2 \bar{u}_i}{\partial x_{i,j,k}^2} - \frac{\partial}{\partial x_i}\left(\rho\overline{u_i'u_j'}\right) \tag{8a}$$

$$\frac{\partial \bar{\phi}_i}{\partial t} + \frac{\partial}{\partial x_i}(\bar{u}_i\bar{\phi}_i) = \frac{\partial}{\partial x_i}\left(\Gamma_T\frac{\partial \bar{\phi}_i}{\partial x_i} - \overline{u_i'\phi_i'}\right) \tag{8b}$$

$$\frac{\nabla^2 \bar{p}_i}{\rho} = \frac{\partial \bar{u}_i}{\partial x_j}\frac{\partial \bar{u}_j}{\partial x_i} + \frac{\partial^2\left(\overline{u_i'u_j'}\right)}{\partial x_i \partial x_j} \tag{9}$$

where, Γ_T is turbulent diffusion. Considering practical turbulent flow, where the Reynold's stress term is modeled as a summation of isotropic stress $(\frac{2}{3}\rho k_{TKE})$ and anisotropic stress (a_{ij}). The anisotropic stress term is responsible for transporting momentum, and isotropic stress term acts as a modified pressure term in mean momentum equation (equation 10).

$$\frac{\bar{D}}{\bar{D}t}(\rho\bar{u}_i) = -\frac{\partial}{\partial x_i}\left(\bar{p}_i + \frac{2}{3}\rho k_{TKE}\right) + \mu\frac{\partial^2 \bar{u}_i}{\partial x_{i,j,k}^2} - \frac{\partial}{\partial x_i}(\rho a_{ij}) \tag{10}$$

Numerically solving this equation requires to model the term a_{ij}, which is represented as a function of strain rate tensor $(\bar{S}_{ij})$ and turbulent viscosity (ν_T) following the Boussinesq's hypothesis and turbulent viscosity hypothesis [8]. Modifying equation 10, arrives equation 11.

$$\frac{\bar{D}}{\bar{D}t}(\rho\bar{u}_i) = -\frac{\partial}{\partial x_i}\left(\bar{p}_i + \frac{2}{3}\rho k_{TKE}\right) + \frac{\partial}{\partial x_i}\left[\rho(\nu + \nu_T)\left(\frac{\partial \bar{u}_j}{\partial x_i} + \frac{\partial \bar{u}_i}{\partial x_j}\right)\right] \tag{11}$$

where, ν is kinematic viscosity and $\mu = \rho\nu$. In an axisymmetric turbulent jet flow, $\left(\vec{\bar{u}_i} \times \vec{x_i}\right) \approx 0$, which reduces to [8],

$$\rho\overline{u_i'u_j'} = -\nu_T\frac{\partial}{\partial x_j}(\rho\bar{u}_i) \tag{12a}$$

$$\overline{u_i'\phi_i'} = -\Gamma_T\frac{\partial}{\partial x_j}(\bar{\phi}_i) \tag{12b}$$

At high Reynold's number and in free stream velocity domain, the ratios $\frac{\nu_T}{\nu}$ and $\frac{\Gamma_T}{\Gamma}$ increases linearly, indicating inferior dominance of molecular viscosity (ν) and diffusivity (Γ) and only holding $\nu_T = f(u_i) = c\sqrt{k}l_m(x_i)$, where, c is a constant, k is turbulent kinetic energy, $l_m(x_i)$ is mixing length, and $\Gamma_T = f(\mathcal{L}) = (u_i')^2\tau_L^*$, where, $\mathcal{L}$ is lengthscale and τ_L^* is Lagrangian integral of timescale $\tau_L \sim \frac{k}{\varepsilon}$, which is a function of turbulent dissipation $\left(\varepsilon = \frac{(u_i')^3}{l_m(x_i)}\right)$ by energy cascade [8]. Modelling $\nu_T = f(k, l_m(x_i))$ and $\tau_L = f(\varepsilon, l_m(x_i))$ requires different assumptions based on the flow topology. The $k-\omega$ SST model is widely used in engineering applications specifically for external flow focusing the study on capturing physics of boundary layer, which mathematically explained by $1 \le y^+ \le 10$ without wall function and with wall function respectively. The $k-\omega$ model is also two equation model closing the model by solving k, turbulence frequency $\left(\omega = \frac{\varepsilon}{k}\right)$ instead of ε and $l_m(x_i) = \frac{\sqrt{k}}{\omega}$.

$$\frac{\bar{D}}{\bar{D}t}(\rho k) = \frac{\partial}{\partial x_i}\cdot\left(\rho\left(\nu + \frac{\nu_T}{(Pr_T)_k}\right)\frac{\partial k}{\partial x_i}\right) + P_T - \left(C_{\varepsilon_2} - 1\right)\rho k\omega \tag{13}$$

$$\frac{\bar{D}}{\bar{D}t}(\rho\omega) = \frac{\partial}{\partial x_i}\cdot\left(\rho\left(\nu + \frac{\nu_T}{(Pr_T)_\omega}\right)\frac{\partial \omega}{\partial x_i}\right) + \left(C_{\varepsilon_1} - 1\right)\frac{\omega P_T}{k} - \left(C_{\varepsilon_2} - 1\right)\rho\omega^2 \tag{14}$$

Menter [9] modified standard $k-\omega$ model capturing near and remote wall effects utilizing a blending function multiplied with ω. This model is known as $k-\omega$ shear stress transport (SST) model. The modified ω transport equation is,

$$\frac{\bar{D}}{\bar{D}t}(\rho\omega) = \frac{\partial}{\partial x_i}\cdot\left(\rho\left(\nu + \frac{\nu_T}{(Pr_T)_\omega}\right)\frac{\partial \omega}{\partial x_i}\right) + \alpha\frac{\omega P_T}{k} - \beta\rho\omega^2 + \frac{2(1-F_b)\rho\nu_T}{(Pr_T)_\omega k}\nabla\omega_i\cdot\nabla k_i \tag{15}$$

where, $\alpha = \left(C_{\varepsilon_1} - 1\right)$, $\beta = \left(C_{\varepsilon_2} - 1\right)$, and F_b is blending factor. This model captures the near and remote wall effect by adjusting $F_b = 0$ on the wall and $F_b = 1$ in free stream region respectively. Table 1 illustrates the model parameters utilized in different k-ω model versions.

Table 1. Details of model constants utilized in $k-\omega$ model ($y^+ = 1.0$)

Parameters	$k-\omega$ SST (near wall)	$k-\omega$ SST (remote from wall)
C_μ	0.09	0.09
α	0.556	0.44
β	0.075	0.0828
$(Pr_T)_k$	1.176	1.0
$(Pr_T)_\omega$	2.0	1.168

Moreover, capturing near wall effects with all URANS approach, enhanced wall treatment approach is utilized allowing separate log law and viscous sublayer modeling approach which enhance the accuracy of prediction irrespective to local grid characteristics [8]. This research utilized two different types of discretization techniques, i.e., tetrahedral mesh and cut-cell mesh, which impose different truncation errors based on the staggered grid approach of finite volume analysis. The tetrahedral mesh is generated for the screws using SCORG (Fig.2(a)) and the rest of fluid domain is generated using ANSYS meshing (Fig.2(b)) with region-based refinement varying mesh cell edge length from 1 mm to 20 mm, as shown in Fig.2. The cut-cell method of mesh generation is part of the CONVERGE CFD code and is a cartesian type grid. Initially a volume is generated such that the surface enclosing the flow domain is a subset of the volume. Next, this volume is discretized with a uniform cartesian grid with grid size 20 mm base grid size, where a Boolean intersection operation is performed keeping only the discretized flow domain enclosed by the surface [10]. However, this algorithm fails at constricted locations, where the cell defeatures and decreases orthogonality significantly.

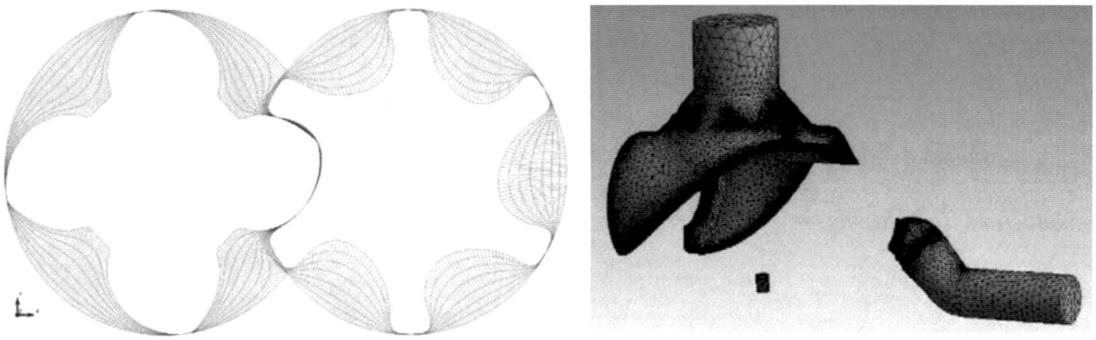

(a) Discretization of rotor using SCORG

(b) Discretization of rest of the fluid domain using ANSYS meshing

Figure 2. Tetrahedral discretization

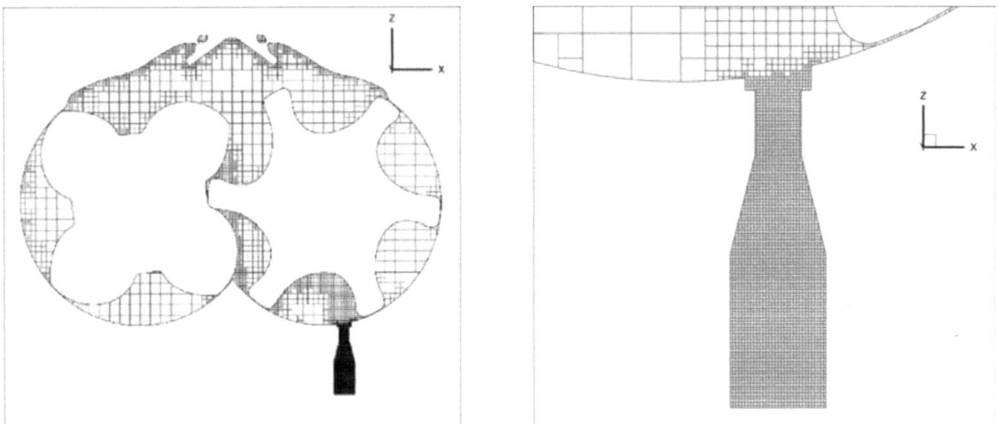

(a) Cut-cell discretization using CONVERGE CFD code

(b) Zoomed view of cut-cell discretization of oil injector using CONVERGE CFD code

Figure 3. Cut-cell discretization

To avoid these defeaturing the mesh is refined at these locations to a refinement level (n) from 3 to 5 minimizing local grid size $\left(\Delta x_i = \frac{\Delta x_{i,base}}{2^n}\right)$ to 2.5 mm to 0.625 mm. Across the multifluid interface, velocity and density gradient normal to the interface is expected to be large, which requires a finer grid size by adaptive mesh refinement (AMR). The mesh is set to maintain a local grid size of 2.5 mm with a refinement level of 3 where $\frac{\partial u_i}{\partial x_i} > 2$ m/s and $\frac{\partial\left(\frac{v_g}{\sum v}\right)}{\partial x_i} > 0.001\%$, allowing a total 1×10^6 number of finite volumes. Figure 3 further illustrates the details of grid implemented in a graphical format.

These truncation errors transport in adjacent finite volumes which possess numerical instabilities. To filter out these instabilities, different numerical algorithms are investigated. The tetrahedral finite volume used incomplete upper and lower triangular (ILU) decomposition and cut-cell method utilized SOR, and the results obtained by these algorithms are compared here after in this work.

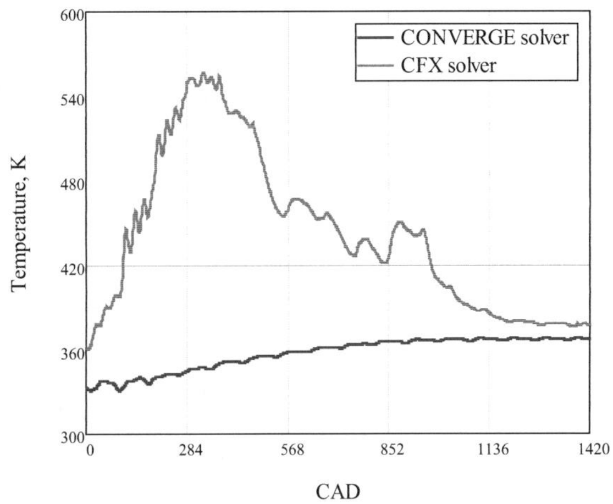

(a) Difference between successive iterations of temperature solution with respect to the crank angle degree (CAD)

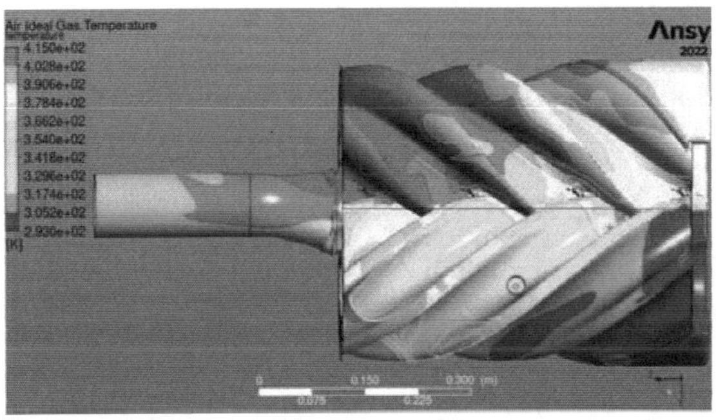

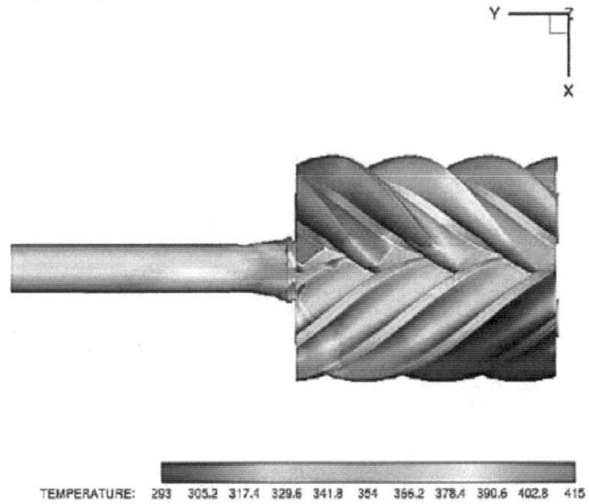

(b) CFX: Contour plot of the temperature distribution over the screws surface

(c) CONVERGE: Contour plot of the temperature distribution over the screws surface

Figure 4. Comparison of temperature solutions between CFX and CONVERGE solvers

The current code with $k - \omega$ SST model required approximately 120 hrs. with CFX and 10 hrs. with CONVERGE to solve 1,420 crank angle degrees (CAD). Both simulations were conducted on a computational environment equipped with two 64-core AMD Epyc 7662 "Rome" CPU processors and 256 GB RAM. For the CFX solver, the two 64-core processors were utilized to achieve the required computational result. Once the CFX simulation was completed, the CONVERGE CFD code was started, also utilizing the two 64-core processors. Both solvers had parallelization enabled, leveraging OpenMP/MPI for reduced simulation time.

3. Results and Discussion

This section presents and analyzes the results from the numerical simulations using the CFX and CONVERGE solvers. The primary objective of this comparative study is to evaluate the performance, stability, and convergence characteristics of these solvers when applied to the transient 3D CFD model of an oil-injected screw compressor. The comparative analysis of the CFX and CONVERGE solvers, illustrated by Fig.4, reveals that while both solvers capture detailed transient phenomena, Fig.4(c and d), the cut-cell method used by the CONVERGE solver is more stable and converges faster, Fig.4(a). The temperature profiles show that the CONVERGE solver stabilizes the temperature distribution efficiently, completing calculations in just 10 hrs., compared to CFX's 120 hrs. Figure 5 showcases the pressure results, highlighting the performance of both CFX and CONVERGE solvers.

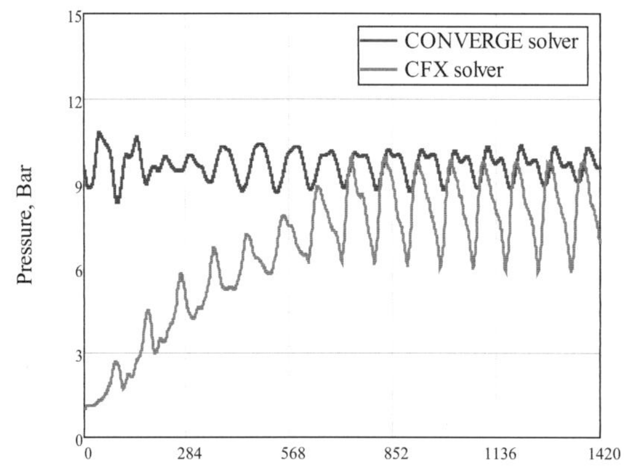

(a) Difference between successive iterations of pressure solution with respect to the crank angle degree (CAD)

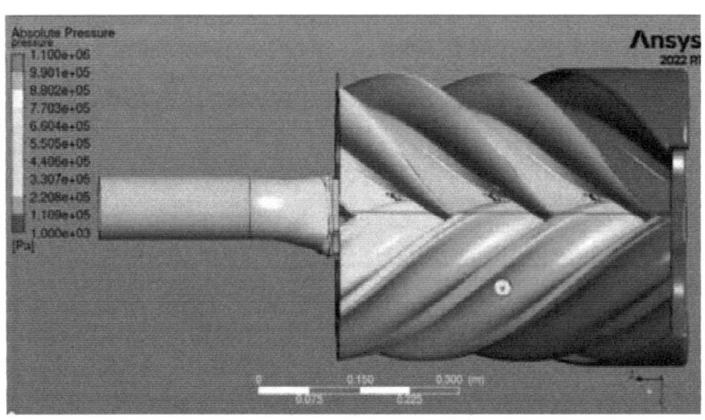

(c) CFX: Contour plot of the pressure distribution over the screws surface

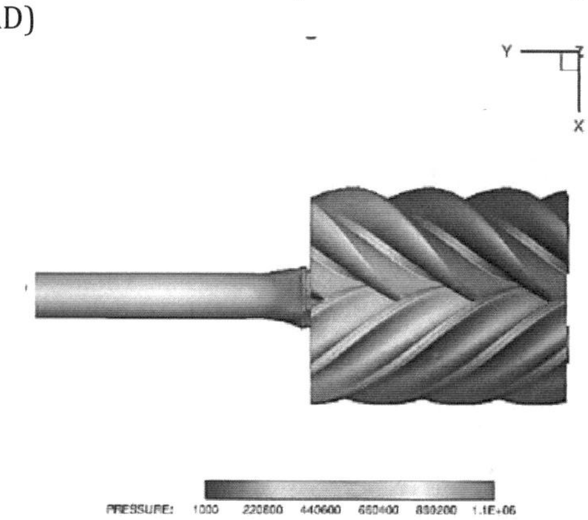

(d) CONVERGE: Contour plot of the pressure distribution over the screws surface

Figure 5. Comparison of pressure solutions between CFX and CONVERGE solvers

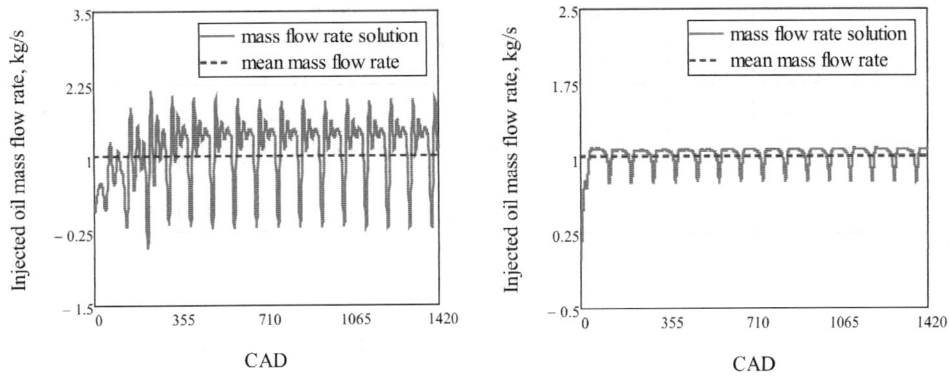

(a) CFX solver (b) CONVERGE solver

Figure 6. Comparison of injected oil mass flow rate solutions between CFX and CONVERGE solvers

The CONVERGE solver, with its cut-cell method, effectively captures detailed transient phenomena and demonstrates enhanced stability and faster convergence. The pressure profiles reveal that the CONVERGE solver achieves efficient stabilization within the same time as the temperature above in contrast to the 120 hours needed by CFX. The mass flow rates of injected and discharged oil have been examined using the CFX and CONVERGE solvers in this study as well. As depicted in Fig. 6 (a), the CFX solver achieved solution stabilization after approximately 355 CAD, whereas the CONVERGE solver stabilized after just 90 CAD, Fig 6 (b). Following this, the arithmetic mean of the mass flow rate was computed for both solvers. A notable variance was observed in the time required for each solver; 120 hours were needed for the CFX, in contrast to only 10 hours for the CONVERGE. Despite these variations in computational time, both solvers converged to the same mean mass flow rate within a maximum error of 2%. Furthermore, Fig. 7 illustrates the solutions for the discharged oil mass flow rates. The CFX solver reached a stable solution after approximately 1136 CAD, with a computational time of 120 hrs. for total 1420 CAD, whereas the CONVERGE solver had not yet stabilized, even after completing 1420 CAD and a computational time of 10 hrs. as illustrated in Figs. 7(a and b).

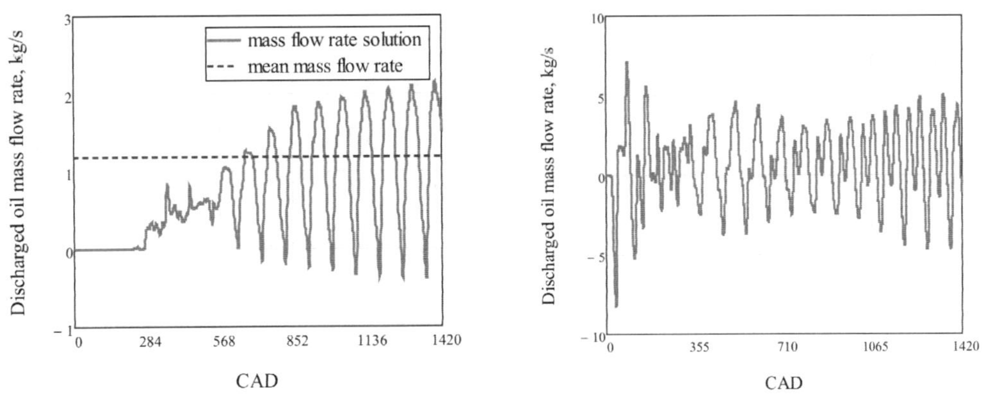

(a) CFX solver (b) CONVERGE solver

Figure 7. Comparison of discharged oil mass flow rate solutions between CFX and CONVERGE solvers

The difference in the stability is due to the perturbations developed by the cut-cell method utilized by CONVERGE. As discussed earlier in Fig. 3, the cut-cell method linearizes the non-linear boundaries of fluid domain, which develop perturbations in the coefficient matrix developing numerical instabilities. These instabilities are magnified only for oil mass flow rate calculation, because of the high shear stress developed on the boundary of the injector and the discharge port due to significant difference in the momentum thickness of the boundary layer. On the other side, CFX utilizes a defeatured tetrahedral discretization, which represents nonlinear boundaries with closer proximity than cut-cell method, decreasing the perturbations and stabilizing the solution. Consequently, to verify that the mass flow rate at the injection port equals that at the discharge port, and that both solvers converge to the same solution, the CONVERGE solver was operated for additional CADs to achieve stabilization. This extended run involved an additional 32 hrs., bringing the total runtime to 42 hrs., with the solver completing 6000 CADs. This is depicted in Fig. 8(a). A stable solution was captured between CADs 3000 and 4000, followed by the calculation of the arithmetic mean of the oil mass flow rate, as shown in Fig. 8(b).

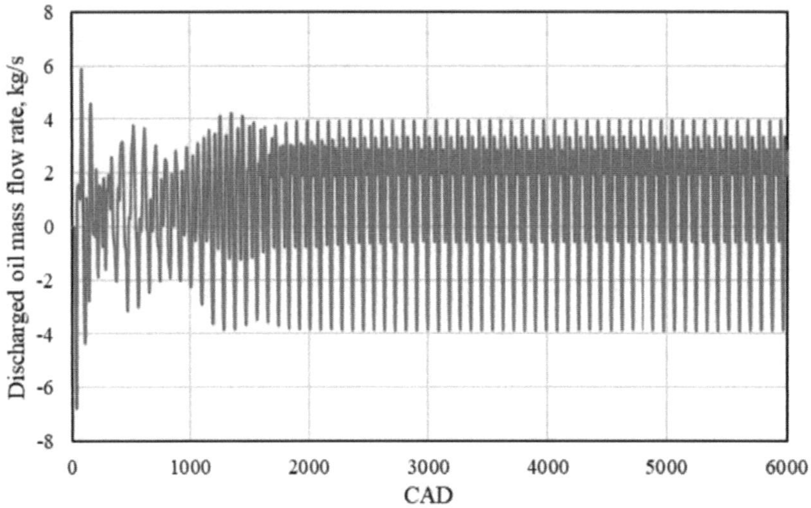

(a) CONVERGE solution for the oil discharged mass flow rate from 0 to 6000 CAD

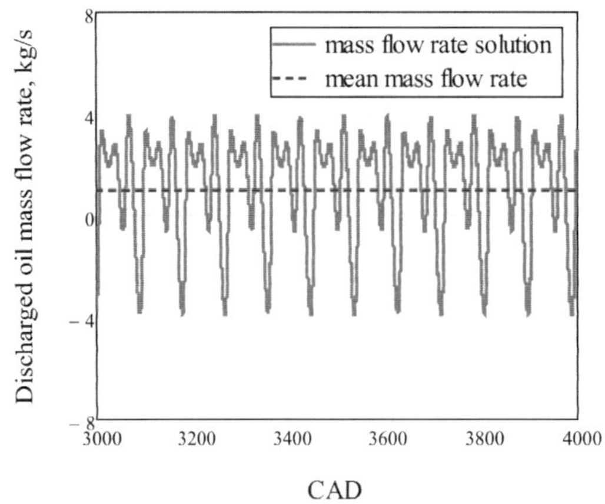

(b) CONVERGE solution for the discharged oil mass flow rate from 3000 to 4000 CAD

Figure 8. Extended solution of the CONVERGE solver for the discharged oil mass flow rate

The numeric outcomes of key performance parameters between the CFX and CONVERGE solvers, including progress (rotations in degrees), air and oil mass flow rates at suction, discharge, injection, and discharge temperature are highlighted in table 2. The results demonstrate that both solvers yield similar outcomes, indicating comparable accuracy in modeling the operative characteristics of an oil-injected screw compressor, thereby reinforcing the consistency of the simulations.

The selection of computational methods significantly impacts both the stability and efficiency of the solution process. In this study, the CONVERGE CFD solver, which incorporates the Successive Over Relaxation (SOR) method, demonstrated faster convergence and numerical stability in the linear systems addressed in this study. Preconditioning was applied to enhance the system's condition number, further improving convergence. The combination of SOR and preconditioning reduced residual errors with each iteration, contributing to the simulation's overall performance. While SOR was essential to these improvements, detailed mathematical analysis of the solver is beyond the scope of this study.

Table 2: highlights of the numeric outcomes of key performance parameters between the CFX and CONVERGE

Parameters	Unit	CFX	CONVERGE
Rotation	degree	1420	6000
Air mass flow at suction	kg/s	0.340	0.314
Air mass flow at discharge		0.349	0.310
Oil mass flow at injection		1.036	1.030
Oil mass flow at discharge		1.022	1.025
Temperature at discharge	K	377.27	364.00

4. Conclusion

This study examined the stability and convergence of preconditioned SOR and incomplete LU decomposition (ILU) iterative linear solvers in the context of an oil-injected screw compressor. The research utilized both a structured grid and a cut-cell grid, and the transient 3D CFD models. Key mathematical models included the unsteady three-dimensional compressible Navier-Stokes equations, continuity, momentum, species transport, energy, and void fraction transport equations. These were implemented with the $k - \omega$ SST turbulence model and the SIMPLE method for pressure-velocity coupling, ensuring accurate simulations of turbulent boundary layers and multifluid interfaces. The comparative analysis of the CFX and CONVERGE solvers revealed that both effectively capture detailed transient phenomena. However, the cut-cell method used by the CONVERGE solver demonstrated enhanced stability and faster convergence, completing simulations in just 42 hrs. compared to the 120 hrs. required by CFX. Despite the significant difference in computational time, both solvers produced approximately similar results for key performance parameters such as air and oil mass flow rates and temperature at discharge. These findings indicate that the CONVERGE solver's cut-cell method offers a more efficient and stable approach for CFD modeling of complex fluid dynamics in screw compressors. This investigation highlights the importance of selecting appropriate numerical methods based on specific project requirements, balancing computational efficiency with simulation accuracy. It should be noted that this study does not promote one solver over the other, and further research is necessary to fully explore the potential and limitations of both solvers in various contexts.

IOP Conf. Series: Materials Science and Engineering **1322** (2024) 012016 doi:10.1088/1757-899X/1322/1/012016

Acknowledgements

The authors show their gratitude towards the Rosen Centre for Advanced Computing (RCAC) at Purdue University for supporting Bell supercomputing facility. Furthermore, the authors are thankful to ANSYS, Inc. and Convergent Science Inc. for providing research and academic licenses and support through their ANSYS Academic Support and CONVERGE Academic Program.

Funding Acknowledgements

The authors would like to acknowledge Hitachi Global Air Power (HGAP) for providing financial and intellectual support.

References

[1] Yang S et al 2021 International Journal of Refrigeration 126 pp 143-154
[2] Andres R, Hesse J, Hetze F and Low D 2018 IOP Conference Series: Materials Science and Engineering 425 012018
[3] Ding H, Jiang Y and Dhar S 2019 IOP Conference Series: Materials Science and Engineering 604 012005
[4] Rane S R, Kovacevic A and Kethidi M 2012 International Compressor Engineering Conference at Purdue Paper 2141
[5] Wang B et al 2024 International Journal of Refrigeration 165 pp 209-222
[6] Wang B et al 2024 International Journal of Refrigeration 158 pp 100-110
[7] Celebi A T et al 2021 *Finite-size effects of diffusion coefficients computed from molecular dynamics: a review of what we have learned so far* Molecular Simulation 47 831
[8] Pope S B 2001 *Turbulent flows* Measurement Science and Technology 12 2020
[9] Menter F R 1994 *Two-equation eddy-viscosity turbulence models for engineering applications* AIAA journal 32 1598
[10] Ghmati R E, Jawad B A and Koutsavdis E 2012 *An investigation of cutcell meshing strategies for accurate aerodynamic performance prediction* SAE Int. J. Passenger Cars-Mechanical Systems 5 369
[11] Rosen Center for Advanced Computing (RCAC) 2024 *Bell* URL: https://www.rcac.purdue.edu/compute/bell
[12] Richards K J, Senecal P K and Pomraning E CONVERGE 3.0 manual
[13] ANSYS Inc. 2009 *ANSYS CFX-Solver Theory Guide Release 12.1.*
[14] Onabid M A 2012 *Solving three-dimensional (3D) Laplace equations by successive over-relaxation method* African Journal of Mathematics and Computer Science Research 5 pp 204-208

IOP Conf. Series: Materials Science and Engineering **1322** (2024) 012017 doi:10.1088/1757-899X/1322/1/012017

Thermodynamic simulation of a water-injected twin-screw steam compressor

M Grieb and A Brümmer

TU Dortmund University, Chair of Fluidics, Dortmund, Germany

E-mail: manuel.grieb@tu-dortmund.de, andreas.bruemmer@tu-dortmund.de

Abstract. Decarbonisation of industrial heating and process steam generation is crucial for the mitigation of advancing climate change. High-temperature heat pumps can meet these demands while effectively integrating with renewable energy sources, outperforming traditional heating and boiling systems, especially when operating with natural refrigerants. Since compressor efficiency is critical for the effectiveness of heat pump systems, particular care is required in compressor design and specification. Detailed thermodynamic simulations of the compressor system are indispensable in this process. The compression of fluids with a negative saturation vapour curve leads to challenging temperature increases during compression, with substantial superheating of the working fluid. In this context, a promising approach is the injection of liquid working fluid into the compressing chambers for internal cooling. This introduces new complexity to compressor simulations for heat-pump applications. In this work, a novel two-phase approach for multi-chamber model simulations of positive displacement compressors is presented. Initially, a quasi-isentropic reference process for vapour compression with internal liquid injection is introduced and efficiency ratings are derived. Furthermore, the simulation process is described in detail, and comprehensive thermodynamic simulations of a water-injected twin-screw steam compressor are carried out.

1. Introduction

Climate change driven by greenhouse gas emissions poses an imminent threat to society, and urgent matters must be addressed for the mitigation of the consequences of global warming. In this context, industry-wide decarbonisation and electrification are pursued. In industrial heating and process steam generation, these challenging demands can be met by utilising high-temperature heat pump systems [1]. The compressor is a key component in every heat pump system, and particular care is required during specification and design of these machines. The single-stage compression of refrigerants with a negative saturation vapour curve leads to large temperature lifts and superheated fluid states on the high-pressure side of the compressor. Besides the high thermal load on the compressor components, superheating is disadvantageous with regard to the design of heat exchangers on the sink side of the heat pump system. In this context, a promising approach is to inject liquid refrigerant into the compressor, which absorbs heat during compression and evaporates during the process. Due to their robustness and proven ability to handle liquid within their working chambers, twin-screw compressors appear to be especially suitable for this application.

In this work, a novel two-phase multi-chamber simulation tool for the thermodynamic calculation of positive displacement machines with internal injection of liquid working fluid is presented. The idealised compression process is described, and efficiency ratings are derived. In addition, comprehensive simulations are carried out for a water-injected twin-screw steam compressor for a high-temperature heat pump system, and the results are presented and discussed.

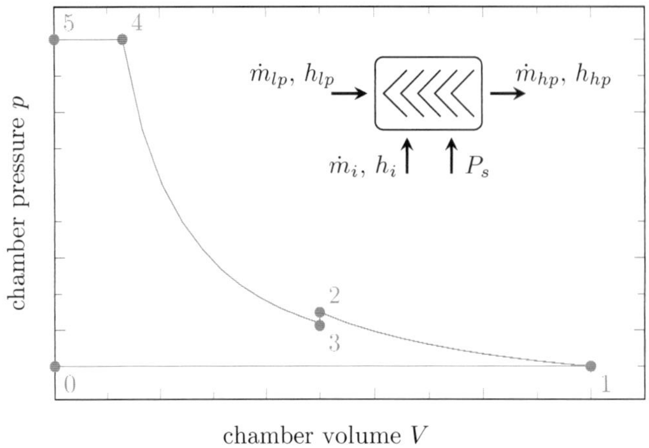

Figure 1. Pressure-volume plot for quasi-isentropic vapour compression cycle with injection of liquid working fluid.

2. Reference process and efficiency ratings for liquid-injected vapour compression

Initially, the reference process for vapour compression with internal injection of liquid working fluid is described, and efficiency ratings and machine coefficients are defined.

For the compression cycle of a twin-screw compressor a commonly used efficiency rating is the inner isentropic efficiency $\eta_{i,s}$, where P_s is the isentropic power, and P_i is the compressor's inner power.

$$\eta_{i,s} = \frac{P_s}{P_i} \tag{1}$$

While isentropic power for dry-running screw machines or for liquid-injected screw machines with purely liquid auxiliary fluid is easily calculated from the gaseous mass flow rate and an isentropic change of state, the calculation of isentropic power for liquid-injected machines with evaporating auxiliary liquids becomes more challenging. Due to the transition from the liquid to the vapour regime within the working chambers, the displaced mass flow rate on the high-pressure side becomes greater than the suction mass flow rate. Hence, the isentropic power of the reference process must account for the injection mass flow rate, possible phase transitions, and for the time at which the liquid is introduced to the working cycle.

Figure 1 shows the pressure-volume plot for a single-fluid quasi-isentropic vapour compression cycle with liquid injection into the enclosed working chamber. During the isobaric suction phase (0-1), vapour enters the working chamber at defined suction pressure. Prior to the injection, an initial compression takes place at constant specific entropy (1-2). The liquid working fluid is instantly injected at constant chamber volume V_i (2-3), which defines the volume ratio of injection $v_{i,i}$, where V_{max} is the compressor's maximal chamber volume.

$$v_{i,i} := \frac{V_{max}}{V_i} \tag{2}$$

For the injection process, thermodynamic equilibrium is assumed. In this respect, it is noted that the isochoric mixing of liquid and vapour cannot take place without entropy production. Hence, the reference process is better described as quasi-isentropic. The subsequent compression of wet working fluid is, again, carried out at constant specific entropy ($s_3 = s_4$). During the discharge phase, the fluid is ideally displaced to the high-pressure side at constant pressure (4-5).

The power P_s of the quasi-isentropic reference process is calculated from the integral energy conservation for the ideal compressor:

$$P_s = \dot{m}_{hp} h_{hp} - \dot{m}_{lp} h_{lp} - \dot{m}_i h_i = \dot{m}_{lp}(h_{hp} - h_{lp}) + \dot{m}_i(h_{hp} - h_i) \tag{3}$$

Here, $\dot{m}$ is the mass flow rate with the indices referring to the low-pressure (lp) and high-pressure (lp) conditions, respectively. The index i describes the properties of the injection, and h is the specific enthalpy. The integral mass conservation equation gives:

$$\dot{m}_{hp} = \dot{m}_{lp} + \dot{m}_i \tag{4}$$

Furthermore, the injection ratio λ_i is defined, where M is the chamber mass or the injection mass, respectively, n is the compressor speed, and z_m is the number of lobes on the main rotor.

$$\lambda_i := \frac{\dot{m}_i}{\dot{m}_{lp}} = \frac{n\, z_m\, M_i}{n\, z_m\, M_1} = \frac{M_i}{M_1} \tag{5}$$

Hence, for known low-pressure and injection parameters, the outlet enthalpy h_{hp} needs to be calculated for the determination of the quasi-isentropic power from equation 3. h_{hp} depends on the timing and magnitude of injection, and it can be calculated from the caloric equation of state when fluid properties at point 4 are sequentially calculated along the compression process 1-2-3-4. For the reference process, fluid properties at point 1 correspond to the low-pressure conditions, while fluid properties at point 4 correspond to the high-pressure conditions.

In the reference process, point 2 is defined by the specific entropy $s_2 = s_1$ and the specific volume v_2:

$$v_2 = \frac{v_1}{v_{i,i}} \tag{6}$$

After injection at constant chamber volume ($V_2 = V_3$), and under consideration that the chamber mass M_2 equals M_1 for the ideal process, v_3 becomes:

$$v_3 = \frac{V_3}{M_3} = \frac{V_2}{M_2 + M_i} = \frac{v_2}{1 + \lambda_i} = \frac{v_1}{v_{i,i}\,(1 + \lambda_i)} \tag{7}$$

The second independent fluid property defining the fluid state at point 3 is the specific energy u_3, which can be calculated from the energy equation for the isochoric injection process 2-3, where changes in total energy E are driven by the injection enthalpy flow. When kinetic energy and gravitational energy are neglected, the total energy corresponds to the internal energy U.

$$U_3 \approx E_3 = E_2 + \int_2^3 \dot{m}_i\, h_i\, \mathrm{d}t = E_2 + M_i\, h_i \tag{8}$$

$$u_3 = \frac{U_3}{M_3} = u_2\, \frac{M_2}{M_2 + M_i} + h_i\, \frac{M_i}{M_2 + M_i} \tag{9}$$

$$u_3 = u_2\, \frac{1}{\lambda_i + 1} + h_i\, \frac{1}{\frac{1}{\lambda_i} + 1} = \frac{u_2 + h_i\, \lambda_i}{\lambda_i + 1} \tag{10}$$

Finally, the specific enthalpy at point 4 (high-pressure side) is calculated from the equation of state with the outlet pressure $p_{hp} = p_4$ and the specific entropy s_4, and the quasi-isentropic power can be calculated from equation 3.

$$s_4 = s_3 = s(u_3, v_3) \tag{11}$$

$$h_{hp} = h_4 = h(p_4, s_4) \tag{12}$$

To rate compressor performance in context of leakage flows and throttling effects, the delivery rate λ is introduced. It is the ratio of actual suction mass flow rate $\dot{m}_{lp}$ to the theoretical suction mass flow rate, which is calculated from the displaced volume flow rate and the low-pressure density ρ_{lp}. It should be noted that this definition of the delivery rate does not account for the injected mass flow rate. The displaced actual mass flow rate on the high-pressure side is increased by the injection mass flow rate.

$$\lambda := \frac{\dot{m}_{lp}}{\dot{m}_{lp,th}} = \frac{\dot{m}_{lp}}{n\, z_m\, V_{max}\, \rho_{lp}} \tag{13}$$

As mentioned before, the injection of liquid into the compressor is associated with entropy production, even for the idealised process. In this context, and under consideration of the thermodynamic evaluation of the heat pump cycle, an investigation of compressor performance regarding exergetic efficiency is expedient. In general, exergy describes the part of energy that can be converted into useful work when reversibly brought to equilibrium with its surrounding environment [2]. Hence, it is a measure of the *quality* of energy in relation to the second law of thermodynamics.

The specific exergy ex carried by a fluid flow is determined as follows. h_e, T_e, and s_e are the specific enthalpy, the temperature, and the specific entropy of the environment. ex becomes zero for $h = h_e$ and $s = s_e$, respectively.

$$ex = h - h_e - T_e\,(s - s_e) \tag{14}$$

Accordingly, the specific anergy b, which is the part of energy that cannot be used in accordance with the second law of thermodynamics, becomes:

$$b = h_e + T_e\,(s - s_e) \tag{15}$$

Both technical work and electrical work do not carry entropy - they consist of pure exergy. Hence, the exergetic balance for the compressor system yields:

$$0 = \dot{m}_{lp}\,ex_{lp} + \dot{m}_i\,ex_i - \dot{m}_{hp}\,ex_{hp} + P - \dot{E}x_l \tag{16}$$

The indexes lp and hp refer to the compressor's low-pressure and high-pressure conditions. Additionally, the exergy loss flow $\dot{E}x_l$ can be expressed in terms of irreversible entropy flow $\dot{S}_{irr}$:

$$\dot{E}x_l = T_e\,\dot{S}_{irr} \tag{17}$$

The exergetic efficiency ζ of a component or system is defined as the ratio of gained exergy flows to invested exergy flows, which results in the following for the liquid-injected compressor:

$$\zeta = \frac{\dot{m}_{hp}\,ex_{hp} - \dot{m}_{lp}\,ex_{lp} - \dot{m}_i\,ex_i}{P} = 1 - \frac{\dot{E}x_l}{P} \tag{18}$$

Here, P is the power added to the system. It is either the effective power $P_e = P_i/\eta_m$ or the electrical power P_{el}, depending on whether the electrical drive is included in the compressor system. In this study, the electric drive is not included in the exergetic ratings.

3. Thermodynamic simulation of liquid-injected compressor

3.1. Multi-chamber model simulation

Within this work, thermodynamic simulations are carried out with a multi-chamber model simulation tool named *Kamel*, which is self-developed at the Chair of Fluidics of TU Dortmund University. The code is written in C++ to make use of the inheritance embedded within the object-oriented programming language. A general description of the simulation tool and its validation against experimental data for a dry-running twin-screw machine is given in [3].

During pre-processing, a non-dimensional chamber model is derived from the compressor geometry. In this process, an abstracted model of fluid capacities and fluid connections is created, whereby capacities (e. g. working chambers, inlet volumes) contain working fluids in a defined thermodynamic state, and connections (e. g. gaps and machine openings) provide fluid flow between those capacities according to a selected flow model. The rotor profile analysis for various rotational angles provides the angle-dependent geometry curves for all capacities (e. g. volume curve of working chamber) and connections (e. g. width curve of housing gap) as well as logical information on the connections between capacities. In the utilised two-chamber model, the course of one main rotor chamber and one gate rotor chamber is considered during runtime. Connections to trailing or leading working chambers are modelled according to the periodicity of screw compressors, whereby dummy capacities are set to the corresponding offset state from the last iteration.

In chamber model simulations, thermodynamic states within all capacities are calculated based on the conservation equations for energy and mass. Since homogeneous states are assumed within the control volumes, and kinetic energy within said volumes is assumed to be small in comparison to internal energy, the momentum equation is not used as a governing equation. The integral mass conservation equation for any control volume within the twin-screw compressor is:

$$\frac{\mathrm{d}M(t)}{\mathrm{d}t} = \sum_j \dot{m}_{r,j} \tag{19}$$

Changes in mass M enclosed in the control volume with respect to time t is given by the sum of mass flows relative to the control volume surface $\dot{m}_r$. The calculation of mass flow rates is carried out in subroutines for any connection. In this study, compressible fluid flow is calculated on the basis of the flow coefficient α, which is defined as the ratio of the actual mass flow rate $\dot{m}_r$ to the theoretical mass flow rate $\dot{m}_{th}$. The theoretical mass flow rate is calculated assuming isentropic flow, with choked flow considered as a possible condition. Homogeneous flow (no slip) in thermodynamic equilibrium is assumed for two-phase fluid flow through connections.

$$\dot{m}_r = \alpha\,\dot{m}_{th} = \alpha\,A\,\rho_{II,s}\,\sqrt{2\,(h_I - h_{II,s})} \tag{20}$$

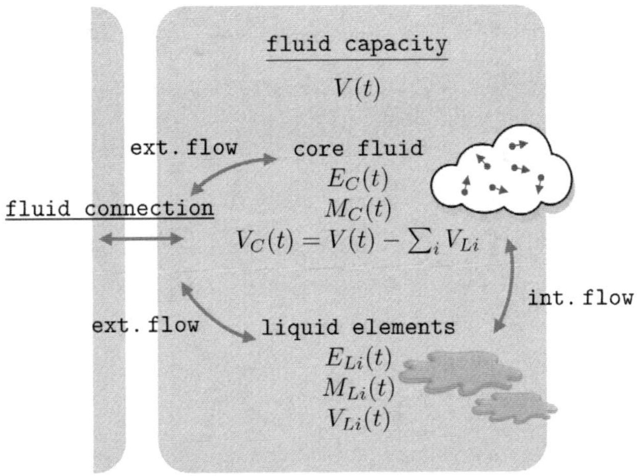

Figure 2. Logical sectioning of fluid capacities in compressible core fluid and list of incompressible liquid elements. Flows between capacites and different fluid regimes is modelled using specific flow models for external and internal flows.

A is the cross-sectional area of the connection (e. g. gap), ρ is the fluid density, and h is the specific enthalpy. The index I describes the fluid states within in the connected control volume at higher pressure, and the index II,s represents the downstream fluid states for the connection in question. For subsonic flow, it is assumed that the static pressure within the downstream chamber is imposed on the smallest cross-section of the connection. The downstream density $\rho_{II,s}$, and the specific enthalpy $h_{II,s}$ are calculated from the equation of state at constant specific entropy s. In case of choked flow through connections, the theoretical mass flow rate is determined for critical states in the smallest cross section. Choked two-phase flow is calculated by locating the maximum of the functional for homogeneous two-phase mass flux in the smallest cross section.

Assuming liquid is injected with a controlled pump, water is injected from a reservoir with a constant mass flow rate over an angular range of 90 °, which equals the maximal extent for injection into the main rotor chamber through a single bore in a 4+6 screw configuration. In this study, injection starts just after the working chambers are disconnected from the inlet port.

The energy equation applied to each control volumes is

$$\frac{\mathrm{d}E(t)}{\mathrm{d}t} = \sum_j h_{t,j}\,\dot{m}_{r,j} - \bar{p}\frac{\mathrm{d}V}{\mathrm{d}t} + \sum_k \dot{Q}_k - \dot{\Phi} \tag{21}$$

where E is the total energy within the control volume V and h_t is the specific total enthalpy. The first sum describes the changes in total energy due to enthalpy flows through the control volume surface. The second expression gives the pressure-volume power, where $\bar{p}$ is the average pressure acting on the control volume surface. $\dot{Q}_k$ are external heat flows, where k is the index of heat transfer connections. Finally, $\dot{\Phi}$ is power induced by tangential shear stresses acting on the moving control volume boundaries. This is often neglected in chamber model simulation tools. The two governing equations are solved during runtime using the forward Euler method, which gives fluid density and specific energy as two independent fluid properties that allow calculating all other relevant fluid properties using an appropriate equation of state. In this work, Pollak's fundamental formulation for the thermodynamic properties of water is implemented [4].

Capacities can contain multiple control volumes for compressible fluid and liquids, where the governing equations are applied to all control volumes, and interaction is modeled using internal connections for heat and mass transfer. Mechanical equilibrium between the phases is assumed at all times, where the pressure of the compressible fluid regime is applied to all liquid phases within the same capacity. Furthermore, in this work, two-phase states within the working chambers are modelled in thermodynamic equilibrium, assuming rapid heat and mass transfer between the injected water and the surrounding steam. This is an approach commonly used in the literature [5, 6]. The logical implementation is depicted in figure 2. In

Table 1. Machine parameters for twin-screw compressor and numerical parameters for thermodynamic simulation.

Manufacturer	Aerzener Maschinenfabrik GmbH
Main rotor number of lobes	$z_m = 4$
Gate rotor number of lobes	$z_g = 6$
Main rotor wrap angle	$\Phi_m = 300\,^\circ$
Length to diameter ratio	$L/D_m = 1.24$
Max. chamber volume	$V_{max} = 0.42 \cdot 10^{-3}\,\mathrm{m}^3$
Built-in volume ratio	$v_i = 4.5$
Volume ratio of injection	$v_{i,i} = 1.003$
Assumed average gap height	$H = 0.1\,\mathrm{mm}$
Flow coefficient inlets/outlets	$\alpha = 0.8$
Average flow coefficient gaps	$\alpha = 0.8$
Flow coefficient blow holes	$\alpha = 0.95$
Angular step size	$\Delta\varphi = 0.1\,^\circ$
Max. number of iterations	$I_{max} = 30$
Fluid model	Helmholtz free energy EOS [4]

principle, fluid capacities can contain an unlimited amount of liquid elements, whereby different types of liquid are implemented (bulk, droplet, film, ...). When applying changes to either the liquid elements or the core fluid, the conservation equations for energy (equation 21) and mass (equation 19) are permanently considered.

3.2. *Compressor characteristics and simulation parameters*

The machine parameters for the compressor investigated in this paper and numerical parameters for the thermodynamic simulation are given in table 1. The data correspond to a compressor prototype currently being manufactured by the project partner Aerzener Maschinenfabrik GmbH. At this point, the investigation is purely theoretical, but the twin-screw steam compressor with water injection is to be investigated shortly on a new test rig at TU Dortmund University. The corresponding R718 heat pump system is currently planned and built at Fraunhofer IEG in Bochum. Machine parameters, such as gap heights and flow coefficients, are to be adapted to the actual parameters in subsequent studies.

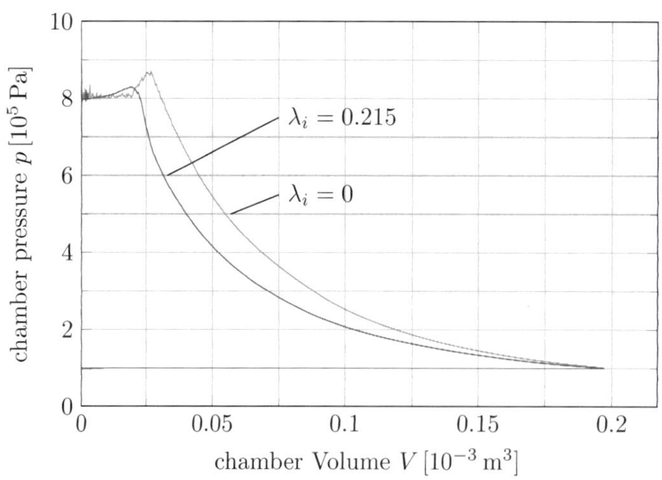

Figure 3. Comparison of main rotor chamber pressure-volume plots for dry compression ($\lambda_i = 0$) and liquid-injected compression ($\lambda_i = 0.215$). $n = 150\,\mathrm{s}^{-1}$, $p_i = 6 \cdot 10^5\,\mathrm{Pa}$, $T_i = 373.15\,\mathrm{K}$, $p_{lp} = 1 \cdot 10^5\,\mathrm{Pa}$, $T_{lp} = 373.15\,\mathrm{K}$, $p_{hp} = 8 \cdot 10^5\,\mathrm{Pa}$.

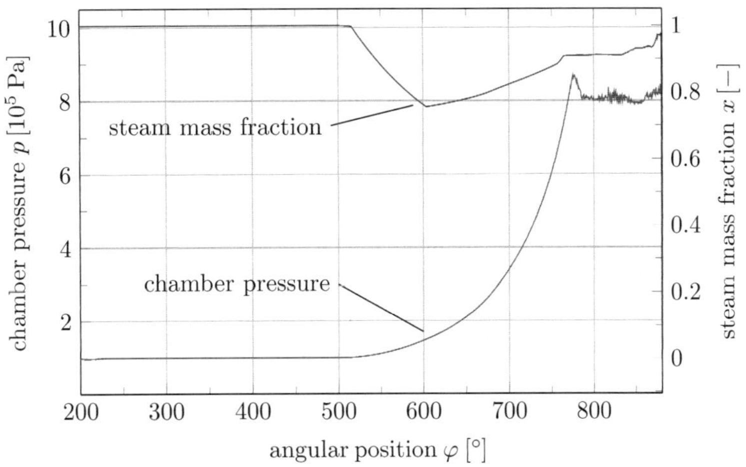

Figure 4. Chamber pressure and steam mass fraction plotted over angular position for main rotor chamber. $\lambda_i = 0.215$, $n = 150\ \text{s}^{-1}$, $p_i = 6 \cdot 10^5$ Pa, $T_i = 373.15$ K, $p_{lp} = 1 \cdot 10^5$ Pa, $T_{lp} = 373.15$ K, $p_{hp} = 8 \cdot 10^5$ Pa.

4. Discussion

Initially, the dry steam compression cycle is compared to a water-injected compression, where the injection ratio is iteratively altered during the calculation to reach the saturation state on the discharge side (steam mass fraction $x_{hp} \approx 1$) for the specified conditions. Figure 3 shows the comparison of pressure-volume plots for dry compression and the liquid-injected process for single-stage compression from 1 to $8 \cdot 10^5$ Pa. The compressor speed is $n = 150\ \text{s}^{-1}$. To achieve saturated state in the compressor outlet port, the injection ratio is only $\lambda_i = 0.215$, which equals a liquid volume fraction of less than 0.03 % at the end of injection. The simulations show a clear reduction in inner compression work for the water-injected steam compression. The absorption of heat from the gaseous phase to the injected mass clearly outweighs the increase in chamber mass and subsequently reduces chamber pressure during compression, thus bringing the process closer to isothermal compression. In addition, the water injection drastically reduces compressor outlet temperature in the simulations. For the operation point referred to in the diagram, outlet temperature is reduced from $T_{hp,dry} = 789.2$ K to $T_{hp,sat} = 443.6$ K. It is a fair assumption that for the targeted single-stage compression ratios in high-temperature R718 heat pump systems, water injection is not an option, but a requirement to facilitate operation.

The corresponding plot of chamber pressure and steam mass fraction over the angular position for the main rotor chamber is given in figure 4. The course of steam quality x is conclusive. Steam quality is gradually reduced during the 90 ° injection period and saturation is reached in the compressor's outlet port. In the calculations, mass and energy flows between the main rotor chamber and the gate rotor chamber are driven by small pressure differences - a behaviour inherent to chamber model simulations, where the velocity distribution within the fluid capacities is not taken into account. It might be expedient to introduce mixing models in future calculations to improve the modelling of liquid distribution when liquid is only injected on one side.

Figure 5 shows a comparison of inner isentropic efficiency and delivery rate for various compressor speeds. Again, the injection rate was adjusted for the water-injected simulations, so that a saturated fluid state is reached in the compressor outlet. Both the isentropic efficiency and the delivery rate are improved when water is injected into the working chambers of the twin-screw compressor. Besides the aforementioned reduction of (inner) pressure-volume work directly increasing the isentropic efficiency, a reduction of internal leakage leads to the improved delivery rate. The strong decrease in delivery rate for low compressor speed results from back flow of hot steam to the suction side, consequently reducing the fluid density within the inlet port. The corresponding injection rate for the water-injected process is given in figure 6. With increasing compressor efficiency, less water needs to be injected relative to the suction mass flow rate to reach saturated fluid states on the compressor's high-pressure side.

Furthermore, figure 6 displays the exergetic compressor efficiency for a variation in machine speed. Temperature and pressure for the surrounding environment are set to $T_e = 288.15$ K and $p_e = 1.013 \cdot 10^5$ Pa, respectively, in this study. Mechanical efficiency is assumed to be $\eta_m = 0.95$ for all simulation points. The absolute values of exergetic efficiency for the steam compressor are high, since, in general, the

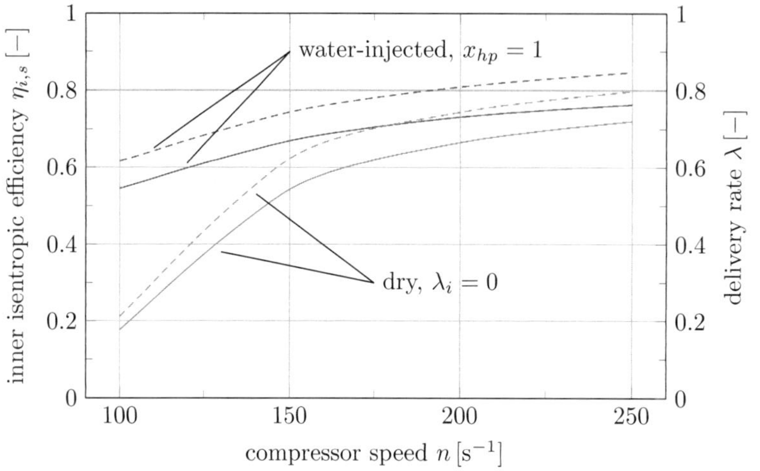

Figure 5. Inner isentropic efficiency (solid line) and delivery rate (dashed line) plotted over compressor speed. $p_i = 6 \cdot 10^5$ Pa, $T_i = 373.15$ K, $p_{lp} = 1 \cdot 10^5$ Pa, $T_{lp} = 373.15$ K, $p_{hp} = 8 \cdot 10^5$ Pa.

anergy content of steam is relatively low. This is due to the high evaporation enthalpy of steam, which adds to the theoretical work potential when brought to equilibrium with the surrounding environment. Despite reaching higher isentropic efficiency throughout the speed range and requiring substantially lower power for steam compression, the exergetic efficiency of the dry compressor exceeds that of the water-injected steam compressor. The exergetic efficiency of the water-injected steam compressor increases with decreasing injection ratio. Hence, from a thermodynamic perspective, liquid injection during vapour compression might bring disadvantages compared to dry compression, even when isentropic compressor efficiency is improved. This, however, is to be put in context with the exergetic efficiencies of the other cycle components and with the overall coefficient of performance of the heat pump system.

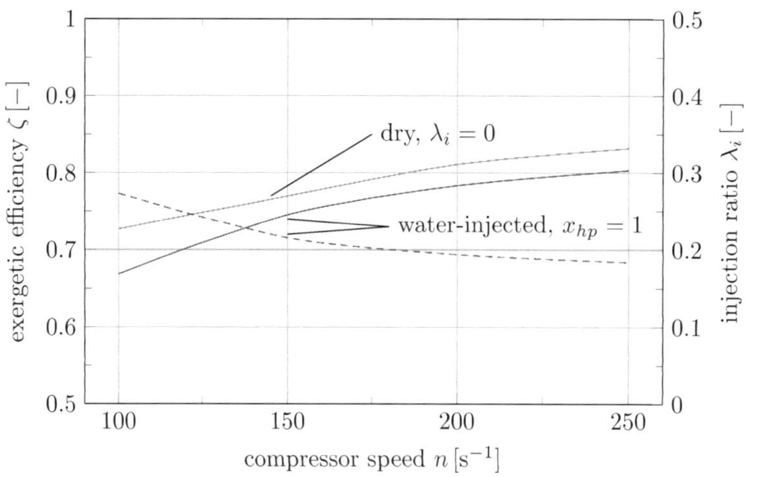

Figure 6. Exergetic efficiency (solid line) and injection rate (dashed line) plotted over compressor speed. $p_i = 6 \cdot 10^5$ Pa, $T_i = 373.15$ K, $p_{lp} = 1 \cdot 10^5$ Pa, $T_{lp} = 373.15$ K, $p_{hp} = 8 \cdot 10^5$ Pa.

5. Conclusion and outlook

This paper presented a detailed thermodynamic simulation framework for a water-injected twin-screw steam compressor, showcasing its potential to improve high-temperature heat pump systems. An ideal quasi-isentropic vapour compression cycle with liquid injection for positive displacement machines is introduced, instructions for calculating isentropic power are provided, and isentropic and exergetic efficiencies for the cycle are defined.

Additionally, a novel two-phase multi-chamber model simulation approach was presented, initially calculating internal flows between vapour and liquid in thermodynamic equilibrium. The simulations demonstrated that the liquid-injected screw compressor is improved in terms of isentropic efficiency, achieving performance closer to isothermal compression. It also showed a moderately higher delivery rate and a drastic reduction in outlet temperature, while, on the downside, reaching lower exergetic efficiency compared to a dry running screw compressor.

Liquid injection in twin-screw steam compressors offers significant potential for enhancing the compressor efficiency and enabling safe operation of high-temperature heat pumps, providing a promising path for more sustainable industrial heating and process steam generation. Future work will focus on the experimental validation of the simulation results to verify the theoretical findings. Additionally, there will be efforts to improve the modelling of phase transitions within the compressor to enhance the accuracy and reliability of the simulations.

Acknowledgements
Supported by: Federal Ministry for Economic Affairs and Climate Action on the basis of a decision by the German Bundestag

References
[1] IPCC 2022 *Climate Change 2022: Mitigation of Climate Change. Contribution of Working Group III to the Sixth Assessment Report of the Intergovernmental Panel on Climate Change* (Cambridge, UK and New York, NY, USA: Cambridge University Press) DOI 10.1017/9781009157926
[2] Dehli M, Doering E and Schedwill H 2022 *Fundamentals of Technical Thermodynamics* (Wiesbaden, GER: Springer) DOI 10.1007/978-3-658-38910-9
[3] Grieb M and Brümmer A 2022 Thermodynamic simulation and experimental validation of an oil-free twin-screw expander *IOP Conf. Ser.: Mater. Sci. Eng.* **1267** DOI 10.1088/1757-899X/1267/1/012015
[4] Pollak R 1974 *Die thermodynamischen Eigenschaften von Wasser - dargestellt durch eine kanonische Zustandsgleichung für die fluiden homogenen und heterogenen Zustände bis 1200 Kelvin und 3000 bar* (Bochum: Ruhr-Universität Bochum)
[5] Tian Y, Geng Y, Yuan H and Zhao Z 2022 Investigation on water injection characteristics and its influence on the performance of twin-screw steam compressor *Energy* **259** DOI 10.1016/j.energy.2022.124886
[6] Yang J, Zhang C, Zhang Z, Yang L and Lin W 2016 Study on mechanical vapor recompression system with wet compression single screw compressor *Applied Thermal Engineering* **103** DOI 10.1016/j.applthermaleng.2016.04.053

Performance Analysis of a Water-Injected Twin-Screw Compressor in a High-Temperature R718 Heat Pump

S Höckenkamp[1], M Grieb[2], M Utri[1] and A Brümmer[2]

[1]Fraunhofer Research Institution for Energy Infrastructure and Geothermal Systems IEG, Bochum, Germany
[2]TU Dortmund University, Chair of Fluidics, Dortmund, Germany

E-mail: simon.hoeckenkamp@ieg.fraunhofer.de

Abstract. High-temperature heat pumps with twin-screw compressors and water (R718) as the working fluid show promising potential for providing industrial heat and process steam at temperature levels of up to 200 °C. This paper presents a simulation approach for a two-layered simulation structure of a water-based high-temperature heat pump system including a twin-screw compressor with water-injection. The heat pump process with two-phase steam compression is investigated with respect to various aspects. In addition to controlling the compressor outlet temperature, the evaporation of the injected liquid leads to an increase in the displaced mass flow rate on the sink side of the heat pump. This effect is analyzed in detail for various system operating points using a thermodynamic simulation framework for the R718-based heat pump cycle and a comprehensive compressor model. The main target of the presented paper is the analysis of system performance and the detailed investigation of the interdependencies. Performance of the twin-screw compressor is determined by means of two-phase chamber model simulations. In the chamber model method thermodynamic properties of both injected liquid and steam are calculated based on the conservation equations of mass and energy for the compression process. Adequate sub-models for internal leakage and heat and mass transfer between vapor and liquid phase are applied. Integral results of the compressor simulations are embedded in the heat pump system simulation using a characteristic map. In conclusion, this paper contributes to the understanding of advanced compression technologies in the field of high-temperature heat pumps, offering a detailed examination of a water-injected steam compressor's applicability and efficiency in a practical system setting.

1. Introduction

The current trend of climate change necessitates urgent actions to reduce greenhouse gas emissions and transition towards sustainable energy systems [1]. One significant contributor to power consumption and CO_2 emissions is the heat demand of industry in the low to medium temperature range [2]. Addressing this demand through innovative technologies is crucial for achieving decarbonisation goals.

Industrial high-temperature heat pumps (HTHPs) present a promising solution to this challenge, potentially filling a substantial gap in decarbonised heat generation [2, 3]. The adoption of low-GWP (Global Warming Potential) and natural refrigerants is becoming increasingly common due to their favorable thermodynamic, ecological, and technical characteristics. Among these, water stands out as an exceptionally suitable refrigerant for high-temperature heat pumps [4]. Its use not only offers thermodynamic advantages but also allows for direct application in process steam lines through mechanical vapor recompression (MVR) technology.

A key component in a HTHP cycle is the compressor. In this context a water-injected screw compressor is implemented, which exhibits robust operating behavior and high efficiency, making it ideal for HTHP and MVR applications [5, 6, 7]. However, the complex thermodynamic effects during operation represent significant challenges for accurate modeling within a heat pump cycle. To address this, a hybrid approach combining detailed compressor simulation with cycle simulation is employed. This approach leverages models that use the internal conditions of the heat pump as input data for compressor parameters. A linear regression is performed to generate a compressor model based on the simulation results of the detailed thermodynamic simulation.

This study aims to utilize the benefits of different simulation approaches to establish a foundation for more complex heat pump models, both, to increase the validity of the cycle simulation and the possibility to implement data from different sources like simulations or experiments. Overall, the simulations are used to analyze the interdependencies between the compressor and the heat pump as a whole. Through this research, the authors seek to contribute to the development of efficient and sustainable industrial heating solutions that align with global decarbonisation efforts.

2. Thermodynamic simulation of the water-injected twin-screw compressor

Within this work, a two-layered simulation structure is utilised for the detailed simulation of heat pump cycle and compressor characteristics. Multi chamber model simulation is used for the prediction of compressor performance and determination of volumetric and isentropic efficiency, which, in turn, serve as input parameters for the functional compressor model within the cycle simulations. In principle, the chamber model method is applicable to all types of displacement compressors.

During pre-processing, a non-dimensional chamber model is derived from the compressor geometry. In this process, an abstracted model of fluid capacities and fluid connections is created, whereby capacities (e. g. working chambers, inlet volumes) contain working fluids in a defined thermodynamic state, and connections (e. g. gaps and machine openings) provide fluid flow between those capacities according to a selected flow model. The rotor profile analysis for various rotational angles provides the angle-dependent geometry curves for all capacities (e. g. volume curve of working chamber) and connections (e. g. width curve of housing gap) as well as logical information on the connections between capacities. In the utilised two-chamber model, the course of one main rotor chamber and one gate rotor chamber is considered during runtime. Connections to trailing or leading working chambers are modelled according to the periodicity of screw compressors, whereby dummy capacities are set to the corresponding offset state from the last iteration. A logical model of the twin-screw compressor in question is depicted in figure 1. To maintain clarity, only selected gap connections are shown.

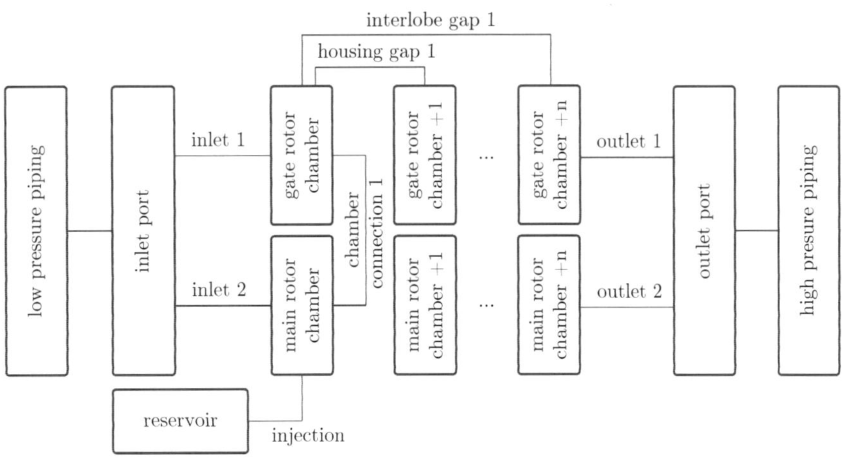

Figure 1. Simplified visualisation of chamber model for thermodynamic simulation of screw compressor.

In chamber model simulations, thermodynamic states within all capacities are calculated based on the conservation equations for energy and mass. Since homogeneous states per fluid phase are assumed within the control volumes, and the kinetic energy within said volumes is considered small compared

IOP Conf. Series: Materials Science and Engineering **1322** (2024) 012018 doi:10.1088/1757-899X/1322/1/012018

to the internal energy, the momentum equation is not used as a governing equation. The integral mass conservation equation for any control volume within the twin-screw compressor is:

$$\frac{\mathrm{d}M(t)}{\mathrm{d}t} = \sum_j \dot{m}_{r,j} \tag{1}$$

Changes in mass M enclosed in the control volume with respect to time t is given by the sum of mass flows relative to the control volume surface $\dot{m}_r$. j is the number of fluid connections. The calculation of mass flow rates is carried out in subroutines for any connection. In this study, compressible fluid flow is calculated on the basis of the flow coefficient α, which is defined as the ratio of the actual mass flow $\dot{m}$ to the theoretical mass flow $\dot{m}_{th}$. The theoretical mass flow rate is calculated assuming isentropic flow, with choked flow considered as a possible condition. Homogeneous flow (no slip) in thermodynamic equilibrium is assumed for two-phase fluid flow through connections.

$$\dot{m}_r = \alpha\, \dot{m}_{th} = \alpha\, A\, \rho_{II,s} \sqrt{2\,(h_I - h_{II,s})} \tag{2}$$

A is the cross-sectional area of the connection (e. g. gap), ρ is the fluid density, and h is the specific enthalpy. The index I denotes the fluid states within in the connected control volume at higher pressure, and the index II,s represents the downstream fluid states for the connection in question. For subsonic flow, it is assumed that the static pressure within the downstream chamber is imposed on the smallest cross-section of the connection. The downstream density $\rho_{II,s}$, and the specific enthalpy $h_{II,s}$ are calculated from the equation of state at constant specific entropy s. Assuming liquid is injected with a controlled pump, water is injected from a reservoir with a constant mass flow rate over an angular range of 90 °, which equals the maximal extent for injection into the main rotor chamber through a single bore in a 4+6 screw configuration.

Table 1. Machine parameters for twin-screw compressor and numerical parameters for thermodynamic simulation.

Manufacturer	Aerzener Maschinenfabrik GmbH
Main rotor number of lobes	$z_m = 4$
Gate rotor number of lobes	$z_g = 6$
Main rotor wrap angle	$\Phi = 300\,°$
Length to diameter ratio	$L/D_m = 1.24$
Max. chamber volume	$V_{max} = 0.42 \cdot 10^{-3}\,\mathrm{m}^3$
Built-in volume ratio	$v_i = 4.5$
Assumed average gap height	$H = 0.1\,\mathrm{mm}$
Flow coefficient inlets/outlets	$\alpha = 0.8$
Average flow coefficient gaps	$\alpha = 0.8$
Flow coefficient blow holes	$\alpha = 0.95$
Angular step size	$\Delta\varphi = 0.1\,°$
Max. number of iterations	$J_{max} = 30$

The energy equation applied to all control volumes is

$$\frac{\mathrm{d}E(t)}{\mathrm{d}t} = \sum_j h_{t,j}\, \dot{m}_{r,j} - \bar{p}\,\frac{\mathrm{d}V}{\mathrm{d}t} + \sum_k \dot{Q}_k - \dot{\Phi} \tag{3}$$

where E is the total energy within the control volume V. $h_t = h + \frac{c^2}{2}$ is the specific total enthalpy, with c being the flow velocity. The first sum describes the changes in total energy due to total enthalpy flows through the control volume surface. Again, j is the number of connections. The second expression gives the pressure-volume power, where $\bar{p}$ is the average pressure acting on the control volume surface. $\dot{Q}_k$

are external heat flows, where k is the index of heat transfer connections. Finally, $\dot{\Phi}$ is power induced by tangential shear stresses acting on the moving control volume boundaries. This is often neglected in chamber model simulation tools. The two governing equations are solved during runtime using the forward Euler method, which gives fluid density and specific energy as two independent fluid properties that allow calculating all other relevant fluid properties using an appropriate equation of state. In this work, Pollak's fundamental formulation for the thermodynamic properties of water is implemented [8].

Capacities can contain multiple control volumes for compressible fluid and liquids, where the governing equations are applied to all control volumes, and interaction is modeled using internal connections for heat and mass transfer. Mechanical equilibrium between the phases is assumed at all times, where the pressure of the compressible fluid regime is applied to all liquid phases within the same capacity. Furthermore, in this work, two-phase states within the working chambers are modelled in thermodynamic equilibrium, assuming rapid heat and mass transfer between the injected water and the surrounding steam. This approach is commonly used in the literature [7, 9]. However, because the high rotational speeds of screw compressors result in a short compression time, this assumption should be evaluated in future work. Further information on the simulation tool and a validation against experimental data is given in [10].

As an integral result of the thermodynamic simulation the compressor's inner power P_i, which is calculated by multiplying the pressure volume work per control volume with the chamber frequency can be obtained. Here, z_m is the number of lobes of the main rotor and l is the number of working chambers.

$$P_i = n\, z_m \sum_l \oint -p\, \mathrm{d}V \tag{4}$$

The low-pressure mass flow rate is calculated by integrating the time-dependent mass flows through the inlets, and the high-pressure mass flow rate is determined accordingly by integrating the flows through the outlets. Naturally, the displaced mass flow on the high-pressure side is increased by the injection mass flow compared to the suction mass flow. Compressor parameters and numerical parameters for this study are given in table 1. Convergence is typically reached after 20 iterations, with relative mass flow rate deviations between calculations of $\delta\dot{m} \approx 10^{-6} \ldots 10^{-5}$. Here, J is the number of iterations, and the relative mass flow rate deviation is calculated from:

$$\delta\dot{m} = \frac{|\dot{m}_J - \dot{m}_{J-1}|}{\dot{m}_J} \tag{5}$$

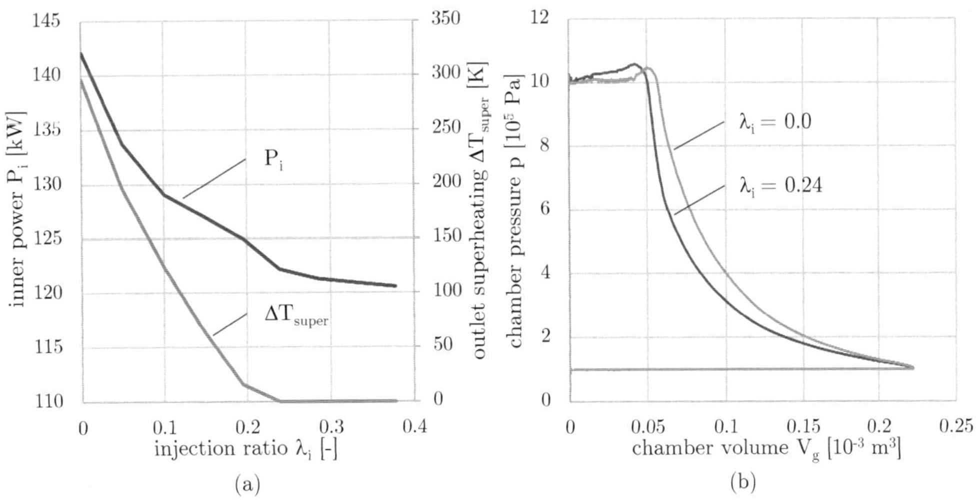

Figure 2. Results of chamber model simulations. (a): inner power and outlet superheating plotted against liquid injection rate. (b): pressure volume diagram for gate rotor chamber for various injection rates. $p_{lp} = 1 \cdot 10^5\,\mathrm{Pa}$, $p_{hp} = 10 \cdot 10^5\,\mathrm{Pa}$, $p_i = 6 \cdot 10^5\,\mathrm{Pa}$, $T_{lp} = 373.15\,\mathrm{K}$, $T_i = 353.15\,\mathrm{K}$, $n = 250\,\mathrm{s}^{-1}$

2.1. Results

In addition to the integral results, such as mass flow rate and inner power, which, in turn, determine volumetric efficiency and isentropic efficiency as input parameters for the heat pump cycle simulation, the thermodynamic simulation provides detailed information on all fluid properties during the compression cycle within the twin-screw compressor. Exemplary results of the compressor simulations are depicted in figure 2 . Comprehensive simulations for various operation points were carried out for the compressor in question, and results were transformed into a multi-dimensional characteristic map that serves as an input for the cycle simulations.

In (a), a variation of the liquid injection rate $\lambda_i = \dot{m}_i/\dot{m}_{lp}$ is presented, where $\dot{m}_i$ is the liquid injection mass flow rate and $\dot{m}_{lp}$ is the suction mass flow rate. In this study, the injection takes place immediately after the working chambers are disconnected from the suction side. Both the compressor's inner power and outlet superheating decrease with the injection rate. For an injection rate of $\lambda_i = 0.24$, the steam exits the compressor in saturated state for the given operation parameters. The inverse relationship between injection mass flow and outlet superheating is due to the liquid absorbing heat during compression, effectively reducing temperature while the injected liquid evaporates. Consequently, the compression process trends towards isothermal compression, which is also manifested in the decreasing power consumption. This effect becomes even more apparent when comparing the pressure volume diagrams in figure 2 (b). Despite compressing a larger fluid mass per cycle, substantially lower pressure-volume work is required for the liquid-injected process. Optimising water injection is hence essential for steam compression and enhances efficiency while simultaneously preventing damage from overheating.

3. Heat pump model

The simplified model of the high-temperature heat pump consists of four basic components: compressor, condenser, expansion valve and evaporator. Figure 3 gives an overview of the cycle and the corresponding pressure-enthalpy (p-h) diagram. The aforementioned screw compressor with liquid water injection is implemented in a functional manner, though not identical to the previously described injection process. The compressor increases the pressure of the wet steam at the inlet (1) to slightly superheated steam at the outlet (2). The amount of injected water (i) is regulated to ensure stable operating conditions. In this model a fictive mixing point (lp,mix) is implemented at the suction port, justified by the assumption of an instantaneous water injection in the enclosed maximum chamber volume just before compression begins. The condenser then transfers the heat of the steam to the heat sink. An adequate degree of subcooling at the outlet of the condenser (3) guarantees the full potential of transferring the latent heat.

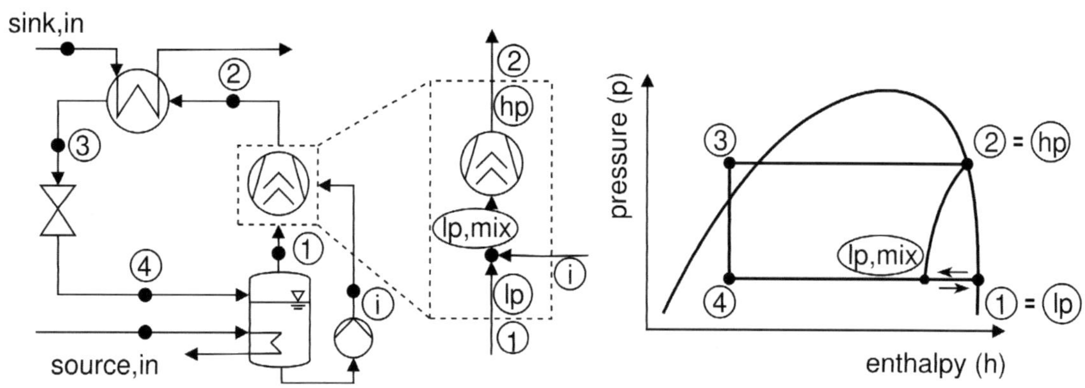

Figure 3. Schematic heat pump cycle and corresponding idealized p-h-diagram

The expansion valve decreases the pressure level and releases wet steam to the evaporator (4). Here, thermodynamic equilibrium is assumed, such that the steam, which passes on to the compressor, is always saturated and the filling has a constant temperature. The injection water is taken from the liquid phase of the evaporator. The ratio of liquid and steam mass flow (λ_i) is used to adjust the steam quality at the start of compression $q_{lp,mix} = (1 + \lambda_i)^{-1}$.

The model used is based on the assumption that the source and the sink stream have the same temperature. In essence, when the source inlet temperature is adjusted, the sink inlet temperature is as well.

3.1. Heat pump parameters

One of the most popular parameters to express the heat pump efficiency is the COP, which is the ratio of usable heat power $\dot{Q}_{23}$ to inserted power consumption $\Sigma_j P_j$. Even though there is a pump used in the model, its power consumption can be neglected due to the injection in the chamber immediately after closure, resulting in no pressure difference. Hence, the only relevant power consumption is the compressor's shaft power P_e.

$$COP = \frac{\dot{Q}_{23}}{\Sigma_j P_j} = \frac{\dot{Q}_{23}}{P_e} \tag{6}$$

The shown heat pump model can be compared with different ideal processes. The most commonly used reference process for heat pumps is probably the Carnot process. The efficiency of a heat pump for a specific operating point can be evaluated by using the Carnot efficiency η_{Carnot}, which is a reasonable indicator for heat pumps with low temperature glide [11]:

$$\eta_{Carnot} = \frac{COP}{COP_{Carnot}} = \frac{COP}{\frac{T_{sink,out}}{T_{sink,out}-T_{source,in}}} \tag{7}$$

The compressor's behaviour is determined by three parameters ($\eta_{vol}, \eta_{i,s}, \eta_{mech}$). Injecting water when the chamber reaches its maximum volume creates wet steam conditions. Since the state point (1) (in figure 3) does not include a liquid phase, it is not representative as a starting point for compression. Thus, the index low pressure after injection (lp,mix) is used where wet steam conditions can be defined and matched with corresponding simulation results from the detailed compressor simulations. The volumetric efficiency η_{vol}, defined as the ratio of the delivered mass flow to theoretically possible mass flow, is expressed by using the low pressure conditions after injection:

$$\eta_{vol} = \frac{\dot{V}}{\dot{V}_{th}} = \frac{\dot{m}}{V_{max} \cdot \rho_{lp,mix} \cdot n \cdot z_m} \tag{8}$$

The inner isentropic efficiency $\eta_{i,s}$ is defined using inlet and outlet enthalpies $h_{lp,mix}$ and h_2 and the outlet enthalpy of an isentropic compression process $h_{2,s}$. In essence, it represents the ratio of isentropic power P_s to inner power P_i, since the mass flow remains constant after the low-pressure mixing point:

$$\eta_{i,s} = \frac{P_s}{P_i} = \frac{h_{2,s} - h_{lp,mix}}{h_2 - h_{lp,mix}} \tag{9}$$

The shaft power P_e can be expressed in terms of the mechanical efficiency η_{mech} which accounts losses in the synchronisation gear, seals and bearings. It is held constant for all operating points at 0.95:

$$\eta_{mech} = \frac{P_i}{P_e} = 0.95 = const. \tag{10}$$

3.2. Compressor model

An essential part of the operating behaviour of the heat pump model is based on the compressor model. The more detailed simulation of a water-injected screw compressor is explained in an earlier chapter. It is essentially the source of data for the compressor model in the heat pump cycle, but the approach shown is applicable regardless of the source of data. The volumetric and inner isentropic efficiencies of the aforementioned simulation data points are not implemented directly. Instead, they are predicted based on a linear regression and incorporated in the compressor model of the heat pump.

The compressor model needs to be robust and responsive in terms of shifting operating points during simulation. Therefore, a linear regression is performed on the chamber model simulation results to extract the functions for calculating the parameters and to define the compressor operation.

The narrow operating range of the compressor within the heat pump cycle with respect to superheating temperature at the outlet $T_{2,super}$ and the mass fraction of injected water $q_{lp,mix}$ make an approach based on a linear regression model sufficient for a reasonable deviation between the chamber model simulation results and the corresponding model. Therefore, all drawn data points are filtered for $T_{2,super} < 10$ K and $q_{lp,mix} > 0.75$.

The compressor model is based on two functions for the volumetric and the inner isentropic efficiencies and dependent on the low pressure $p_{lp,mix}$, the low pressure quality $q_{lp,mix}$, the rotational speed n and the high pressure p_2. The index $pred$ indicates that the calculated values differ from those obtained in the original chamber model simulation:

$$\eta_{vol,pred} = \beta_{10} + \beta_{11} \cdot p_{lp,mix}^{\delta_{11}} + \beta_{12} \cdot q_{lp,mix}^{\delta_{12}} + \beta_{13} \cdot n^{\delta_{13}} + \beta_{14} \cdot p_2^{\delta_{14}} \tag{11}$$

$$\eta_{i,s,pred} = \beta_{20} + \beta_{21} \cdot p_{lp,mix}^{\delta_{21}} + \beta_{22} \cdot q_{lp,mix}^{\delta_{22}} + \beta_{23} \cdot n^{\delta_{23}} + \beta_{24} \cdot p_2^{\delta_{24}} \tag{12}$$

with $\beta_{ij} \in \mathbb{R}$ and $\delta_{ij} \in [-2, 2]$ and for $i \in \{1, 2\}$ and $j \in \{1, 2, 3, 4\}$.

As shown before, the linear regression is performed with a rudimentary transformation ("power transformation") [12]. The linear regression was performed using *scikit-learn 1.2.2*. The exponents δ_{ij} were optimised with respect to least squares errors in an usual range of [-2,2] [12]. The used library was *scipy 1.12.0*.

The residual plot in figure 4 gives an overview of the discrepancy between the original data and the simplified model by using linear regression. The efficiencies were normalized for confidentiality reasons ($\eta_{norm} = \eta/\eta_{max}$). While the volumetric efficiency seems to deliver reliable results (mean error: 0.94%, 95th percentile error: 2.5%), the inner isentropic efficiency requires more caution when implemented in simulations (mean error: 7.5%, 95th percentile error: 5.5%). Due to some outliers, the 95th percentile error is smaller than the mean. With the implementation of more complex models with additional (dimensionless) parameters the prediction quality increases. But so does the risk of overfitting, since the number of used data points was with 60 rather low.

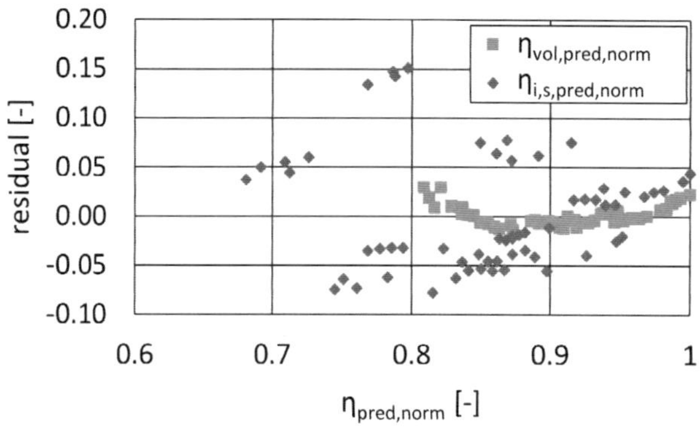

Figure 4. Residual plot of predicted efficiencies for given data points

As the residual plot indicates, the variance is non-constant for the aforementioned data points. Although the model requires further refinement, it performs better than using constant efficiencies. In comparison to the arithmetic mean of all 60 data points, the model reduces the error for both, volumetric efficiency (mean error: 4.9%, 95th percentile error: 13 %) and inner isentropic efficiency (mean error: 11%, 95th percentile error: 7.1%).

4. Heat pump cycle simulation

4.1. Boundary conditions

The thermodynamic simulation is performed with *Dymola 2023* (Dassault Systèmes©) with *TIL libraries (3.13.0)* (TLK Thermo GmbH). Table 2 gives an overview of implemented settings and boundary conditions, which were used in the following chapters. Both heat exchangers are defined by a pinch point temperature difference ΔT_{pinch} of 2 K. The enthalpy is assumed to stay constant via the expansion valve.

The injected compressor, as key element of the heat pump, needs multiple parameters for a complete definition. The volumetric and isentropic efficiency are calculated using functions mentioned above 11 and 12, in which losses of the compressor are included. The mechanical efficiency is assumed to be constant at 0.95. The maximum delivered volume flow is expressed as the conveyed volume per revolution ($V_{rot} = V_{max} \cdot z_m$) and the rotational speed (n), which are kept constant for all simulations. It is accordingly adjusted when the volumetric efficiency is taken into account.

Hydraulic and heat losses are neglected in several components of the heat pump (piping, heat exchangers). The injection pump is necessary to guarantee a fluid flow for the liquid injection and does not consume power, since there is almost no need to increase the pressure and hydraulic losses are neglected.

Table 2. Definition of key components in the heat pump model

Component	Parameters
Compressor	$T_{2,super} = 1$ K
	$V_{rot} = 1.65$ l
	$n = 250$ Hz
	$\eta_{vol} = f(p_{lp.mix}, q_{lp,mix}, n, p_2)$ (see equation 11)
	$\eta_s = \eta_{i,s} = f(p_{lp,mix}, q_{lp,mix}, n, p_2)$ (see equation 12)
	$\eta_{mech} = 0.95$
Condenser	$\Delta T_{pinch} = 2$ K
	$T_{3,sub} = 10$ K
Evaporator	$\Delta T_{pinch} = 2$ K
	$q_1 = 1$
Expansion valve	$h_3 = h_4$

Sink and source flows may differ in their mass flows but have the same inlet temperature, as shown in figure 3. The sink outlet temperature is a boundary condition of the simulation. In consequence, the mass flow is adjusted accordingly.

4.2. Simulation results

Each simulation result represents one specific operating point and the simulated time was set such that the convergence is guaranteed. The characteristics of the process can most easily be analysed in the pressure-enthalpy-diagram, shown for an exemplary operating point in figure 5. The heat pump raises the temperature from 100 °C on source side to 180 °C on the sink side. Due to necessity of a driving temperature difference and the pinch point temperature difference in the heat exchangers, the compressor needs to lift a lower temperature ($\vartheta_{lp,mix} = 91.5$ °C, $p_{lp,mix} = 0.74$ bar) to a higher level than the sink temperature ($\vartheta_2 = 181.9$ °C, $p_2 = 10.5$ bar). Consequently, the compressor needs to overcome a higher pressure ratio ($\frac{p_2}{p_{lp,mix}} = 14.1$).

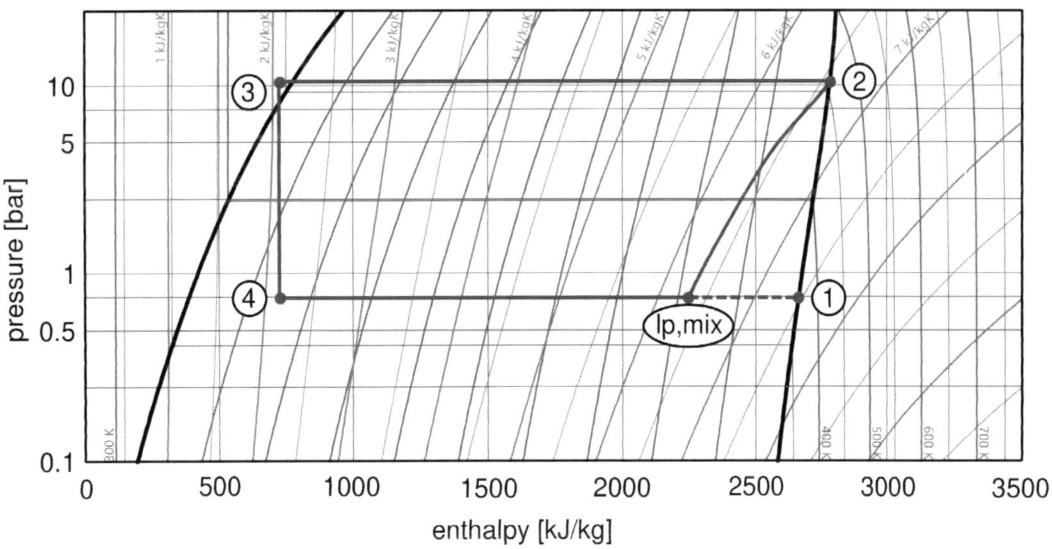

Figure 5. Characteristic heat pump operating point in p-h-diagram ($\vartheta_{source,in} = 100$°C, $\dot{m}_{source} = 10$ kg/s, $\vartheta_{sink,out} = 180$°C)

With the assumption of saturated steam conditions at the inlet and the outlet of the compressor, a lift of 100 °C to 180 °C has to be performed with a screw compressor ($\frac{p_2}{p_{lp,mix}} = 9.9$). In accordance with the temperature difference within the heat pump cycle, the same compressor might fail to achieve that temperature lift.

The injection of water at the start of the compression shifts the compression process in wet steam conditions, figure 5. The amount of water is adjusted, such that the outlet of the compressor is slightly superheated by 1 K. Without the use of injection water the outlet temperature would increase dramatically. For example, an isentropic compression starting in (1) would result in an outlet temperature of 405 °C. Besides the reduction of superheating, the injection has multiple beneficial effects like the reduction of power consumption and the reduction of gas leakage inside of the compressor.

4.3. Parameter study

In the scope of the performed parameter study simulations had to be executed, which go beyond scope of the compressor training data. Sink temperatures above 180 °C and source temperatures less than 90 °C are listed for the sake of completeness, but have to be checked in further studies and experiments.

Figure 6 gives an overview of the heat pump behavior for different sink an source temperatures for constant source mass flow of 10 kg/s and rotational speed of 250 Hz. The pressure ratio above 12 is greyed out to indicate the boundaries of the simulations.

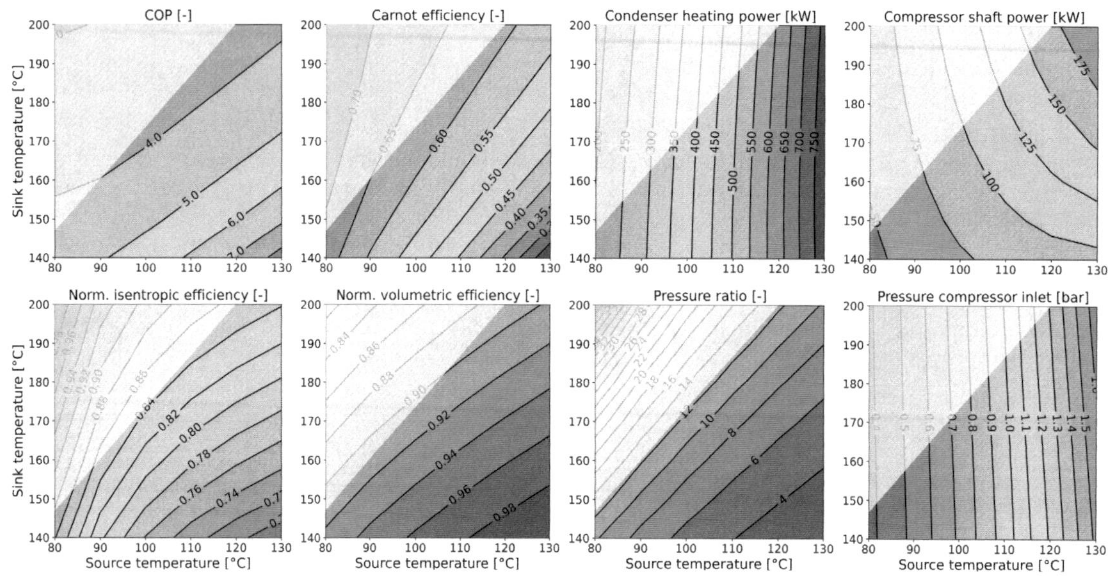

Figure 6. Operating range of high-temperature heat pump for source and sink temperature range ($\dot{m}_{source}$ =10 kg/s =*const.*, n =250 Hz =*const.*)

The COP increases with high source temperatures and low sink temperatures, figure 6, like the definition of the Carnot-COP (equation 7) indicates as well. The condenser power increases moderately while the compressor power grows proportionally with the source temperature. In consequence, the ratio of both parameters, i.e. the COP, increases. The compressor power increases with growing sink temperature and reduces the COP accordingly. The shown pressure ratio gives the corresponding information; the power consumption of the compressor increases with condensing pressure, while there is almost no effect on the heat flow in the condenser.

Interestingly, the Carnot efficiency decreases with increasing COP, as shown in figure 6. The aforementioned effect of the temperature difference in the heat exchangers becomes more pronounced with a decreasing temperature lift of the heat pump. The inlet pressure of the compressor is a good indicator of how much additional pressure lift is necessary. Therefore, in these simulations the pressure ratio of the compressor is always higher than the one of the ideal Carnot-based heat pump.

The efficiencies seem to follow a similar trend; a low temperature lift supports the volumetric efficiency and harms the isentropic efficiency. Again, the pressure ratio is a main driver for the behavior of the compressor. High pressure ratios favor leakage through gaps harm the volumetric efficiency. The isentropic efficiency on the other side increases with an increasing pressure ratio. This makes sense up to the point the design pressure ratio is achieved (approx. 7.5), throttling effects during discharging and leakages might increase this limit. Beyond this limit a further increase of the inner isentropic efficiency should be viewed with skepticism. Therefore, the greyed out area should not only be considered as the assumed technical limit of the compressor, but rather the validity limit of the compressor model within the heat pump.

5. Conclusion

This study presented a chamber model simulation of a water-injected screw compressor, enabling precise prediction of the compressor's performance for specific operating points. A linear regression approach was performed to integrate simulation data from the detailed chamber model into simplified models for heat pump cycle simulation. This method proved to be a robust and responsive solution, resulting in reasonable errors when compared to models assuming constant efficiencies.

Additionally, simulations were conducted for multiple operating points in a grid, where both source and sink temperatures were varied. The analysis revealed that the temperature difference between the heat exchanger sides is a critical factor influencing the COP and compressor behavior.

Future improvements will address the complex interdependencies of multi-phase thermodynamics in the chamber model simulation and the overall heat pump model. Enhancing the compressor efficiency model will be a major focus with the use of dimensionless parameters and additional data points as a foundation. Furthermore, the compressor model will be refined to account for liquid injection at various stages during the compression process. Experimental data will provide significant insights for refining both the chamber model and the cycle simulation. For this purpose, a detailed experimental investigation of the screw-compressor in question will be carried out shortly at TU Dortmund University. In addition, a comprehensive test rig for the entire high-temperature heat pump system is currently build up at Fraunhofer IEG in Bochum. Ongoing updates on the project's progress will be provided, with findings from experiments and simulations continually integrated into new compressor stages developed by Aerzener Maschinenfabrik GmbH.

Acknowledgements

Supported by: Federal Ministry for Economic Affairs and Climate Action on the basis of a decision by the German Bundestag

References

[1] IPCC 2023 AR6 Synthesis Report: Climate Change 2023 Tech. rep. Intergovernmental Panel on Climate Change Geneva, Switzerland DOI 10.59327/IPCC/AR6-9789291691647

[2] Wolf S 2017 *Integration von Wärmepumpen in industrielle Produktionssysteme* Phd thesis University of Stuttgart URL `https://elib.uni-stuttgart.de/bitstream/11682/9610/7/Dissertation_Integration_von_Waermepumpen_in_industrielle_Produktionssysteme_von_Stefan_Wolf.pdf`

[3] Agora Energiewende and Fraunhofer IEG 2023 The roll-out of large-scale heat pumps in Germany Tech. rep. Agora Energiewende Berlin DOI 10.24406/publica-1440

[4] Pachai A C, Normann J, Arpagaus C and Hafner A 2021 Screening of future-proof working fluids for industrial high-temperature heat pumps up to 250 °C *Proceedings of the 9th IIR Conference on Ammonia and CO2 Refrigeration Technologies* 0007 DOI 10.18462/IIR.NH3-CO2.2021.0007

[5] Chamoun M, Rulliere R, Haberschill P and Peureux J L 2014 Experimental and numerical investigations of a new high temperature heat pump for industrial heat recovery using water as refrigerant *International Journal of Refrigeration* **44** 177–188 DOI 10.1016/j.ijrefrig.2014.04.019

[6] Wu D, Jiang J, Hu B and Wang R 2020 Experimental investigation on the performance of a very high temperature heat pump with water refrigerant *Energy* **190** 116427 DOI 10.1016/j.energy.2019.116427

[7] Tian Y, Geng Y, Yuan H and Zhao Z 2022 Investigation on water injection characteristics and its influence on the performance of twin-screw steam compressor *Energy* **259** DOI 10.1016/j.energy.2022.124886

[8] Pollak R 1974 *Die thermodynamischen Eigenschaften von Wasser - dargestellt durch eine kanonische Zustandsgleichung für die fluiden homogenen und heterogenen Zustände bis 1200 Kelvin und 3000 bar* (Bochum: Ruhr-Universität Bochum)

[9] Yang J, Zhang C, Zhang Z, Yang L and Lin W 2016 Study on mechanical vapor recompression system with wet compression single screw compressor *Applied Thermal Engineering* **103** DOI 10.1016/j.applthermaleng.2016.04.053

[10] Grieb M and Brümmer A 2022 Thermodynamic simulation and experimental validation of an oil-free twin-screw expander *IOP Conf. Ser.: Mater. Sci. Eng.* **1267** DOI 10.1088/1757-899X/1267/1/012015

[11] Arpagaus C, Bless F, Uhlmann M, Schiffmann J and Bertsch S S 2018 High temperature heat pumps: Market overview, state of the art, research status, refrigerants, and application potentials *Energy* **152** 985–1010 DOI 10.1016/j.energy.2018.03.166

[12] Weisberg S 2005 *Applied Linear Regression* 3rd ed Wiley series in probability and statistics (Hoboken, N.J: Wiley-Interscience) ISBN 978-0-471-66379-9

IOP Conf. Series: Materials Science and Engineering **1322** (2024) 012019 doi:10.1088/1757-899X/1322/1/012019

Combined Rotor Rack Generation for Twin Screw Vacuum Pump Rotor Profile Design

Y Lu* and A Kovacevic

Department of Engineering, City, University of London, London, UK

E-mail: yang.lu.8@city.ac.uk

Abstract. Rotor profile design is a critical factor influencing the efficiency and functionality of twin screw vacuum pumps. This paper delves into a specific aspect of rotor profile design using the rack generation method. The paper utilises an algorithm of rack generation tailored specifically to twin screw vacuum pump applications. The utilization of straight lines to form the rack ensures the generation of involute profiles on the rotors. Additionally, rack-generated rotors are typically easier to manufacture. Geometrical parameters, such as displacement and meshing line length across three rack generated rotor profiles are explored. Furthermore, chamber modelling is employed for the development of rotor profiles, augmenting the efficacy and reliability of the rack generation process. By focusing on rack generation, this paper offers valuable insights into rotor profile design, contributing to a deeper understanding of the methodologies driving innovation in twin screw vacuum pump technology.

1. Introduction

A twin screw vacuum pump is a positive displacement machine where the main rotor and gate rotor rotate in opposite directions along parallel axes within a casing. To enhance pumping speed and minimize leakage from discharge to suction, the pump features a high area utilization coefficient and wrap angle. Variable pitch or conical rotors are also incorporated to increase the volume index, accommodating high-pressure ratio working conditions [1]. The performance of a twin screw vacuum pump is significantly influenced by the rotor profile, which can be designed based on a primary curve defined on either the main rotor, gate rotor, or rack [2]. Rack is a rotor with infinite radius can be used as a location of the primary arc [3]. Rack has been used for rotor profile design for twin screw compressor [4].In addition, rotor profile can also be designed based on geometrical characteristic sealing line [5-7]. Zaytsev [8] derived an analytical method for generating the profile of twin screw compressor rotors from a prescribed meshing line. Stosic [9] has developed a series of rotor profiles for vacuum and multiphase pumps. Performance calculations were conducted, and the results demonstrated a strong alignment with published experimental data. Three types of rotor profiles are commonly utilized in twin-screw vacuum pumps: Quimby, square-cut, and hybrid profiles as shown in Figure 1. In order to simplify the manufacturing process, both the main and gate rotors are designed to share the same rotor profile. The Quimby rotor profile is particularly prevalent due to its ability to align the cusp points of the main and gate rotors precisely with those of the casing. This alignment effectively prevents the formation of blowhole, thereby substantially reducing potential leakage issues. For twin screw vacuum pump rotor profile design, rack can be two straight line and with optimisation of the

IOP Conf. Series: Materials Science and Engineering **1322** (2024) 012019 doi:10.1088/1757-899X/1322/1/012019

connection between straight line. The presented method aids in developing rotors with a reduced contact line length and no blowhole areas, thereby enhancing compressor volumetric efficiency.

Figure 1. Three types of rotor profile.

2. Rotor profile design

In this section, a comprehensive introduction to the rack-based rotor profile generation method for the Quimby rotor profile was presented. Three fundamental design parameters: the base circle radius, addendum, and dedendum, were firstly defined. From these three fundamental design parameters, six additional parameters integral to the construction of the rotor profile were derived. The mathematical equations that describe the rack geometry were articulated, emphasizing the geometric relationships crucial for accurate profile generation. Subsequently, the algorithmic process that transforms the rack profile into the rotor profile was detailed, providing a rigorous step-by-step description of the transformation methodology.

2.1. Rotor design parameters

Quimby rotor profile consists of four segments which are outer arc, transition curve, inner arc and epitrochoid as shown in Figure 2. For the combined rotor-rack generation method, epitrochoid is a pre-defined curve and not integrated with rack. Only outer arc, transition curve and inner arc are generated from rack. The shortened rack profile for visualization is also shown in Figure 2. The basic rack comprises three straight segments: segment AB, segment BC, and segment CD. Segment AB is utilized to generate the inner arc of the main rotor and the outer arc of the gate rotor. Segment BC is responsible for generating the transition curve, while segment CD is used to generate the outer arc of the main rotor and the inner arc of the gate rotor. Length of AB is equation to length of CD since inner arc is meshing with outer circle. A connection curve can be added between segments AB and BC to create a smoother transition, thereby simplifying the manufacturing process. However, this approach is not discussed in this paper.

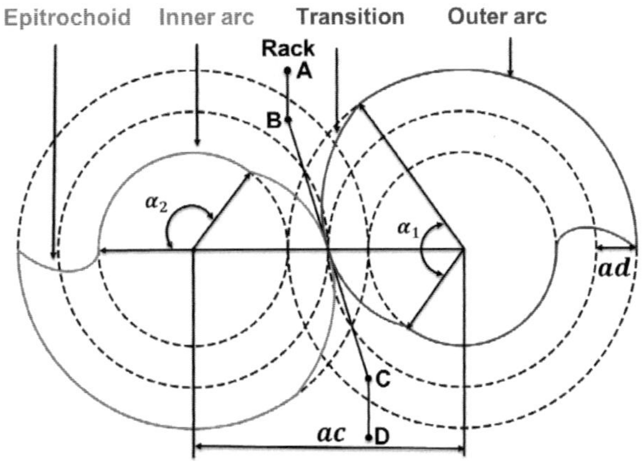

Figure 2. Consist of rotor profile and rack.

The three design parameters of the rack, as defined in Table 1, are the axis distance, addendum, and the angle of the transition curve. For the same lobe combination rotor profile, the addendum is equal to the dedendum.

Table 1. Design parameters.

Parameters	Symbol	Unit	Example
Axis distance	ac	*[mm]*	100
Addendum	ad	*[mm]*	15
Angle of transition curve	α_1	*[/]*	$1.2 \cdot pi$

Six design parameters—pitch radius, outer radius, inner radius, angle of the arc, and length of the rack—can be derived from three primary design parameters shown in Table 2. The expressions for these derived parameters are also provided in Table 2. Since the inner arc meshes with the outer arc, the angle of the inner arc is equal to that of the outer arc.

Table 2. Derived design parameters.

Parameters	Symbol	Unit	Expression	Example
Pitch radius	rp	*[mm]*	$0.5 \cdot ac$	50
Outer radius	ro	*[mm]*	$rp + ad$	65
Inner radius	ri	*[mm]*	$rp - ad$	35
Angle of arc	α_2	*[/]*	$0.5 \cdot (2 \cdot pi - \alpha_1)$	$0.4 \cdot pi$
Rack vertical length	L	*[mm]*	$2 \cdot pi \cdot rp$	314
Length of AB	Lab	*[mm]*	$\alpha_2 \cdot rp$	63

2.2 Equations of rack

The coordinate system is illustrated in Figure 3. Three fixed coordinate systems $X_1O_1Y_1$, $X_2O_2Y_2$ and $X_3O_3Y_3$ are defined for main rotor, gate rotor and rack individually. Two local coordinate systems $X_{01}Y_{01}$ and $X_{02}Y_{02}$ connecting with main rotor and gate rotor rotate anticlockwise and clockwise respectively. The motion parameter θ is same for main rotor and gate rotor.

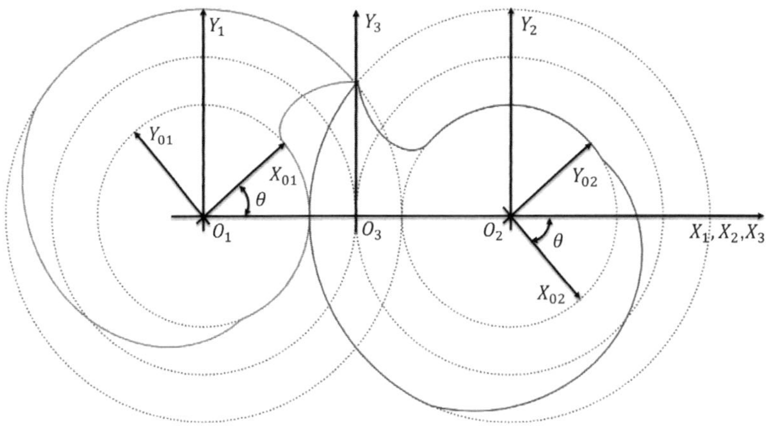

Figure 3. Coordinate system.

As mentioned in the section 2.1, there are three segments of rack which are defined in the rack coordinate system $X_3O_3Y_3$. The equations of three segments are expressed with defined parameters are shown in this section.

Segment AB is defined in equation 1 as:

$$\begin{cases} x_{AB} = -ad \\ 0.5 \cdot L \leq y_{AB} \leq 0.5 \cdot L - Lab \end{cases} \tag{1}$$

Segment BC is defined in equation 2 as:

$$\begin{cases} -ad \leq x_{BC} \leq ad \\ y_{BC} = -\dfrac{0.5 \cdot L - Lab}{ad} \cdot x_{BC} \end{cases} \tag{2}$$

Segment CD is defined

$$\begin{cases} x_{CD} = ad \\ -0.5 \cdot L - Lab \leq y_{CD} \leq 0.5 \cdot L \end{cases} \tag{3}$$

2.3 Transformation from rack to rotor profile

The procedure of rack based rotor profile generation for twin screw compressor is well explained by Stosic [10]. If the primary curves are given on the rack, their coordinates x_{0r} and y_{0r}, as well as their first derivatives dy_{0r}/dx_{0r} should be known and the generated curves will be calculated at the rotor as:

$$\begin{cases} x_{01} = x_{0r} \cdot \cos\theta - (y_{0r} - rp \cdot \theta) \cdot sin\theta \\ y_{01} = x_{0r} \cdot \sin\theta + (y_{0r} - rp \cdot \theta) \cdot \cos\theta \end{cases} \tag{4}$$

With meshing condition, θ can be solved analytically with the equation 5:

$$\frac{dy_{0r}}{dx_{0r}} \cdot (rp \cdot \theta - y_{0r}) - (rp - x_{0r}) = 0 \tag{5}$$

With the coordinates of generated curve and meshing angle, the contact line can also be calculated with the equation 6:

$$\begin{cases} x_{0c} = x_{01} \cdot \cos\theta - y_{01} \cdot sin\theta + ac \\ y_{0c} = x_{01} \cdot sin\theta + y_{01} \cdot cos\theta \end{cases} \tag{6}$$

2.4 Equations of epitrochoid

An epitrochoid is a geometric curve traced by a point fixed on a circle rolling around the outside of another fixed circle. The definition of epitrochoid curve is shown in equation 7. Parameter t represents the relative position between rolling circle and fixed circle.

$$\begin{cases} x_{01} = ro \cdot \sin(2 \cdot t) - \mathrm{ac} \cdot \sin(t) \\ y_{01} = \mathrm{ac} \cdot \cos(t) - ro \cdot \cos(2 \cdot t) \end{cases} \tag{7}$$

After the transformation process from the rack to the rotor profile, the rack-generated curve comprising the inner arc, transition curve, and outer arc can be obtained. By rotating the epitrochoid, it can be aligned and connected with the rack-generated rotor profile, resulting in the complete rotor profile

IOP Conf. Series: Materials Science and Engineering **1322** (2024) 012019 doi:10.1088/1757-899X/1322/1/012019

3. Geometry analysis of rotor profile

In this section, three different rotor profiles with different angle of transition curves are presented. The geometrical characteristics such as displacement and meshing line length are compared.

3.1. Geometry of rotors

The length of segment AB on rack determined the angle of the arc on the rotor profile which will affect the tip leakage length. Three different angle of transition α_1 are shown in Table 3. Other rotor design parameters such as addendum and pitch radius are shown in Table 1 and Table 2.

Table 3. Three cases parameters.

Parameters	Case 1	Case 2	Case 3
Angle of transition curve	$0.8 \cdot pi$	$1.0 \cdot pi$	$1.2 \cdot pi$
Angle of arc curve	$0.6 \cdot pi$	$0.5 \cdot pi$	$0.4 \cdot pi$

Figure 4 (a) shows the rotor profile and meshing line of three cases. A meshing line is formed between two rotors showing where two rotors contact. The epitrochoid is same. With the increasing of the angle of transition curve from $0.8 \cdot pi$ to $1.2 \cdot pi$, angle of arc curve reduced from $0.6 \cdot pi$ to $0.4 \cdot pi$. Transition curve of three cases intersect at same point on pitch circle. Figure 4 (b) shows the meshing line of three cases. Longer transition curve produced shorter sealing line length. Case 3 has the shortest meshing line comparing to Case 1 and Case2.

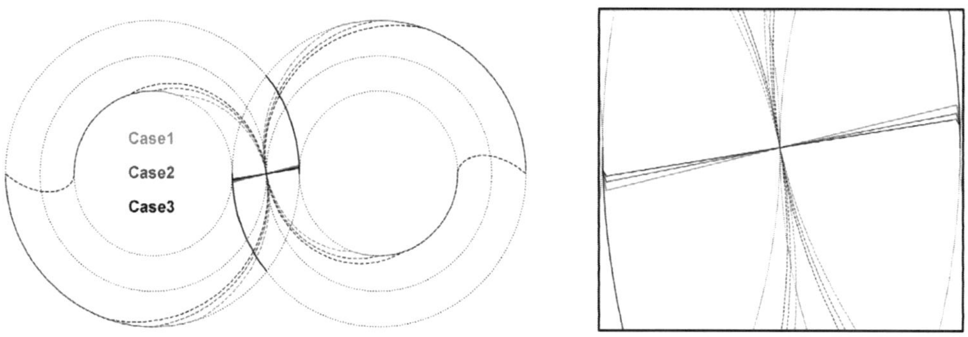

(a) Rotor profile (b) Meshing line

Figure 4. Comparison of cases.

3.2. Geometrical characteristics

To compare the geometrical characteristics, same rotor configurations are applied to three cases as shown in Table4. Rotor has constant pitch with axial discharge port. The designed volume index is 3. Rotor length is 350 mm and wrap angle is $4.0 \cdot pi$.

Table 4. Rotor configuration.

Parameters	Unit	Example
Rotor length	*[mm]*	350
Wrap angle	*[/]*	$4.0 \cdot pi$
Volume index	*[/]*	3

With the design parameters, displacement and meshing line length are compared for three cases. Displacement is calculated by integrate the cross-section area between two rotors and casing along the length of rotor. As shown in Table 5, Case 3 has largest displacement and shortest length of meshing line. Comparing Case 3 with Case 1, displacement increased by 1.6 % and length of meshing line reduced by 2.0 %. By increasing the angle of transition curve, profile would have advantage of larger displacement and shorter meshing line.

Table 5. Geometrical characteristic comparison.

Parameters	Unit	Case 1	Case 2	Case 3
Displacement	*[l/rev]*	*0.741*	*0.749*	0.753
Length of meshing line	*[mm]*	126.9	*125.3*	124.3

4. Performance analysis

The performance of the twin screw vacuum pumps with three rotor profiles are compared in this section. The calculation process of geometric parameters such as chamber volume, leakage area and discharge port area has been integrated to a standard software package SCORG [11]. The developed mathematic chamber model originally developed for twin screw compressor is used for the calculation of the volume flow rate and power consumption.

The working conditions for performance comparison are shown in Table 6. Working medium is air. Suction pressure is 0.1 *bar* and discharge pressure is 1 *bar*. Suction temperature is 293 *K*. This working condition makes sure the continuum flow inside the working chamber and the mathematic model developed for twin screw compressor is still valid for twin screw vacuum pumps.

Table 6. Working conditions.

Parameters	Unit	Example
Speed	*[rpm]*	8000
Suction pressure	*[bar]*	0.1
Suction temperature	*[K]*	293
Discharge pressure	*[bar]*	1

The volume flow rate, indicated power and specific power are compared for three cases and the results are shown in Table 7. Volume flow rate is calculated from inlet. By comparing Case 3 with Case 1, volume flow rate shows increasing of 1.93 % while power reducing of 1.95 %. Specific power reduced 3.97 %. It shows the performance improvement by increasing angle of transition curve.

Table 7. Geometrical characters comparison.

Parameters	Unit	Case 1	Case 2	Case 3
Volume flow rate	$[m^3/min]$	5.70	*5.77*	5.81
Power	*[kW]*	7.19	*7.13*	7.05
Specific power	$kW/m^3/min$	1.26	*1.23*	1.21

5. Conclusions

This paper presents details of combined rotor rack generation method for designing rotor profiles in twin screw vacuum pumps. Using this approach, three distinct rotor profiles were developed, each featuring different transition curve angles. These profiles were evaluated based on displacement and meshing line length, while volume flow rate, power consumption, and specific power were analyzed under identical working conditions. The findings indicate that this method offers significant flexibility in rotor profile design. An increase in the transition curve angle leads to profiles with larger displacement, shorter meshing lines, higher volume flow rates, and reduced power consumption and specific power.

Nomenlature

ac	*Axis distance*
ad	*Addendum*
α_1	*Angle of transition curve*
α_2	*Angle of inner or outer arc*
θ	*Meshing angle*
rp	*Pitch radius*
ro	*Outer radius*
ri	*Inner radius*
L	*Length of rack*
L_{AB}	*Length of AB*

References

[1] Li D, He Z, Wang C, Guo Y, Wei W, Lin D and Xing Z 2021 Design methodology and performance analysis of conical rotors for dry screw vacuum pumps *Vacuum* **185**

[2] Stosic N and Hanjalic K 1996 General Method for Screw Compressor Profile Generation. In: *International Compressor Engineering Conference at Purdue,* (West Lafayette, Indiana, USA

[3] Stos̆ic´ N and Hanjalic´ K 1997 Development and Optimization of Screw Machines With a Simulation Model—Part I: Profile Generation *Journal of Fluids Engineering* **119** 659-63

[4] Stosic N, Rane S, Kovacevic A and Smith I K 2016 Screw Rotor Profiles of Variable Lead Vacuum and Multiphase Machines and Their Calculation Models *Volume 6A: Energy*

[5] Wu Y-R and Fong Z-H 2008 Improved rotor profiling based on the arbitrary sealing line for twin-screw compressors *Mechanism and Machine Theory* **43** 695-711

[6] Nguyen T L and Wu Y R 2019 A Novel Rotor Profile Generation Method for Claw-Type Vacuum Pumps Based on Sealing Line *Applied Mechanics and Materials* **894** 34-40

[7] Hoang M-T and Tran T-V 2022 Methodology for generating a general profile of asymmetrical rotors of twin-screw vacuum pumps with point meshing feature by a pre-defined sealing line *Proceedings of the Institution of Mechanical Engineers, Part B: Journal of Engineering Manufacture* **237** 1582-95

[8] Zaytsev D and Infante Ferreira C A 2005 Profile generation method for twin screw compressor rotors based on the meshing line *International Journal of Refrigeration* **28** 744-55

[9] N. Stosic I K S, A. Kovacevic, E. Mujic 2006 Vacuum and Multiphase Screw Pump Rotor Profiles and Their Calculation Models. In: *Schraubenmaschinen 2006,* (Dortmund, Germany pp 53–66

[10] Stosic N, Smith I K and Kovacevic A 2005 *Screw Compressor Geometry Mathematical Modeling and Performance Calculation*: Springer)

[11] PDMAnalysis 2024 SCORG Help Manual

IOP Conf. Series: Materials Science and Engineering **1322** (2024) 012020 doi:10.1088/1757-899X/1322/1/012020

Design of toothed belt driven screw vacuum pumps

R Müller*, **T Dreifert and A Hellmig**

Leybold GmbH, Cologne, Germany

*E-mail: roland.a.mueller@atlascopco.com

Abstract. Most screw vacuum pumps use oil-lubricated gears for timing of the rotors. An alternative is a toothed belt to drive and synchronise the screw rotors. In combination with grease-lubricated bearings this drive concept allows the design of completely oil-free pumps. However, belts are less accurate in terms of timing, hence the design of efficient pumps is challenging. In this paper, the development of a new series of belt driven dry screw vacuum pumps is discussed. It covers the overall pump concept including cooling system, material selection and noise damping. Particular attention is paid to the design of the screw rotors with respect to the properties of the belt drive. It is shown that the disadvantage of the inaccurate timing and the need for wide rotor to rotor clearance can be overcome with a high number of wraps. Furthermore, it gives insight to applications and practical field experience.

1. Introduction

Twin shaft pumps are widely used as dry vacuum pumps such as Roots, claw, or screw pumps. All mechanisms need precise timing of both rotors for contact-free operation. In most pumps this is achieved by a pair of gears, which typically need oil lubrification [1]. An alternative is an individual drive for both shaft and a control device to synchronise the rotor electronically [2].

Yet another option is the use of a toothed belt as described in [3]. The belt runs on three pulleys: one on each pump shaft and a third one on the motor shaft. In this case, the belt drive synchronises the pump shafts and at the same time provides a speed transmission to the motor drive. This makes the up speeding of the pump shafts simple as no additional gear or inverter driven motor is needed.

Furthermore, the belt drive does not require any lubrication, it can run dry on the pulleys. The combination of a dry-running belt drive and grease lubricated bearings allows the design of a completely oil-free pump. This is from an application point of view desirable as a contamination of the process gas can be excluded. Additionally, it simplifies the design of the shaft seal between the gear box and the pump chamber as it does not need to seal against oil. Here, a relatively simple labyrinth seal is sufficient to protect the gearbox from particles in the process gas flow.

Despite the advantages of the belt drive no pumps with this concept have appeared in the market until recently. The main reasons are the inaccurate timing in comparison to gears and the limited life of the belt. However, due to technical improvements in belt technology both precision and lifetime have improved significantly. This makes belts in general feasible for twin shaft pumps. The issues of rotor timing and life of the belt must be addressed in the design of the pump. One example for such a pump is the "Varodry" pump series produced by Leybold GmbH, which is discussed in the following sections.

2. Basic design of pump

The "Varodry" pump series are screw pumps with a pumping speed of 65 -200 m^3/h. The rotors are simply supported and have a belt drive at the exhaust side. As in figure 1, the belt drive provides an up speeding of the rotors to approx. twice the motor speed. Greased lubricated bearings are used at both inlet and exhaust side, hence no oil is needed in the pump. The bearings are grease lubricated for service life. The seals are a combination of piston rings and simple labyrinths. If needed, purge gas can be supplied to the seals to protect the bearings from process gas. The belt section is covered with an enclosure. It protects the belt from particles coming from the environment and reduces the noise level of the pump. In case of a belt failure, the additional fail-safe gears ensure a safe run-down of the pump. During normal pump operation, the fail-safe gears are not in contact und run with a small clearance. For belt service the enclosures can easily be disassembled, and the belt be replaced.

Figure 1. Basic concept of belt driven twin shaft machine. The belt synchronises the two shaft (left hand side) and provides an up speeding of the pump shafts.

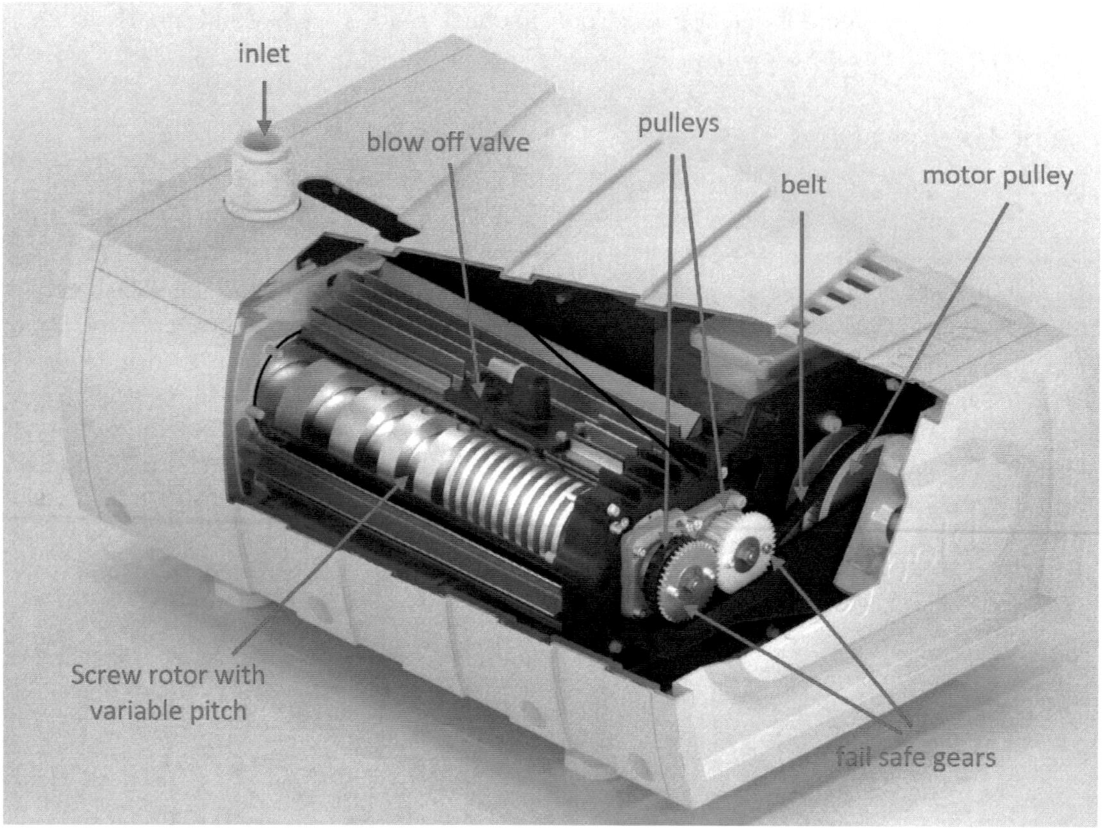

Figure 2. Structure of belt driven screw vacuum pump "Varodry"

The pump is fully air cooled. A fan is mounted on the motor shaft, and the air flow is guided inside the enclosure to the finned pump housing.

The pump housing and the rotors are made from aluminium alloys. It has the advantage of a high heat conductivity, favourable machining properties and good corrosion resistance. In addition, the rotors are anodised which improves their robustness. For modularity, the pump housing is made from an extruded profile. This allows simple variation of the housing length for different pump sizes.

All rotors have a built-in volume ratio. To avoid too high over compression and power consumption a "Blow-Off" valve is installed. It is a simple gravity actuated ball valve that guides the process gas into the exhaust plenum of the pump. For pumping vapours, the pump can be equipped with gas ballast.

The silencer of the pump is positioned underneath the pump housing. For cost and corrosion reasons, it is made from plastic. It consists of two chambers which sizes are adapted the exhaust gas pulsation frequency.

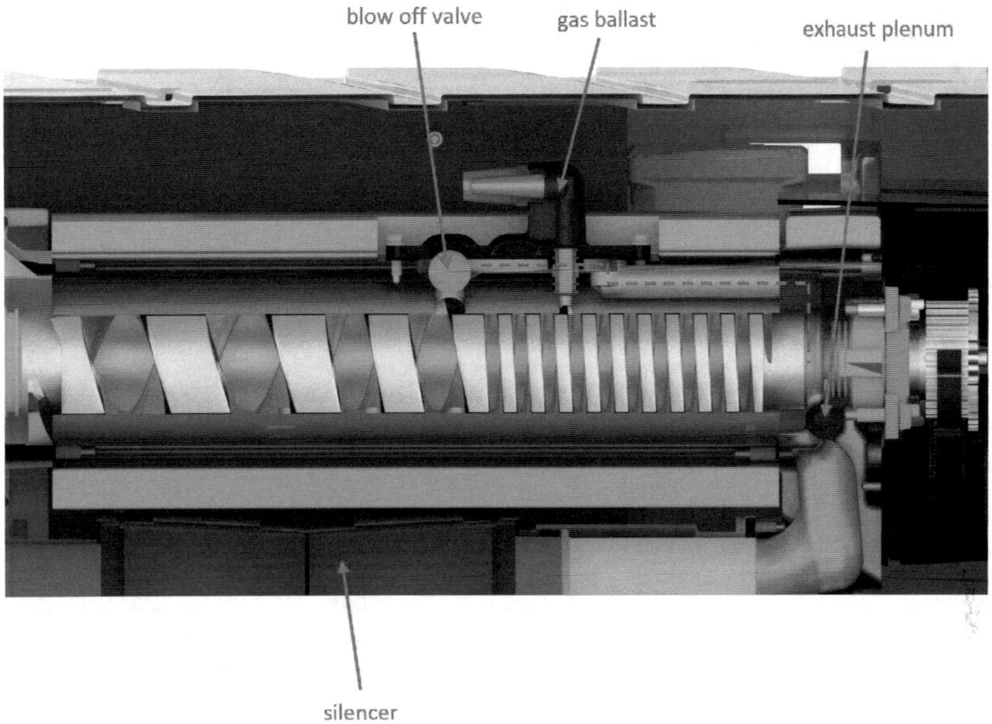

Figure 3. Cross section of the pump

3. Rotor design

As discussed above, the timing of the belt drive is less accurate than the conventional solution using gears. Tests have shown that with this kind of belt drive an angular error of up to +/-1° can be expected [4]. This is significantly more compared to conventional timing gears and hence the rotors must have sufficiently wide flank-to-flank clearances to avoid rotor contact during operation.

The rotors have a non-constant pitch to realise the desired volume ratio. The flank clearance is adapted to the local pitch to achieve the desired angular play:

$$C_{flank} = \frac{p}{360} \cdot \Delta\alpha \tag{1}$$

p is the local pitch (in mm/revolution) and $\Delta\alpha$ the targeted angular play (1° in this case). Due to the dependency on the local pitch, the flank clearance is not constant along the rotor. At the inlet side with high pitch, the flank clearance is much wider than at the exhaust side with low pitch.

The wide flank clearances cause relatively high back leakage. In principle, this reduces the volumetric efficiency and gives a higher ultimate pressure. To compensate for this, a high number of rotor wraps is needed. For efficient operation in the fine vacuum region, a high number of small exhaust wraps is needed.

In screw vacuum pumps two basic types of profile are established [5]. One is the symmetrical cycloid profile, the other one is the non-symmetrical "Quimby" type profile (see figure 4).

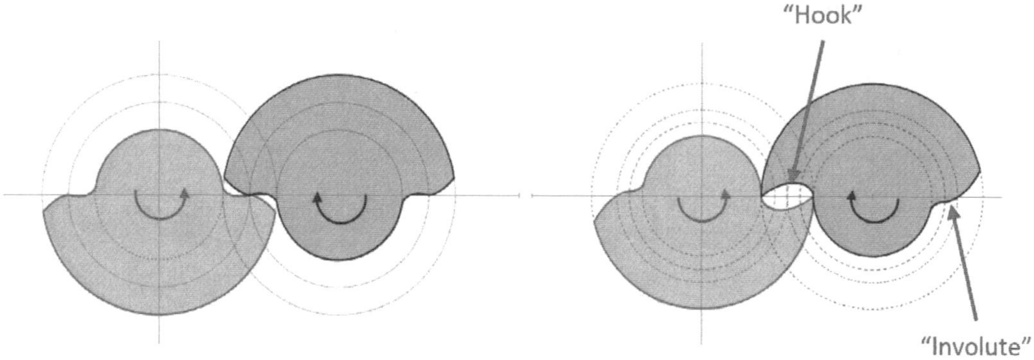

Figure 4. Symmetrical cycloid profile (left), Quimby profile (right)

The Quimby profile connects the working chambers of both rotors via a relatively wide blow hole, so it forms pairs of working chambers with approximately the same pressure. These pairs are well sealed against the upstream pair of working cells as there is virtually no blow hole between the pairs. This property makes the Quimby profile well suited for highly efficient pumps. The drawback is the undercut of the "Hook" flank, which difficult to machine.

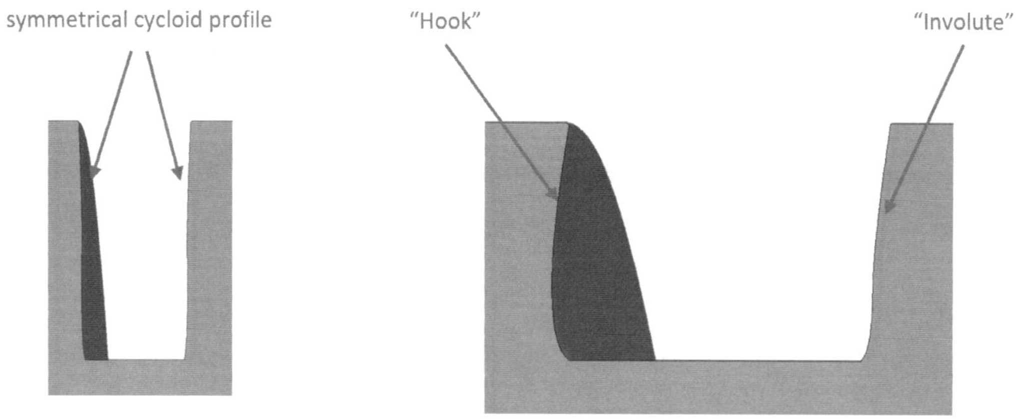

Figure 5. Axial cross section of cycloid profile (left) and Quimby profile (right)

The symmetrical profile forms blow holes between all working chambers, which makes the profile less efficient. The size of the blow holes depends on the pitch; for small pitch the blow holes are small. An advantage of this type of profile is the symmetry and the small undercut, which simplifies machining. In case of a constant pitch, both flanks have the same shape. This is advantageous for manufacturing as both flanks can be made with a form cutter which moves through the middle of the groove and cuts both flanks at the same time. This allows cost effective machining of a high number of wraps, particularly if combined with aluminium as the material. All these properties make the symmetrical profile suitable for the exhaust side [6,7,8].

On the inlet side the symmetrical profile would create high back leakage due to large blow holes, hence the Quimby profile is more suitable for the inlet side. The advantages of both types can be combined into a single rotor if the symmetrical profile is used for the exhaust side and the Quimby profile for the inlet side. The rotor consists of multiple sections with constant, but different pitches, and the exhaust side has the symmetrical profile whereas the inlet side has the Quimby profile. At the transition points of the sections the rotors do not mesh, and the tool run out creates a recess. As this is small and causes a leakage only when engaging with the other rotor, it hardly affects the overall pump performance.

4. Pump performance

Figure 6 shows the effective pumping speed curves of the "VARODRY 200" and the "VARODRY 65". Both pumps reach and ultimate pressure of 0.01 mbar. The peak speed is reached at approximately 5 mbar inlet pressure. Due to the internal compression and back leakage, the pumping speed drops from the peak and reaches a minimum at approximately 250 mbar inlet pressure. At this point the blow-off opens, and the pumping speed rises again.

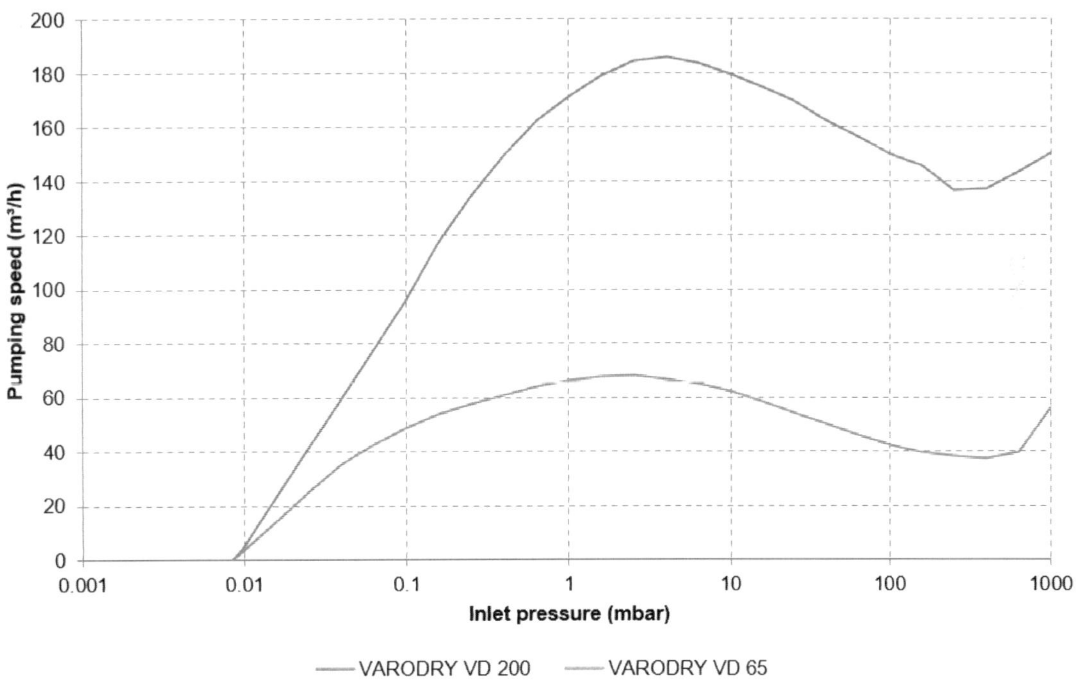

Figure 6. Pumping speed curves of VARODRY 200 and VARODRY 65

The power curves of both pumps are shown in figure 7. Both pumps reach their maximum power at approximately 250 mbar inlet pressure. For higher inlet pressure, the power drops as over compression is reduced by the blow-off valve. At low inlet pressure, the effect of the build-in volume ratio becomes apparent. The VD200 has a higher volume ratio compared to the VD65 which leads to relatively low power consumption at low pressure.

This becomes more obvious by comparing the "Pumping Efficiency" of both pumps. It is defined by the ratio of pumping speed to power consumption:

$$PE = \frac{S}{P} \quad (2)$$

Simply speaking, the value states how much pumping speed is generated per input power, conveniently expressed in the unit $(m^3/h)/kW$. A high value means high pumping speed for a given power, i.e. an efficient pump. The pumping efficiency depends on the inlet pressure and is shown in the curves in figure 8 for both pumps. Due to the higher volume ratio the VD200 reaches a higher peak efficiency compared to the VD65. Also, in the atmospheric decade up to approx. 800 mbar inlet pressure it is more efficient than the VD65 which can be explained by the higher volumetric efficiency in this pressure range.

In addition, the pumping efficiency curve of a typical oil-sealed rotary vane pump of the 200 m^3/h class is plotted. Except for the atmospheric decade, it is less efficient than the Varodry 200 pump due to friction losses.

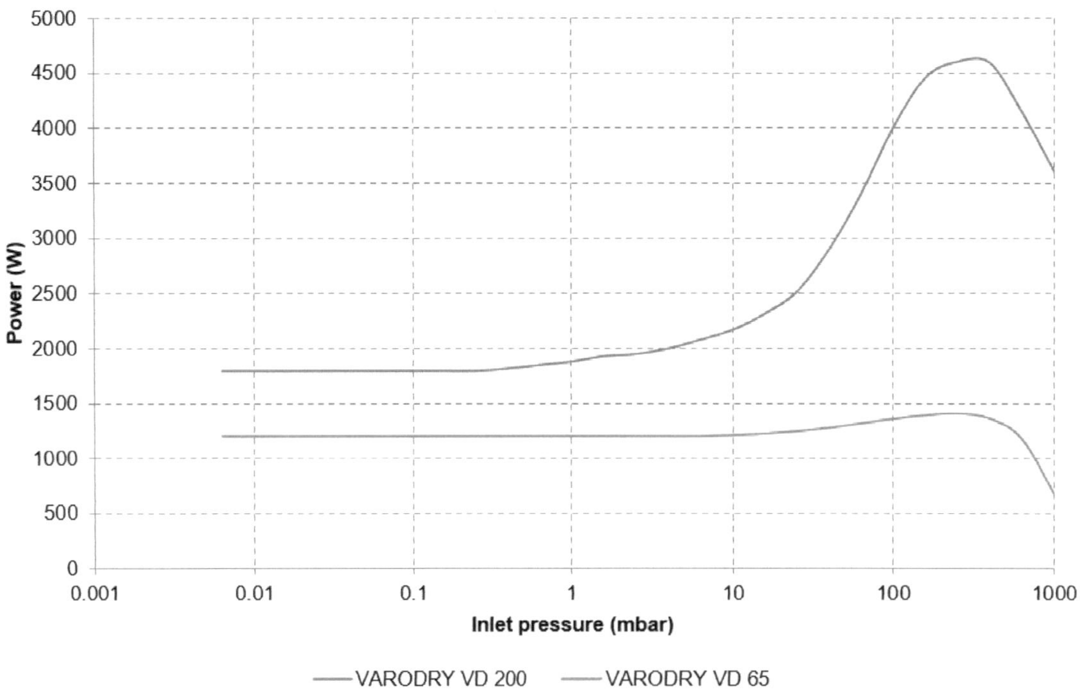

Figure 7. Power curves of VARODRY 200 and VARODRY 65

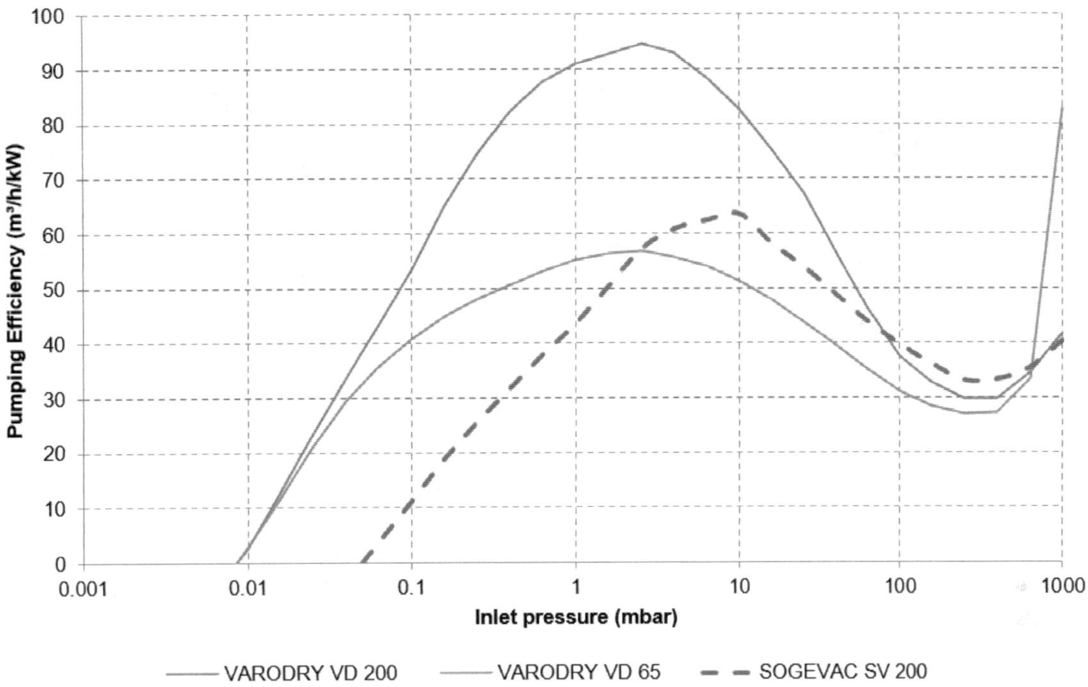

Figure 8. Pumping efficiency curves of VARODRY 200, VARODRY 65 and SOGEVAC SV200 (oil sealed rotary vane pump)

5. Application examples

The belt-driven pumps are used in a variety of industrial applications. One example is coating of electrical components with Parylene. Parylene is a transparent polymer that protects the components from outside influence. In the coating process the Parylene vapour desublimates on the substrate at low pressure. Residuals of the vapour enter the vacuum pump and must be handled by it. A dry screw pump is particularly suited as it can transport the vapour without contaminating any oil.

Lithium-Ion battery production involves many steps where vacuum is needed. Examples are leak detection and electrode production. The process steps involve the handling of water vapour and aggressive solvents by the pump. The dry and corrosion-protected design makes the pump well suited for this type of application.

The pumps are also used in food industry, e.g. for freeze drying. In this process, food is dried at low temperature and low pressure, and it releases large amount of water vapour. Along with the water vapour organic acids enter the pump. Commonly, liquid ring or rotary vane pumps are used for this type of application. A dry solution offers the advantage of no contamination of water or oil. Despite higher installation costs, the dry solution is therefore cost competitive. A costumer installation is shown in figure 9, where two VARODRY200 pumps successfully replaced a liquid ring pump.

Figure 9. Food drying installation with two 200 m^3/h belt driven screw pumps

Yet another food application is the harvesting of maple sap. In this process, the sap is taken from the maple trees via a vacuum line. The oil free design of the pumps ensures food-safety. The pumps are only operated during the harvesting period of approx. 2 month per year. Even after the long shutdown time the pumps can restart without any maintenance.

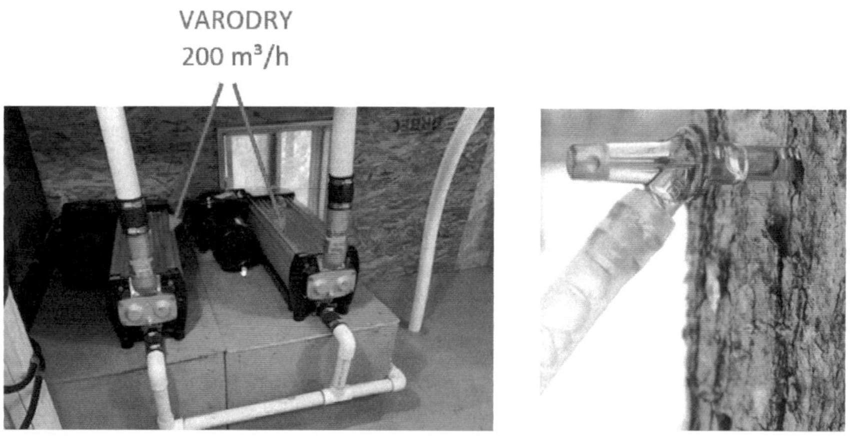

Figure 10. Maple sap harvesting, pump installation (left), tree connection (right, from [9])

6. Conclusion

The VARODRY pump series proves that belt drives are generally suited for industrial screw vacuum pumps. The design of the pump must be adopted to the properties of the belt, in particular the issue of inaccurate timing of both rotors has to be addressed. The discussed rotor design overcomes this issue and provides high pumping speed and low ultimate pressure.

In the following table the differences to conventional pump in the pumping speed range are highlighted:

Table 1. Comparison of belt driven and conventional pumps.

	VARODRY	Conventional pumps [10,11,12]
Rotor timing	Belt	Oil lubricated gears
Bearing lubrication	Grease	Oil
Rotor, stator material	Aluminium	Gray cast iron
Cooling system	Air cooling	Air / water cooling
Maintenance	6000 h belt change	5000 h – 9000 h oil change
Ultimate pressure	0.01 mbar	0.05-0.1 mbar

The oil-free design makes the pumps well suited for industrial applications in which oil contamination must be avoided. Additionally, both aluminium rotor and aluminium stator provide good corrosion resistance against water and many solvents, which makes the pump suitable for drying applications. On top of that the air cooling and its high energy efficiency makes the pump attractive and very cost competitive. By design, the pump can easily be serviced, and the belt replacement can be done by the customer.

References

[1] Jousten K 2018 *Handbuch Vakuumtechnik* 12. Auflage Springer Vieweg
[2] Flowserve SIHI Germany GmbH 2016 *SIHI Dry Produktbroschüre*
[3] Steffens R 2005 *Patent specification DE 103 34 481 A1*
[4] Dreifert T et al 2016 *Patent specification DE 20 2016 005 208 U1*
[5] Dreifert T and Müller R 2014 *Screw vacuum pumps* Proceeding of ICSM
[6] Huck C et al 2014 *Thermodynamical operation behaviour of screw vacuum pump with Cycloid- and Quimby-toothed rotors* Proceedings of ICSM
[7] Malov D et al 2020 *Analysis or Profile of Screw Vacuum Pumps* ICSAS Proceedings
[8] Dreifert T et al 2016 *Patent specification DE 10 2016 216 279 A1*
[9] CDL Catalog 2024
[10] Leybold GmbH 2023 *LEYVAC LV80 LV140 Operating Instructions*
[11] Dr.-Ing. K. Busch GmbH 2022 *Instruction Manual COBRA NC 0100-0300 B*
[12] Gardner Denver Schopfheim GmbH 2011 *S-VSI 100 S-VSI 300 Product leaflet*

IOP Conf. Series: Materials Science and Engineering **1322** (2024) 012021 doi:10.1088/1757-899X/1322/1/012021

Investigations to reduce rarefied gap flows within positive displacement vacuum pumps by utilising surface structures

S Brock[1*], H Pleskun[1], G Polus[2], J Saelzer[2], D Biermann[2] and A Brümmer[1]

[1]Chair of Fluidics, TU Dortmund University, Dortmund, Germany
[2]Institute of Machining Technology, TU Dortmund University, Dortmund, Germany

*E-mail: sven.brock@tu-dortmund.de

Abstract. The ultimate pressure and efficiency of positive displacement vacuum pumps is limited by the backflow through the gaps within the machines. These gap flows are typically reduced by decreasing the gap height, which is limited for reasons of operational safety. Therefore, this study concentrates on the reduction of rarefied gap flows by means of suitable surface structures whose dimensions are small compared to the gap height without simultaneously reducing the minimum gap height. For this purpose, a method is presented to efficiently calculate the impact of the surface structure on these gap flows. Based on this method, design parameters for the choice of a suitable surface structure are discussed. Furthermore, a possible transfer for the application on the rotors is shown including a proposal for the machining process.

1. Introduction

Numerous industrial processes require vacuum technology, whereby vacuum pumps are often required to provide clean vacuum, meaning without contamination from auxiliary products, or to be resistant to process-related contamination. Both of these requirements are met by rotary positive displacement vacuum pumps, resulting in screw vacuum pumps becoming established in the fine vacuum range and multi-stage Roots pumps in the high vacuum range.[1, 2, 3]

Screw vacuum pumps consist of two helical rotors that mesh with each other and are surrounded by a housing. Due to the large pitch angle, many working chambers are created between the teeth and housing, in which the fluid is transported axially and compressed from the low-pressure side to the high-pressure side by the rotation of the rotors. The working chambers are connected to each other and to the high-pressure and low-pressure ports by gaps, whereby the housing gap has by far the greatest influence due to its large width. Due to the geometry of the housing gap the rarefied gas flow that occurs can be abstracted as a superposition of a pressure-driven flow and a drag flow in the plane-parallel gap. [4, 5, 6, 7, 8]

These gap connections cause backflow, which can have a strong influence on the operating behaviour. This results in energy losses due to the re-expansion and also reduces the pumping speed and ultimate pressure [9, 10]. However, the backflows also reduce the overcompression at high intake pressures, a condition that leads to large losses and high thermal stress, particularly during start-up [10, 11].

Every fluid is composed of countless particles in chaotic motion, constantly colliding with one another and with the boundary walls. For rarefied flows, wall interactions predominate, which can be described well by diffuse scattering in technical applications.[12] Due to multiple interactions with the roughness peaks, the particle is scattered from the wall with a random speed and direction, but on average exactly normal to the wall with a speed proportional to the square root of the wall temperature. Figure 1a shows the resulting cosine distribution (red) for a rarefied flow in the plane-parallel gap. In case of a moving lower wall, the distribution function is distorted compared to that of the static upper wall by the influence of the wall velocity c_W.[13, 14]

If a structured surface is used as shown in figure 1b, the particles are often thrown back in the opposite direction when interacting with one of the flanks. If, due to a fluid flow, more particles interact with one flank than with the other, the particles overall tend to be scattered against the flow direction. This causes the flow rate $\overline{\mathbf{c}}$ to be lower than in the case of smooth walls. Investigations by Sazhin show that this behaviour can reduce the mass flow rate of a rarefied pressure-driven flow through a plane-parallel gap with sawtooth structures by up to 25 %. This effect is caused by wall interactions, thus is diminishing with an increase in gas rarefaction parameter until it vanishes when approaching continuum flows.[15]

This article will therefore examine the influence of such structures on the gap flow in more detail, especially considering the impact of the geometry and extending this approach to Couette flows. A novel wall reflection model for structured surfaces enables the effective calculation of such flows with the Direct Simulation Monte Carlo (DSMC) method or Monte Carlo (MC) method. After validation different geometries for surface structures are compared and a possible implementation in rotary positive displacement vacuum pumps is discussed.

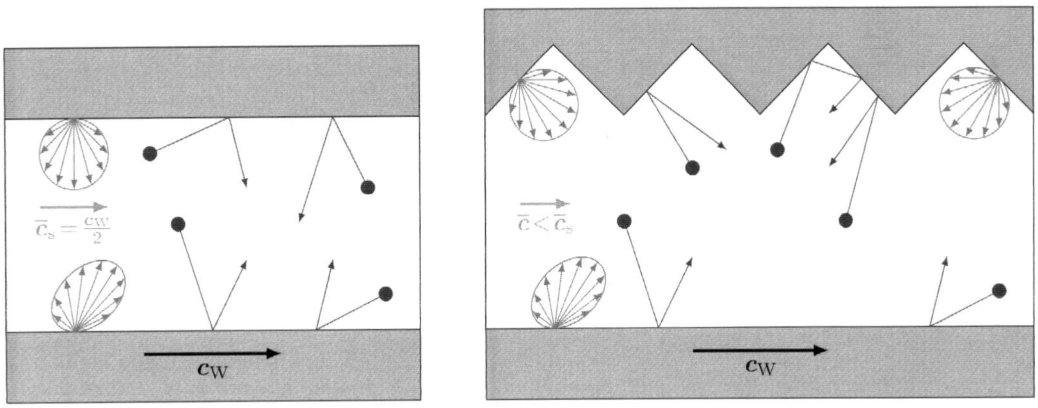

(a) smooth walls **(b)** one wall with surface structures

Figure 1. Rarefied Couette flow in plane-parallel gap, considering individual particle movement with diffuse scattering. The flow rate $\overline{\mathbf{c}}$ with surface structures (1b) is lower than the flow rate $\overline{\mathbf{c}}_s$ with smooth walls (1a), because particles tend to be scattered by the surface structures in opposite direction to the fluid flow.

2. Simulation of rarefied gap flows

In the following two common simulation approaches for rarefied flows will be addressed and used to calculate the Couette and Poiseuille flow in the plane-parallel gap with and without surface structures. While the Couette flow is often discussed in the infinitely long and wide gap for generalisation, it is necessary to use a finite length for the Poiseuille flow due to the pressure gradient. Therefore, different simulation domains are used for these flow types, both of which conform to the following definition according to figure 2a.

Here, only plane-parallel gaps are considered, whereby the plate spacing represents the gap height H and the flow occurs along the gap length L. The gap width $W \to \infty$ is infinite, so that the flow is invariant to this direction and can therefore also be described in two dimensions.

All surface structures shall be two dimensional as well as periodic and initially the ideal sawtooth profile, which is shown in figure 2b, will be examined. The orientation of the two flanks can be described by the angles α, β, where $\alpha + \beta$ form the opening angle of the sawtooth and thus a smooth wall is described by $\alpha = \beta = 90°$. The structure height h expresses the depth of the grove. The tip is the edge that protrudes into the main flow and the root describes the deepest point of the profile in the wall. The gap height H is not influenced by the surface structures, as it is measured between the structure tips and

IOP Conf. Series: Materials Science and Engineering **1322** (2024) 012021 doi:10.1088/1757-899X/1322/1/012021

the opposing wall. For the structure to have an influence on the flow, it must be positioned perpendicular to the direction of flow, in other words along the width.

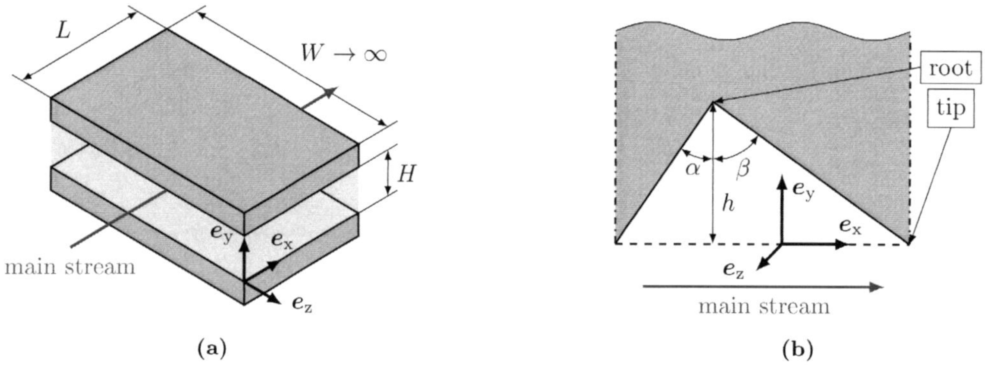

Figure 2. Definition of (2a) plane-parallel gap and (2b) ideal sawtooth profile.

2.1. *DSMC method*

The DSMC method is a possible way to simulate gas flows by solving the Boltzmann equation. Its main features are the representation of many real particles, meaning atoms or molecules, by one simulated particle and the decoupling of particle movements and intermolecular collisions. Different effects of the fluid flow, for example wall interactions, intermolecular collisions or chemical reactions, can be implemented by various statistical models.[13]

This work employs the solver *dsmcFoam+* of the extension *MicroNanoFlow* from the University of Edinburgh for the open-source project *OpenFOAM* [16, 17]. The No-Time-Counter and Acceptance-Rejection methods are used for determining collision partners and the Variable-Hard-Sphere model is applied to calculate the velocity vectors for each binary, intermolecular collision [13]. Wall interactions utilise the fully diffuse reflection model [14].

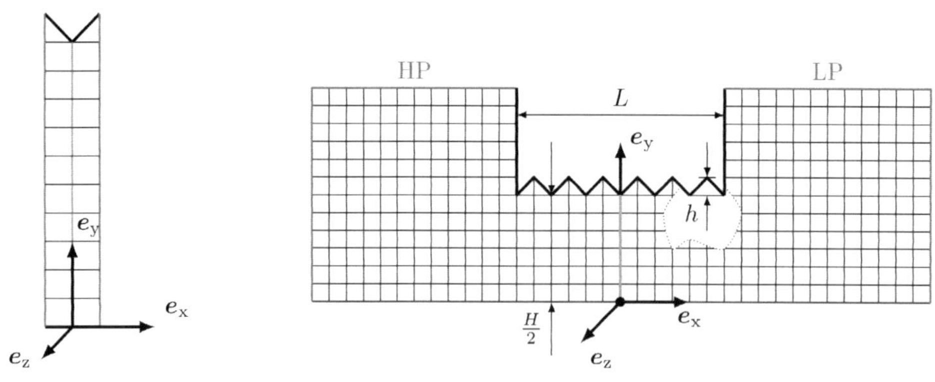

(a) Couette flow (b) Poiseuille flow

Figure 3. Meshed domains of (3a) Couette and (3b) Poiseuille flow with geometrically resolved surface structures for DSMC simulations. The domains use walls (thick), reference planes (green), symmetry planes (blue), cycling boundary conditions (red) and pressure boundary conditions (magenta) respectively and are both one cell wide (not displayed) in direction $\boldsymbol{e}_z$.

The plane-parallel gap of infinite width is simulated using one cell in lateral direction with symmetric boundary conditions. As shown in figure 3a, the Couette flow can utilise cyclic boundary conditions to simulate an infinitely long gap with two cells, while this is not applicable for the Poiseuille flow because of the underlying pressure gradient from high pressure (HP) to low pressure (LP). Therefore, according to figure 3b, the whole channel length is meshed and pressure chambers are added to take inflow and outflow losses into account. The Poiseuille flow uses surface structures on both plates of the gap so that another symmetry plane can simplify the domain height-wise. In contrast the Couette flow must employ a structured and a smooth plate, since for two plates with the same reflection behaviour the flow velocity is always equal to the mean of both wall velocities [18, 19]. In both cases, the mass flow rate $\dot{m}$ through a reference plane is measured as difference of the particle transfer per time step Δt, where $\sum_i m_{+,i}$ and $\sum_i m_{-,i}$ represent the total mass of all particles crossing the plane in positive or negative directions respectively.

$$\dot{m} = \frac{\sum_i m_{+,i} - \sum_i m_{-,i}}{\Delta t} \tag{1}$$

2.2. *Monte Carlo method for rarefied Poiseuille flow*

The MC method is a very basic tool to simulate free molecular flows through a component connecting two chambers, e.g. a tube or gap, and therefore neglects intermolecular collisions. Particles are initialised randomly at the entry of the gap and their linear trajectories including wall reflections are calculated until they leave through either entry or exit. This is repeated in opposite direction and used to calculate the transition probabilities P_{01}, P_{10}. Furthermore, for equal temperatures T in both chambers and equal cross-sectional areas A at entry and exit the mass flow rate $\dot{m}$ can be calculated utilising the most probable particle velocity c_{m} and both high and low pressures p_{HP}, p_{LP}. [3]

$$\dot{m} = \frac{A}{\sqrt{\pi} \cdot c_{\mathrm{m}}} \cdot (p_{\mathrm{HP}} \cdot P_{01} - p_{\mathrm{LP}} \cdot P_{10}) \, , \qquad c_{\mathrm{m}} = \sqrt{2 \cdot R_{\mathrm{s}} \cdot T} \tag{2}$$

The Poiseuille flow within an infinitely wide gap is essentially two dimensional and can be simulated using the domain shown in figure 4a. In contrast to the DSMC method, no mesh is required, since neither macroscopic state variables are of interest, nor intermolecular collisions are carried out.

For initialisation the particles are given a uniformly distributed random position on the entry. The initial velocity is determined using the following probability density functions f for its components c_{n}, c_{t1}, c_{t2} in normal and tangential directions respectively, which need to be transformed into the global coordinate system $(\boldsymbol{e}_{\mathrm{x}}, \boldsymbol{e}_{\mathrm{y}}, \boldsymbol{e}_{\mathrm{z}})$ depending on the orientation. The probability density functions are derived from the Maxwell-Boltzmann-distribution, i.e. the velocity is component-wise normal distributed, but must point into the channel. This also describes the fully diffuse scattering, so that these equations are also applied in this case: [3, 13, 14]

$$f_{\mathrm{n}}(c_{\mathrm{n}}) = \frac{2\, c_{\mathrm{n}}}{c_{\mathrm{m}}^2} \cdot \exp\left(-\frac{c_{\mathrm{n}}^2}{c_{\mathrm{m}}^2}\right) , \quad f_{\mathrm{t1}}(c_{\mathrm{t1}}) = \frac{1}{\sqrt{\pi}\, c_{\mathrm{m}}} \cdot \exp\left(-\frac{c_{\mathrm{t1}}^2}{c_{\mathrm{m}}^2}\right) , \quad f_{\mathrm{t2}}(c_{\mathrm{t2}}) = \frac{1}{\sqrt{\pi}\, c_{\mathrm{m}}} \cdot \exp\left(-\frac{c_{\mathrm{t2}}^2}{c_{\mathrm{m}}^2}\right) \tag{3}$$

To generate random velocity components complying with the probability density functions of equation 3 the inverse transformation sampling method can be used to derive a explicit term. This utilises a generator rand() for uniformly distributed numbers between 0 and 1. Alternatively a normally distributed random number generator norm() can be used for c_{t1} and c_{t2}: [14, 20]

$$c_{\mathrm{n}} = c_{\mathrm{m}} \cdot \sqrt{-\ln(\mathrm{rand}())} \, , \qquad c_{\mathrm{t1}} = \frac{c_{\mathrm{m}}}{\sqrt{2}} \cdot \mathrm{norm}() \, , \qquad c_{\mathrm{t2}} = \frac{c_{\mathrm{m}}}{\sqrt{2}} \cdot \mathrm{norm}() \tag{4}$$

This simulation approach is performed by a custom code in Matlab.

2.3. *Monte Carlo method for rarefied Couette flow*

For the Couette flow further simplifications of the MC method shall be made to simulate an infinitely long and wide gap. Here the particle is scattered repeatedly between two opposing parallel plates. While one of the plates is moving with velocity $\boldsymbol{c}_{\mathrm{W}}$, the other has surface structures, whereby the groves are perpendicular to the direction of movement of the first plate.

The flow velocity can now be calculated by using a weighted average of the individual particle velocities after each wall reflection. Fast particles reach the opposing plate in short time, meaning they have less impact on the flow velocity, and the reverse is true for slow particles. Thus, the weights g_i for this average should be the corresponding time between to wall collisions, which can be calculated from the gap height and the velocity c_{y} along that direction.

$$\overline{\boldsymbol{c}} = \frac{\sum_i g_i \cdot \boldsymbol{c}_i}{\sum_i g_i} \, , \qquad g_i = \Delta t_i = \frac{H}{|c_{i,\mathrm{y}}|} \tag{5}$$

This approach is part of current investigations and will be examined in more detail in pending works.

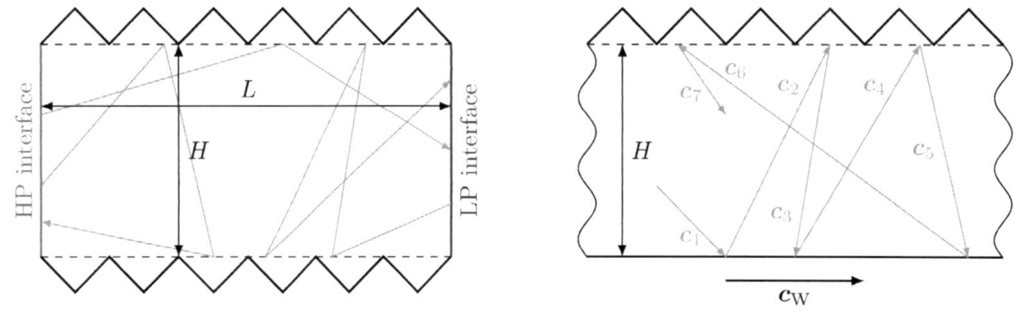

(a) Poiseuille flow

(b) Couette flow

Figure 4. Domain and working principle of different MC methods for (4a) Poiseuille and (4b) Couette flow. The mass flow rate is calculated using transition probabilities for the Poiseuille flow, while it is derived as weighted average of all particle velocities for the Couette flow.

3. Wall reflection model for surface structures

As shown before, it is possible to mesh and simulate geometries with structured surfaces. However, a special wall reflection model will now be shown, whereby the structured wall can be represented in the simulation by a smooth surface. When a particle enters the structure, meaning the collision with the theoretical smooth surface, the model takes the particle movements and interactions within the structure into account by applying a subroutine. This approach significantly simplifies the meshing of the main flow, as the cell size is no longer determined by the dimensions of the surface structure. Coarser meshing has fewer collisions cells and, in accordance with best practice guidelines, fewer particles and larger time steps can be used [21]. This wall reflection model neglects intermolecular collisions within the structure, because the local gas rarefaction parameter and rate of intermolecular collisions are lower due to the small dimensions of the structure.

When a particle is colliding with the theoretically smooth wall at position $\boldsymbol{x}$ with the velocity $\boldsymbol{c}$ in the absolute coordinate system, it enters the structure and is simulated using the wall reflection model. In reality, the particle remains in the structure for a certain amount of time Δt_{coll} and exits the structure at another point $\boldsymbol{x}^*$ with a new velocity $\boldsymbol{c}^*$. Here, the dwell time and the displacement $\boldsymbol{x}^* - \boldsymbol{x}$ shall be ignored so that the particle gains a new velocity instantaneously upon impact with the theoretical wall.

The key requirement of this new wall reflection model is that the surface structure must be periodic and two-dimensional to reduce the workload. As the example in figure 5b shows, the relative wall coordinate system is defined by the normal vector $\boldsymbol{e}_{\mathrm{y}'}$ of the wall and the tangential vector $\boldsymbol{e}_{\mathrm{z}'}$ along which the structure is invariant. The remaining vector $\boldsymbol{e}_{\mathrm{x}'}$ is obtained by the cross-product, resulting in the orthonormal basis $(\boldsymbol{e}_{\mathrm{x}'}, \boldsymbol{e}_{\mathrm{y}'}, \boldsymbol{e}_{\mathrm{z}'})$. The geometry of the structure can be described by a 2D point cloud of I points $\boldsymbol{P}_i$ using the relative coordinate system. In the following the z′-direction is only taken into account for the sake of completeness, thus this component of the interpolation points is set to zero. Due to the periodicity, it is required that the first and last points are located on the entry/exit plane with $y' = 0$, and the points $\boldsymbol{P}_i$ should also be sorted along the direction $\boldsymbol{e}_{\mathrm{x}'}$. This also results in the structure width w along the periodicity and the structure depth h:

$$i \in [1, I] \in \mathbb{N}: \qquad \boldsymbol{P}_i = (x'_i \quad y'_i \quad z'_i)^{\mathrm{T}}, \qquad x'_i \leq x'_{i+1}, \qquad y'_1 = y'_I = z'_i = 0 \tag{6}$$

$$w = x'_I - x'_1, \qquad h = \max_i (y'_i) \tag{7}$$

Upon entry the particle must be transformed into the relative wall coordinate system, including a change of observer due to a potential wall movement. The velocity $\boldsymbol{c}^{(0)}$ is obtained by subtracting the wall velocity $\boldsymbol{c}_{\mathrm{W}}$ and then multiplying it by the inverse of the transformation matrix $\boldsymbol{T}$. This matrix is constructed column by column using the unit vectors of the relative coordinate system and, due to orthogonality, its inverse is equal to its transpose. The entry position is not determined precisely, but chosen at random with uniform distribution on the entry plane, as the displacement of the particle is neglected anyway when it exits the structure.

$$\boldsymbol{T} = (\boldsymbol{e}_{\mathrm{x}'} \quad \boldsymbol{e}_{\mathrm{y}'} \quad \boldsymbol{e}_{\mathrm{z}'}), \qquad \boldsymbol{T}^{-1} = \boldsymbol{T}^{\mathrm{T}} \tag{8}$$

$$\boldsymbol{c}^{(0)} = \begin{pmatrix} u'^{(0)} & v'^{(0)} & w'^{(0)} \end{pmatrix}^{\mathrm{T}} = \boldsymbol{T}^{-1} \cdot (\boldsymbol{c} - \boldsymbol{c}_{\mathrm{W}}) \tag{9}$$

$$\boldsymbol{x}^{(0)} = \begin{pmatrix} x'^{(0)} & y'^{(0)} & z'^{(0)} \end{pmatrix}^{\mathrm{T}} = (\mathrm{rand}() \cdot w + x'_1 \quad 0 \quad 0)^{\mathrm{T}} \tag{10}$$

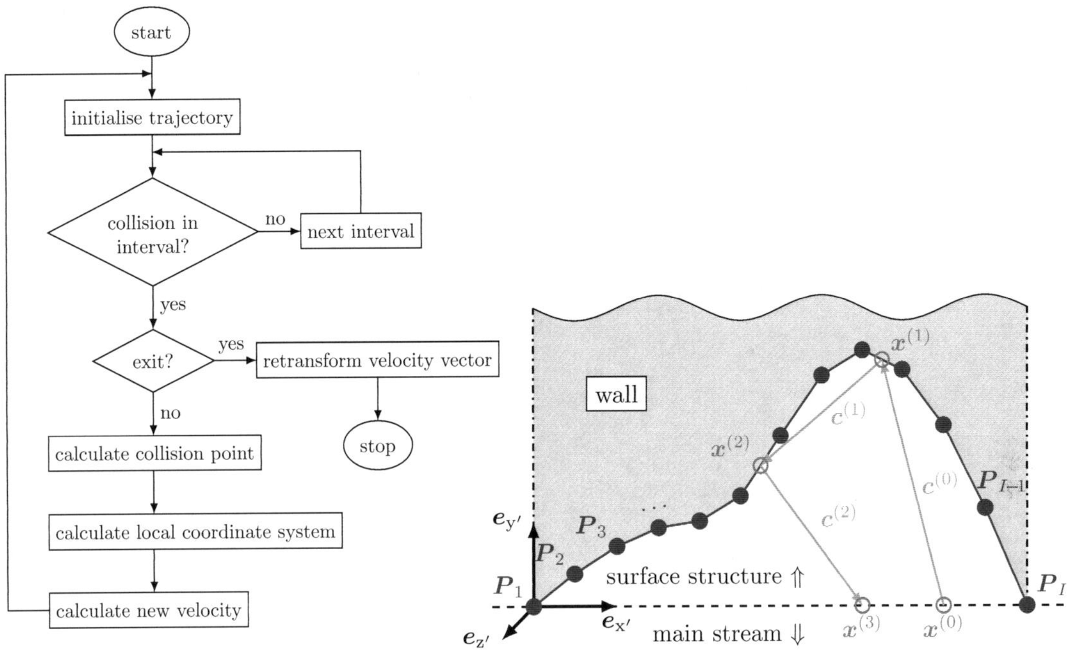

(a) program schedule (b) graphical example

Figure 5. Algorithm of the subroutine as (5a) program schedule and via (5b) graphical example: After initialisation, the movement of the particle is considered according to the linear trajectories and wall interactions until it leaves the structure.

These initial position and velocity can now be used to repeatedly calculate a collision point within the structure and a new particle velocity by utilising the diffuse reflection. This is done until the particle exits the structure through the entry/exit-plane at $y' = 0$. This is illustrated in figure 5b for an exemplary trajectory and the corresponding program schedule is displayed in figure 5a. The collision point is determined by the linear trajectory $f(x')$ and $g(x')$ of the particle, whereby the collision with the wall or the exit from the structure is checked iteratively for each interpolation interval $[x'_i, x'_{i+1}]$ in order to take undercuts into account.

$$f^{(k)}(x) = \frac{v'^{(k)}}{u'^{(k)}} \cdot \left(x - x'^{(k)}\right) + y'^{(k)}, \qquad g^{(k)}(x) = \frac{w'^{(k)}}{u'^{(k)}} \cdot \left(x - x'^{(k)}\right) + z'^{(k)} \tag{11}$$

$$i \to i + \mathrm{sgn}\left(u'^{(k)}\right): \quad \left(y'_i - f^{(k)}(x'_i)\right) \cdot \left(y'_{i+1} - f^{(k)}(x'_{i+1})\right) < 0 \quad \Leftrightarrow \quad \text{wall collision} \tag{12}$$

$$i \to i + \mathrm{sgn}\left(u'^{(k)}\right): \quad f^{(k)}(x'_i) \cdot f^{(k)}\left(x'_{i+1}\right) < 0 \quad \Leftrightarrow \quad \text{exiting structure} \tag{13}$$

In case of a collision with the wall within a specific interval, the collision point $\boldsymbol{x}^{(k+1)}$ is calculated as the intersection of the trajectory function $f^{(k)}(x)$ and the interpolation $p_i(x)$, as shown in figure 6. The new velocity vector $\boldsymbol{c}^{(k+1)}$ can be derived through diffuse scattering via equation 3. For this, a new relative coordinate system $(\boldsymbol{e}_{\mathrm{t}}, \boldsymbol{e}_{\mathrm{n}}, \boldsymbol{e}_{\mathrm{z}'})$ needs to be formulated using the gradient of the interpolation polynomial:

$$p_i(x) = \left(\frac{\Delta y}{\Delta x}\right)_i \cdot (x - x'_i) + y'_i, \qquad \left(\frac{\Delta y}{\Delta x}\right)_i = \frac{y'_{i+1} - y'_i}{x'_{i+1} - x'_i} \tag{14}$$

$$f^{(k)}\left(x'^{(k+1)}\right) = p_i\left(x'^{(k+1)}\right), \qquad \boldsymbol{x}^{(k+1)} = \left(x'^{(k+1)} \quad f^{(k)}\left(x'^{(k+1)}\right) \quad g^{(k)}\left(x'^{(k+1)}\right)\right)^{\mathrm{T}} \tag{15}$$

$$\boldsymbol{e}_{\mathrm{t}} = \frac{1}{\sqrt{1+\left(\frac{\Delta y}{\Delta x}\right)_i^2}} \cdot \left(-1 \quad -\left(\frac{\Delta y}{\Delta x}\right)_i \quad 0\right)^{\mathrm{T}}, \qquad \boldsymbol{e}_{\mathrm{n}} = \frac{1}{\sqrt{1+\left(\frac{\Delta y}{\Delta x}\right)_i^2}} \cdot \left(\left(\frac{\Delta y}{\Delta x}\right)_i \quad -1 \quad 0\right)^{\mathrm{T}} \tag{16}$$

As soon as the particle in the subroutine leaves the surface structure via the exit plane, its last velocity $\boldsymbol{c}^{(K)}$ is transferred to the original particle. For this, the velocity vector $\boldsymbol{c}^*$ after the interaction with the structure is obtained by back transformation into the absolute system:

$$\boldsymbol{c}^* = \boldsymbol{T} \cdot \boldsymbol{c}^{(K)} + \boldsymbol{c}_{\mathrm{W}} \tag{17}$$

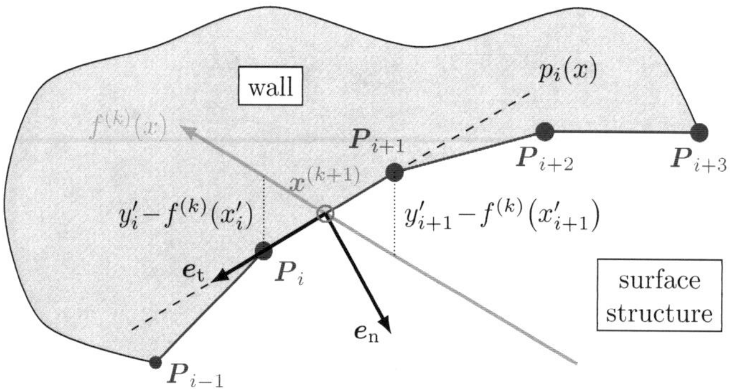

Figure 6. Calculating collision point $\boldsymbol{x}^{(k+1)}$ as intersection of two linear functions $f^{(k)}(x) = p_i(x)$ after iteratively checking for collision within each interpolation interval $[x_i, x_{i+1}]$. Coordinate system $(\boldsymbol{e}_{\mathrm{t}}, \boldsymbol{e}_{\mathrm{n}}, \boldsymbol{e}_{\mathrm{z'}})$ is constructed from slope of corresponding interpolation polynom $p_i(x)$.

4. Results

All simulations use air as a mixture of 77.7 % nitrogen and 22.3 % oxygen, which results in a specific gas constant $R_{\mathrm{s}} = 287.058\,\mathrm{J/(kg\,K)}$. Both molecules are bi-atomic but for the temperature of $T = 293.15\,\mathrm{K}$ vibrational degrees of freedom are neglected. The gap height $H = 1 \cdot 10^{-3}\,\mathrm{m}$ serves a the characteristic length and the structure depth h is defined by the fraction $h/H = 0.1$. For the Poiseuille flow the low pressure is set to $p_{\mathrm{LP}} = 0$, while the high pressure p_{HP} can be derived as a function of gas rarefaction, temperature and material parameters, namely the gas viscosity $\mu = 1.814 \cdot 10^{-5}\,\mathrm{Pa\,s}$ and the most probable particle velocity c_{m} respectively at that temperature. The pressure p_{C} for the Couette flows shall equal the high pressure of the Poiseuille flow.

$$p_{\mathrm{C}} = p_{\mathrm{HP}} = \frac{\delta \cdot \mu \cdot c_{\mathrm{m}}}{H}, \qquad c_{\mathrm{m}} = \sqrt{2 \cdot R_{\mathrm{s}} \cdot T} \tag{18}$$

To investigate the effect of the reduction of gap flows, the normalised mass flow rate $\dot{m}^*$ is utilised, which compares the mass flow rate $\dot{m}$ using surface structures with that for smooth walls $\dot{m}_{\mathrm{s}}$ for equal gap heights H.

$$\dot{m}^* = \frac{\dot{m}}{\dot{m}_{\mathrm{s}}} \tag{19}$$

While the mass flow rate for the Poiseuille flow with smooth walls must also be calculated using simulations in order to take inlet/outlet losses into account, an analytical solution can be used for the Couette flow in the plane-parallel gap. For two infinite plates with the same temperature and accommodation factors the fluid velocity is equal to the average speed of both plates, so that the following applies in the case of one static plate and two smooth walls: [18, 19]

$$\overline{\boldsymbol{c}}_{\mathrm{s,C}} = \frac{\boldsymbol{c}_W}{2}, \qquad \dot{m}_{\mathrm{s,C}} = \rho \cdot A \cdot \overline{\boldsymbol{c}}_{\mathrm{s,C}} \tag{20}$$

Because of this, it is assumed that for Couette flow the effect of surface structures cancels, when both surfaces are equipped with them, resulting in equation 20. Thus for Couette flow, only the static wall has a structured surface, while for the Poiseuille flow both surfaces are structured to maximize the observable effect.

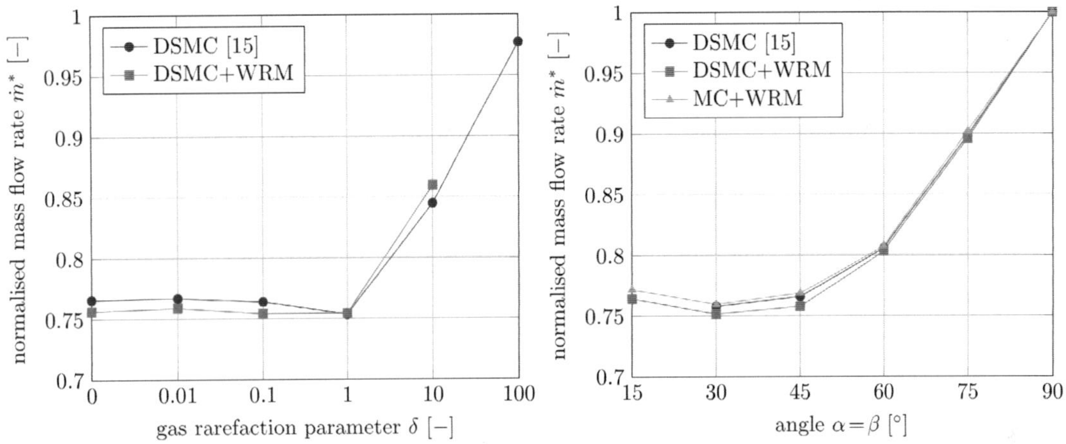

(a) Poiseuille: $L/H = 10$, $\alpha = \beta = 45°$ **(b)** Poiseuille: $L/H = 10$, $\delta = 0$

(c) Couette: $c_W = 30\,\mathrm{m/s}$, $\alpha = \beta = 45°$ **(d)** Couette: $c_W = 30\,\mathrm{m/s}$, $\delta = 0$

Figure 7. Validation of new wall reflection model (WRM) for DSMC method and MC method by comparing results for Poiseuille and Couette flow. Variation of (7a), (7c) gas rarefaction and (7b), (7d) sawtooth angles. Data available at [22].

To validate the novel wall reflection model (WRM) for structured surfaces, it is implemented into the DSMC method and the MC method and known results shall be reproduced. Thus, the Poiseuille flow with ratio of length to height $L/H = 10$ and the Couette flow are simulated using ideal sawtooth profiles. Figures 7a and 7c show results for angles $\alpha = \beta = 45°$ with varying gas rarefaction. The MC method cannot be used here, because intermolecular collisions are neglected. It can be seen that the WRM is suitable for simulating flows with surface structures for various gas rarefaction parameters, since the deviation between the models is at most 2 %. For the Poiseuille flow with increasing gas rarefaction an initial sharp decrease in normalised mass flow rate can be seen, which reaches a plateau for $\delta \leq 1$, while the Couette flow decreases more gradual and reaches a limit for free molecular flow. Additionally the effect of surface structures on the Poiseuille flow is present with higher gas rarefaction parameters than for the Couette flow, as it already vanishes at $\delta = 10$ for the latter. Since the free molecular flow requires comparatively little computational effort and the surface structures show great benefit here, various symmetrical sawtooth profiles are also simulated for $\delta = 0$, the results of which are shown in figures 7b and 7d. For the Poiseuille flow both approaches using the WRM, but especially the MC method, show great similarity with previous findings by Sazhin, as the deviation is smaller than 1.5 % [15]. A initially sharp drop in the gap flow can be observed with decreasing the angles and a minimum of 76 % can be seen for $\alpha = \beta = 30°$. The results for the Couette flow replicate this behaviour and comparing the WRM in DSMC and MC simulations to standard DSMC simulations shows a maximum deviation of 5 %.

These findings validate the new WRM, which is now used to further investigate the effect different geometries of surface structures have on the reduction of gap mass flow rates.

4.1. Optimisation of sawtooth angles

The following investigations concentrate on the free molecular flow $\delta = 0$, because it is easier to simulate and the effect only slightly varies for $\delta < 1$, and thus is suited the best for optimising the geometry of the surface structures. Additionally, this enables the use of the MC method, which is advantageous due to its ease of use and short computation time.

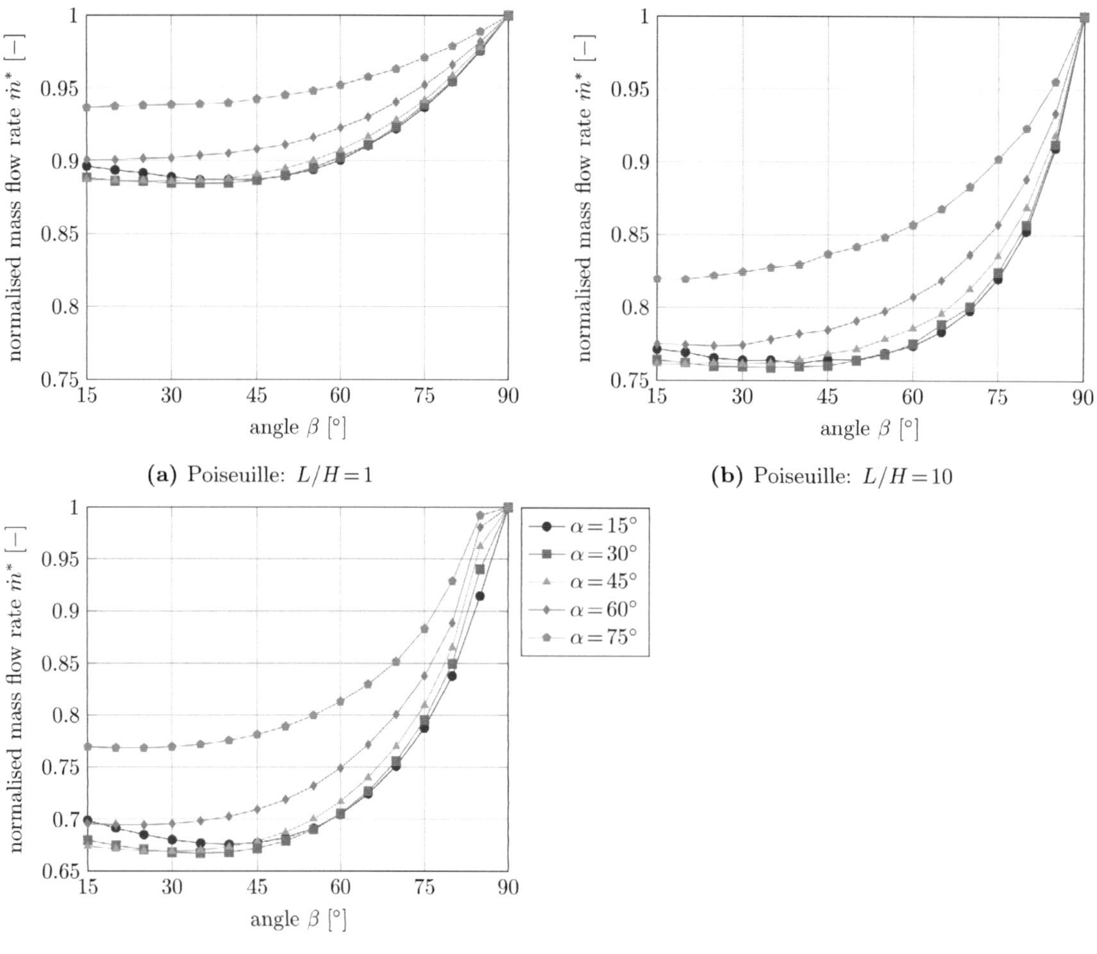

(a) Poiseuille: $L/H = 1$

(b) Poiseuille: $L/H = 10$

(c) Couette: $c_W = 30\,\text{m/s}$

Figure 8. Normalised mass flow rate of (8a), (8b) Poiseuille and (8c) Couette flow plotted against the angles of the ideal sawtooth profile defined in figure 2b. Data available at [22].

For the Poiseuille flow, longer gaps can utilise the effect of surface structures to reduce gap mass flow rates better. Figures 8a and 8b can be compared for this purpose, whereby the normalised mass flow rate for $L/H = 1$ is reduced to about 88 % and that for $L/H = 10$ reaches up to 76 %. The influence of

IOP Conf. Series: Materials Science and Engineering **1322** (2024) 012021 doi:10.1088/1757-899X/1322/1/012021

the angles α and β, however, is similar for both Poiseuille flow regardless of gap length and Couette flow, as seen in figure 8c. Decreasing either angle initially improves performance significantly, but this then stagnates and for very small angles the normalised mass flow rate can even slightly increase. Additionally, it must be emphasised that the results are almost at the optimum over large ranges, from which they deviate by only about 1 % in the range $\alpha, \beta \in [20-50]^\circ$. Therefore, there is a large selection of very good sawtooth profiles that can be selected according to other aspects, such as manufacturability.

4.2. Effect of rounding the sawtooth edges

As discussed before the ideal sawtooth profile offers many combinations of angles with nearly optimal reduction of gap flows. To improve manufacturing and to give insight into the requirements for tool design, the edges of the sawtooth profile are to be rounded with a radius r. For now, a radius will only be applied to the root of the profile, as rounding the tip seems to harm performance [23]. To reduce complexity, only symmetric profiles will be investigated for root rounding. Figure 9d showcases how the rounded profile is derived from a ideal sawtooth profile.

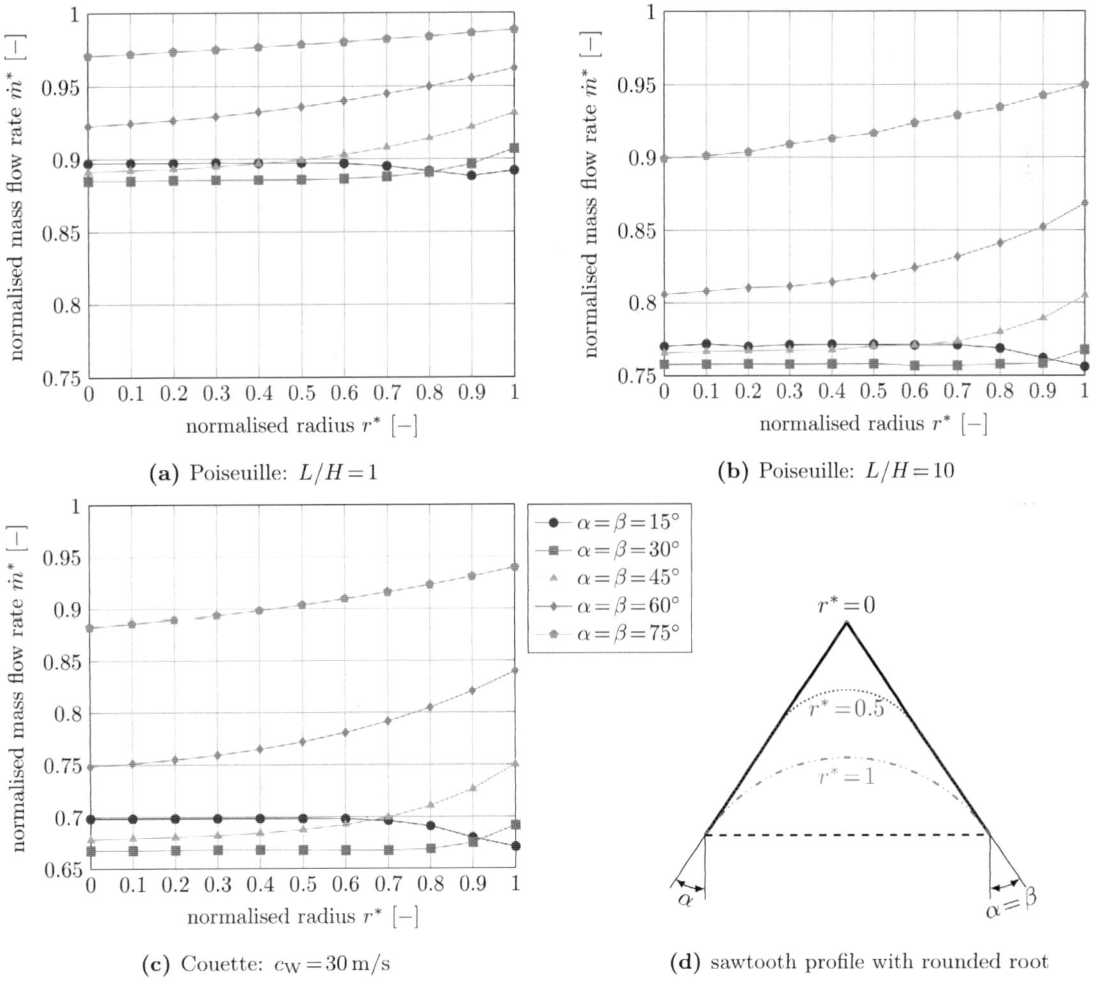

(a) Poiseuille: $L/H=1$

(b) Poiseuille: $L/H=10$

(c) Couette: $c_W=30\,\text{m/s}$

(d) sawtooth profile with rounded root

Figure 9. Normalised mass flow rate of (9a), (9b) Poiseuille and (9c) Couette flow plotted against the normalised radius for different angles of the (9d) symmetrical, rounded sawtooth profile. Data available at [22].

For better comparability, the normalised radius r^* is used, which relates the actual radius r with the maximal radius $r_{\max}$ at which the profile is completely rounded to form a circle segment. The maximum radius can be calculated by:

$$r^* = \frac{r}{r_{\max}}, \qquad r_{\max} = h \cdot \frac{\tan(\alpha)}{\cos(\alpha)} \tag{21}$$

Considering the results shown in figure 9, rounding the sawtooth root has very similar effects for both Couette and Poiseuille flow. For large angles, the relationship between normalised mass flow rate and normalised radius is almost linear and this becomes increasingly progressive as the angle decreases. For very small angles, conversely a slight improvement in performance can be seen with the use of larger radii. The best results can be observed for $\alpha = \beta = 30°$ and here rounding the root with $r^* < 0.5$ even has no influence on the normalised mass flow rate. But again, there is a large number of suitable profiles with less than 3 % deviation from the optimum, so that the focus can be put on manufacturability rather than optimising the geometry for performance.

4.3. Movement of structured surface

For now, the Couette flow is only investigated for a static structured surface and a moving smooth wall. While this case leads to a reduction in gap flow, it can be argued conversely that the movement of the structured wall increases the mass flow rate. More precisely, a static observer B, where the structured wall moves in $-x'$-direction, experiences the same as an observer C, which moves together with the smooth wall in $+x'$-direction. As illustrated in figure 10b, the latter can be derived from the initial case A of a static structured wall and a moving smooth wall by changing the reference system. Thus, the flow velocities and normalised mass flow rates for cases A and B can be related to each other:

$$\bar{c}_{\mathrm{x,B}} = \bar{c}_{\mathrm{x,A}} - c_{\mathrm{W}} = \dot{m}^*_{\mathrm{A}} \cdot \frac{c_{\mathrm{W}}}{2} - c_{\mathrm{W}} \qquad \Leftrightarrow \qquad \dot{m}^*_{\mathrm{B}} = \frac{\bar{c}_{\mathrm{x,B}}}{-\frac{1}{2}c_{\mathrm{W}}} = 2 - \dot{m}^*_{\mathrm{A}} \tag{22}$$

To prove these theoretical considerations, for various symmetrical, ideal sawtooth profiles the smooth wall is moved as before and then the structured wall is moved in opposite direction. The results for this can be found in figure 10a. It can be clearly seen that the movement of the structured surface leads to an increase in the gap mass flow rate and also that equation 22 is fulfilled.

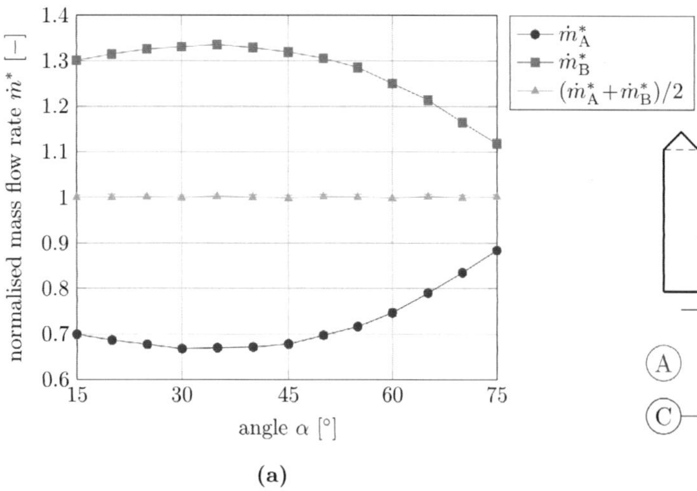

(a)

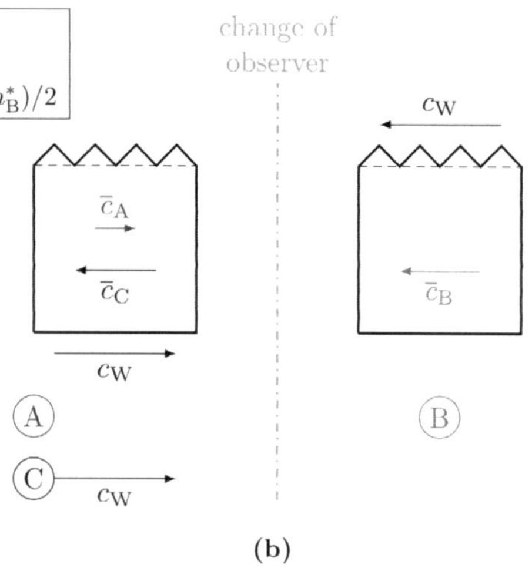

(b)

Figure 10. (10a) Normalised mass flow rate of Couette flow and evaluation of equation 22 against the angle of the symmetrical, ideal sawtooth profile. (10b) Change of observer to derive results for case B with $-x'$ movement of structured surface from case A with $+x'$ movement of the smooth wall. $c_{\mathrm{W}} = 30\,\mathrm{m/s}$. Data available at [22].

5. Application in rotary positive displacement vacuum pumps

The previous results will now be used to investigate how structured surfaces may be purposefully used in rotary positive displacement vacuum pumps to reduce backflows. The focus will initially be on the Roots vacuum pump due to its simple geometry. As with all rotary positive displacement machines the housing gap is located radially between the tip of the tooth and the housing. Figure 11a illustrates the housing gap in the axial profile plane in a rotor-fixed manner, so that the housing wall moves purely tangentially from the high-pressure port to the low-pressure port. The distance between both components represents the gap height, which is variable along the gap length due to the rotor geometry. The gap length extends in tangential direction and the gap width is the dimension of the rotor in axial direction. As discussed before, the flow occurs along the gap length and for very wide gaps, as here, is invariant to the width. These geometric circumstances mean that the surface structures must be arranged as grooves in the axial direction.

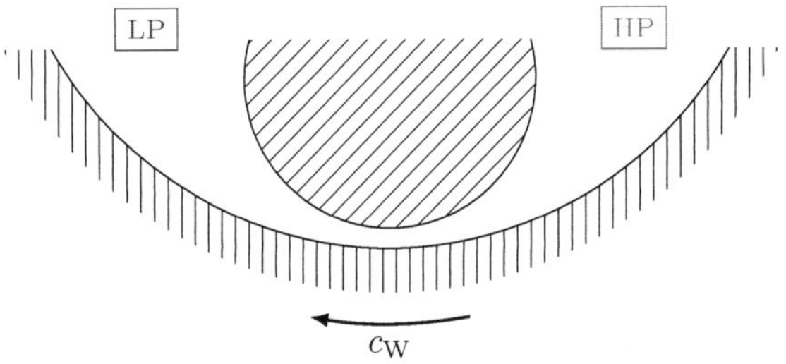

(a) roots vacuum pump

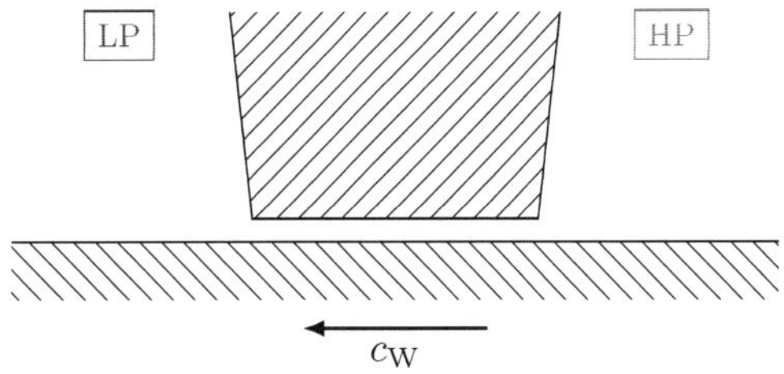

(b) screw spindle vacuum pump [1, 5, 8]

Figure 11. Rotor-fixed housing gap displayed (11a) for roots vacuum pump and (11b) untwisted for screw spindle vacuum pump.

The backflow through the housing gap of Roots vacuum pumps therefore occurs as a superposition of pressure and drag driven flow. While it would be beneficial with regard to the Poiseuille flow to provide both the tooth tip and the housing wall with surface structures, this would nullify the effect for the Couette flow. As shown above, the use of structures on moving walls leads to an increase in the mass flow rate of the Couette flow, which means that the structures must be applied to the rotor tooth. This is particularly advantageous for production, as the structures can be machined onto an outer contour instead of the inner surfaces of the housing with limited access. Furthermore, it may be sufficient to apply the structures only in a small area near the narrowest cross-section, compared to machining the whole inner contour of the housing. A planing process is suitable for this and is currently in development to enable the production of the necessary micro-structures.

In screw spindle vacuum pumps, on the other hand, the housing gap runs helically through the machine and forms a twisted but rectangular channel with a constant gap height. As a result, the gap here has a large length to height ratio, which is particularly the case at the low-pressure side of machines with variable lead. In addition, due to the rotor geometry and movement, a secondary Couette flow occurs along the gap width, but this has hardly any influence on the mass flow rate through the gap [24]. Figure 11b can therefore be used as a principle sketch of the gap flow, which can again be understood as a superposition of Couette and Poiseuille flow from a working chamber with higher pressure to one with lower pressure [1, 5, 8]. Therefore, surface structures should also be applied to the rotor tip, for which a turning process would be suitable due to the helical orientation of the rotor tip.

6. Conclusion and outlook

A novel wall reflection model for structured surfaces is presented for the simulation of rarefied gas flows, which reduces the effort for defining the domain as well as for the simulation itself. The trajectory of the particles within the structure is simulated by considering linear movement, neglecting intermolecular collisions and utilising conventional wall reflection models, such as diffuse scattering, to calculate the velocity vector after this interaction. The model is implemented in a DSMC method and a MC method and validated by comparison with known simulation results.

The influence of surface structures on rarefied gap flows is investigated in more detail and extended from the Poiseuille to the Couette flow. Particular attention is paid to the influence of the geometry of the structures, whereby not only the ideal sawtooth profiles but also those with rounded root are considered. It is found that a reduction in the gap mass flow rate of up to 33 % is possible for Couette flows and up to 25 % for the Poiseuille flow in a long gap with $L/H = 10$. However, a large number of profiles can achieve nearly optimal results, so that the focus can be laid on manufacturability.

Finally, the application in rotary displacement vacuum pumps such as the Roots or screw spindle vacuum pumps are discussed. The housing gap is very promising due to its great impact on backflows. In the future, in addition to the experimental investigation of the fluid mechanics and production of these surface structures, the potential benefit on the operating behaviour on vacuum pumps will be assessed.

Acknowledgment
Funded by the Deutsche Forschungsgemeinschaft (DFG, German Research Foundation). Gefördert durch die Deutsche Forschungsgemeinschaft (DFG) – Projektnummer 513663608.

The authors gratefully acknowledge the computation time provided on the Linux HPC cluster at Technical University Dortmund (LiDO3), partially funded in the course of the Large-Scale Equipment Initiative by the Deutsche Forschungsgemeinschaft (DFG, German Research Foundation) as project 271512359.

Nomenclature

Variables			Operators/Indices/Abbreviations	
Symbol	Unit	Explanation	Symbol	Explanation
A	$[\mathrm{m}^2]$	cross-sectional area	**bold**	tensor/matrix/vector
$\boldsymbol{c}$	[m/s]	velocity vector (of particle)	$\bar{\bullet}$	average/mean
$\bar{\boldsymbol{c}}$	[m/s]	flow rate	$\dot{\bullet}$	time derivative
c_m	[m/s]	most probable particle velocity	$\bullet^*$	after collision or normalised
$\boldsymbol{c}_\mathrm{W}$	[m/s]	velocity vector of wall	$\bullet'$	image of transformation
$\boldsymbol{e}$	[−]	unit vector	$\bullet^\mathrm{T}$	transpose of matrix
f	[−]	probability density function	$\bullet^{-1}$	inverse of matrix
g	[s]	weighting factor	$\bullet^{(k)}$	iteration
f, g	[m]	particle trajectory functions	Δ	difference/increment
H, L, W	[m]	height/length/width of gap	norm()	normal distributed random numbers
h, w	[m]	height/width of structure		
i, I	[−]	running index and its maximum	rand()	uniformly distributed random numbers between 0 and 1
k, K	[−]	iterator and last iteration		
m	[kg]	mass	sgn()	sign function
$\dot{m}$	[kg/s]	mass flow rate	01	from HP to LP
P	[−]	probability	10	from LP to HP
$\boldsymbol{P}_i$	[m]	point of point cloud	C	Couette
p	[Pa]	pressure	char	characteristic
p	[m]	interpolation polynomial	coll	collision
r	[m]	radius	DSMC	Direct Simulation Monte Carlo
R_s	[J/(kg K)]	specific gas constant	HP	high pressure
T	[K]	temperature	LP	low pressure
$\boldsymbol{T}$	[−]	transformation matrix	m	mean
t	[s]	time	MC	Monte Carlo method
$\boldsymbol{x}$	[m]	positional vector of particle	n	normal
x, y, z	[m]	components of position	s	smooth
u, v, w	[m/s]	components of velocity	t	tangential
α, β	[°]	angles of sawtooth profile	W	Wall
δ	[−]	gas rarefaction parameter	WRM	wall reflection model
ρ	$[\mathrm{kg/m^3}]$	density		

References

[1] Wenderott D 2001 *Spaltströmungen im Vakuum* Dissertation TU Dortmund Düsseldorf ISBN 978-3-18-342307-1
[2] Jorisch W 2008 Trockenlaufende Vakuumpumpen für die Prozess-Industrie *Vakuum in Forschung und Praxis* **20** 33–39 DOI 10.1002/vipr.200800344

[3] Jousten K 2018 *Handbuch Vakuumtechnik* 12th ed (Wiesbaden: Springer Vieweg) DOI 10.1007/978-3-658-13386-3
[4] Pfaller D 2012 *Einfluss der Verdrängergeometrie auf die Auslegung von Schraubenspindel-Vakuumpumpen* Dissertation TU Dortmund Dortmund ISBN 978-3-8440-1291-0
[5] Müller R 2013 *Spaltströmung mit Wärmeübertragung in Vakuumpumpen* Dissertation TU Kaiserslautern Kaiserslautern URN urn:nbn:de:hbz:386-kluedo-36582
[6] Huck C, Pleskun H and Brümmer A 2018 Measurement and simulation of rarefied Couette Poiseuille flow with variable cross section *J. Vac. Sci. Technol. A* **36** DOI 10.1116/1.5024899
[7] Huck C 2020 *Verdünnte Strömung in Spalten von Vakuumpumpen mit bewegter Berandung* Dissertation TU Dortmund Dortmund DOI 10.17877/DE290R-22546
[8] Jünemann T and Brümmer A 2020 One-dimensional simulation of Couette–Poiseuille flow with variable cross section for the full range of gas rarefaction *J. Vac. Sci. Technol. B* **38** DOI 10.1116/6.0000207
[9] Dreifert T and Müller R 2014 Screw vacuum pumps - The state of the art *9th International Conference on Screw Machines 2014* (VDI-Berichte 2228) pp 29–42 URL `https://ft.mb.tu-dortmund.de/storages/ft-mb/r/Publikationen/ICSM/2014/T._Dreifert__R._Mueller_2014_-_Screw_vacuum_pumps.pdf`
[10] Brock S, Pleskun H, Polus G, Saelzer J, Biermann D and Brümmer A 2024 Potential of surface structures for the reduction of vacuum gap flows *Process Technology & Components* **2024** 46–51 URL `https://www.harnisch-digital.de/puk-process-technology-components-2024/68664363`
[11] Kösters H and Eickhoff J 2006 Trockene Schraubenvakuumpumpe mit hoher innerer Verdichtung *Schraubenmaschinen 2006* (VDI-Berichte 1932) pp 423–428 URL `https://ft.mb.tu-dortmund.de/storages/ft-mb/r/Publikationen/ICSM/2006/Koesters__Eickhoff_2006_-_Trockene_Schraubenvakuumpumpe_mit_hoher_innerer.pdf`
[12] Sharipov F 2016 *Rarefied Gas Dynamics* (Weinheim: Wiley-VCH) DOI 10.1002/9783527685523
[13] Bird G A 1994 *Molecular Gas Dynamics and the Direct Simulation of Gas Flows* (Oxford: Clarendon Press) DOI 10.1093/oso/9780198561958.001.0001
[14] Shen C 2005 *Rarefied Gas Dynamics* (Berlin Heidelberg: Springer) DOI 10.1007/b138784
[15] Sazhin O 2020 Rarefied gas flow through a rough channel into a vacuum *Microfluid. Nanofluidics* **24** 27 DOI 10.1007/s10404-020-2330-y
[16] White C 2013 *Benchmarking, Development and Application of an Open Source DSMC Solver* Ph.D. thesis University of Strathclyde Glasgow DOI 10.48730/jsh2-je35
[17] White C, Borg M K, Scanlon T J, Longshaw S M, John B, Emerson D R and Reese J M 2018 dsmcFoam+: An OpenFOAM based direct simulation Monte Carlo solver *Comput. Phys. Comm.* **224** 22–43 DOI 10.1016/j.cpc.2017.09.030
[18] Pleskun H, Bode T and Brümmer A 2022 Couette flow in a rectangular channel in the whole range of the gas rarefaction *Phys. Fluids* **34** DOI 10.1063/5.0082940
[19] Pleskun H and Brümmer A 2022 Gas–surface interactions of a Couette–Poiseuille flow in a rectangular channel *Phys. Fluids* **34** DOI 10.1063/5.0099256
[20] Überhuber C 1995 *Computer-Numerik 2* (Berlin Heidelberg: Springer) DOI 10.1007/978-3-642-57794-9
[21] Bird G A 2007 Sophisticated DSMC Notes prepared for a short course at the DSMC07 meeting URL `http://gab.com.au/DSMC07notes.pdf`
[22] Brock S, Pleskun H, Polus G, Saelzer J, Biermann D and Brümmer A 2024 Supplementary data for: Investigations to reduce rarefied gap flows within positive displacement vacuum pumps by utilising surface structures DOI 10.17877/tudodata-2024-lz9qq0vr
[23] Polus G, Saelzer J, Brock S, Pleskun H, Biermann D and Brümmer A 2024 Experimental and simulative investigations of burr formation in planing of AISI 1045 *Int. J. Adv. Manuf. Technol.* DOI10.1007/s00170-024-14150-1
[24] Pleskun H and Brümmer A 2021 Schlussbericht. Verdünnte Spaltströmungen in trockenlaufenden Vakuumpumpen mit bewegter Berandung beliebiger Richtungen. TU Dortmund. IGF-Vorhaben Nr. 19859 N/1

AUTHOR INDEX